Selected Titles in This Series

13 **Paul C. Shields,** The ergodic theory of discrete sample paths, 1996
12 **N. V. Krylov,** Lectures on elliptic and parabolic equations in Hölder spaces, 1996
11 **Jacques Dixmier,** Enveloping algebras, 1996 Printing
10 **Barry Simon,** Representations of finite and compact groups, 1996
9 **Dino Lorenzini,** An invitation to arithmetic geometry, 1996
8 **Winfried Just and Martin Weese,** Discovering modern set theory. I: The basics, 1996
7 **Gerald J. Janusz,** Algebraic number fields, second edition, 1996
6 **Jens Carsten Jantzen,** Lectures on quantum groups, 1996
5 **Rick Miranda,** Algebraic curves and Riemann surfaces, 1995
4 **Russell A. Gordon,** The integrals of Lebesgue, Denjoy, Perron, and Henstock, 1994
3 **William W. Adams and Philippe Loustaunau,** An introduction to Gröbner bases, 1994
2 **Jack Graver, Brigitte Servatius, and Herman Servatius,** Combinatorial rigidity, 1993
1 **Ethan Akin,** The general topology of dynamical systems, 1993

The Ergodic Theory of Discrete Sample Paths

The Ergodic Theory of Discrete Sample Paths

Paul C. Shields

Graduate Studies
in Mathematics

Volume 13

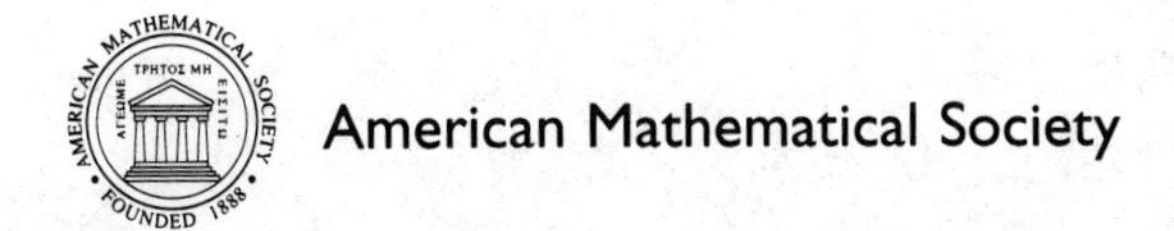

1991 *Mathematics Subject Classification.* Primary 28D20, 28D05, 94A17;
Secondary 60F05, 60G17, 94A24.

ABSTRACT. This book is about finite-alphabet stationary processes, which are important in physics, engineering, and data compression. The book is designed for use in graduate courses, seminars or self study for students or faculty with some background in measure theory and probability theory.

Library of Congress Cataloging-in-Publication Data

Shields, Paul C.
The ergodic theory of discrete sample paths / Paul C. Shields.
p. cm. — (Graduate studies in mathematics, ISSN 1065-7339; v. 13)
Includes bibliographical references and index.
ISBN 0-8218-0477-4 (alk. paper)
1. Ergodic theory. 2. Measure-preserving transformations. 3. Stochastic processes. I. Title. II. Series.
QA313.S55 1996
519.2′32—dc20 96-20186
CIP

---

10 9 8 7 6 5 4 3 2 1 01 00 99 98 97 96

Contents

Preface

This book is about finite-alphabet stationary processes, which are important in physics, engineering, and data compression. The book is designed for use in graduate courses, seminars or self study for students or faculty with some background in measure theory and probability theory. The focus is on the combinatorial properties of typical finite sample paths drawn from a stationary, ergodic process. A primary goal, only partially realized, is to develop a theory based directly on sample path arguments, with minimal appeals to the probability formalism. A secondary goal is to give a careful presentation of the many models for stationary finite-alphabet processes that have been developed in probability theory, ergodic theory, and information theory.

The two basic tools for a sample path theory are a packing lemma, which shows how "almost" packings of integer intervals can be extracted from coverings by overlapping subintervals, and a counting lemma, which bounds the number of n-sequences that can be partitioned into long blocks subject to the condition that most of them are drawn from collections of known size. These two simple ideas, introduced by Ornstein and Weiss in 1980, immediately yield the two fundamental theorems of ergodic theory, namely, the ergodic theorem of Birkhoff and the entropy theorem of Shannon, McMillan, and Breiman. The packing and counting ideas yield more than these two classical results, however, for in combination with the ergodic and entropy theorems and further simple combinatorial ideas they provide powerful tools for the study of sample paths. Much of Chapter I and all of Chapter II are devoted to the development of these ideas.

The classical process models are based on independence ideas and include the i.i.d. processes, Markov chains, instantaneous functions of Markov chains, and renewal and regenerative processes. An important and simple class of such models is the class of concatenated-block processes, that is, the processes obtained by independently concatenating fixed-length blocks according to some block distribution and randomizing the start. Related models are obtained by block coding and randomizing the start, or by stationary coding, an extension of the instantaneous function concept which allows the function to depend on both past and future. All these models and more are introduced in the first two sections of Chapter I. Further models, including the weak Bernoulli processes and the important class of stationary codings of i.i.d. processes, are discussed in Chapter III and Chapter IV.

Of particular note in the discussion of process models is how ergodic theorists think of a stationary process, namely, as a measure-preserving transformation on a probability space, together with a partition of the space. This point of view, introduced in Section I.2, leads directly to Kakutani's simple geometric representation of a process in terms of a recurrent event, a representation that not only simplifies the discussion of stationary renewal and regenerative processes but generalizes these concepts to the case where times between recurrences are not assumed to be independent, but only stationary. A further generalization, given in Section I.10, leads to a powerful method for constructing examples known as cutting and stacking.

The book has four chapters. The first chapter, which is half the book, is devoted to the basic tools, including the Kolmogorov and ergodic theory models for a process, the ergodic theorem and its connection with empirical distributions, the entropy theorem and its interpretations, a method for converting block codes to stationary codes, the weak topology and the even more important $\bar{d}$-metric topology, and the cutting and stacking method.

Properties related to entropy which hold for every ergodic process are discussed in Chapter II. These include entropy as the almost-sure bound on per-symbol compression, Ziv's proof of asymptotic optimality of the Lempel-Ziv algorithm via his interesting concept of individual sequence entropy, the relation between entropy and partitions of sample paths into fixed-length blocks, or partitions into distinct blocks, or partitions into repeated blocks, and the connection between entropy and recurrence times and entropy and the growth of prefix trees.

Properties related to entropy which hold only for restricted classes of processes are discussed in Chapter III, including rates of convergence for frequencies and entropy, the estimation of joint distributions in both the variational metric and the $\bar{d}$-metric, a connection between entropy and $\bar{d}$-neighborhoods, and a connection between entropy and waiting times.

Several characterizations of the class of stationary codings of i.i.d. processes are given in Chapter IV, including the almost block-independence, finitely determined, very weak Bernoulli, and blowing-up characterizations. Some of these date back to the original work of Ornstein and others on the isomorphism problem for Bernoulli shifts, although the almost block-independence and blowing-up ideas are more recent.

This book is an outgrowth of the lectures I gave each fall in Budapest from 1989 through 1994, both as special lectures and seminars at the Mathematics Institute of the Hungarian Academy of Sciences and as courses given in the Probability Department of Eötvös Loránd University. In addition, lectures on parts of the penultimate draft of the book were presented at the Technical University of Delft in the fall of 1995. The audiences included ergodic theorists, information theorists, and probabilists, as well as combinatorialists and people from engineering and other mathematics disciplines, ranging from undergraduate and graduate students through post-docs and junior faculty to senior professors and researchers.

Many standard topics from ergodic theory are omitted or given only cursory treatment, in part because the book is already too long and in part because they are not close to the central focus of this book. These topics include topological dynamics, smooth dynamics, random fields, K-processes, combinatorial number theory, general ergodic theorems, and continuous time and/or space theory. Likewise little or nothing is said about such standard information theory topics as rate distortion theory, divergence-rate theory, channel theory, redundancy, algebraic coding, and multi-user theory.

Some specific stylistic guidelines were followed in writing this book. Proofs are sketched first, then given in complete detail. With a few exceptions, the sections in Chapters II, III, and IV are approximately independent of each other, conditioned on the material in Chapter I. Theorems and lemmas are given names that include some information about content, for example, the entropy theorem rather than the Shannon-McMillan-Breiman theorem. Likewise, suggestive names are used for concepts, such as building blocks (a name proposed by Zoli Györfi) and column structures (as opposed to gadgets.) Also numbered displays are often (informally) given names similar to those used for LaTeX labels. Only those references that seem directly related to the topics discussed are included. Exercises that extend the ideas are given at the end of most sections; these range in difficulty from quite easy to quite hard.

I am indebted to many people for assistance with this project. Imre Csiszár and Katalin Marton not only attended most of my lectures but critically read parts of the manuscript at all stages of its development and discussed many aspects of the book with me. Much of the material in Chapter III as well as the blowing-up ideas in Chapter IV are the result of joint work with Kati. It was Imre's suggestion that led me to include the discussions of renewal and regenerative processes, and his criticisms led to many revisions of the cutting and stacking discussion. In addition I had numerous helpful conversations with Benjy Weiss, Don Ornstein, Aaron Wyner, and Jacob Ziv. Others who contributed ideas and/or read parts of the manuscript include Gábor Tusnády, Bob Burton, Jacek Serafin, Dave Neuhoff, György Michaletzky, Gusztáv Morvai, and Nancy Morrison. Last, but far from least, I am much indebted to my two Toledo graduate students, Shaogang Xu and Xuehong Li, who learned ergodic theory by carefully reading almost all of the manuscript at each stage of its development, in the process discovering numerous errors and poor ways of saying things.

No project such as this can be free from errors and incompleteness. A list of errata as well as a forum for discussion will be available on the Internet at the following web address.

http://www.math.utoledo.edu/~pshields/ergodic.html

I am grateful to the Mathematics Institute of the Hungarian Academy for providing me with many years of space in a comfortable and stimulating environment as well as to the Institute and to the Probability Department of Eötvös Loránd University for the many lecture opportunities. My initial lectures in 1989 were supported by a Fulbright lectureship. Much of the work for this project was supported by NSF grants DMS-8742630 and DMS-9024240 and by a joint NSF-Hungarian Academy grant MTA-NSF project 37.

This book is dedicated to my son, Jeffrey.

Paul C. Shields
Toledo, Ohio
March 22, 1996

Chapter I

Basic concepts.

Section I.1 Stationary processes.

A (discrete-time, stochastic) process is a sequence $X_1, X_2, \ldots, X_n, \ldots$ of random variables defined on a probability space (X, Σ, μ). The process has *alphabet* A if the range of each X_i is contained in A. In this book the focus is on finite-alphabet processes, so, unless stated otherwise, "process" means a discrete-time finite-alphabet process. Also, unless it is clear from the context or explicitly stated stated otherwise, "measure" will mean "probability measure" and "function" will mean "measurable function" with respect to some appropriate σ-algebra on a probability space.

The cardinality of a finite set A is denoted by $|A|$. The sequence $a_m, a_{m+1}, \ldots, a_n$, where each $a_i \in A$, is denoted by a_m^n. The set of all such a_m^n is denoted by A_m^n, except for $m = 1$, when A^n is used.

The *k-th order joint distribution* of the process $\{X_k\}$ is the measure μ_k on A^k defined by the formula

$$\mu_k(a_1^k) = \text{Prob}(X_1^k = a_1^k), \quad a_1^k \in A^k.$$

When no confusion will result the subscript k on μ_k may be omitted. The set of joint distributions $\{\mu_k\colon k \geq 1\}$ is called the *distribution* of the process. The distribution of a process can, of course, also be defined by specifying the start distribution, μ_1, and the successive conditional distributions

$$\mu(a_k|a_1^{k-1}) = \text{Prob}(X_k = a_k|X_1^{k-1} = a_1^{k-1}) = \frac{\mu_k(a_1^k)}{\mu_{k-1}(a_1^{k-1})}.$$

The distribution of a process is thus a family of probability distributions, one for each k. The sequence cannot be completely arbitrary, however, for implicit in the definition of process is that the following *consistency* condition must hold for each $k \geq 1$,

$$\mu_k(a_1^k) = \sum_{a_{k+1}} \mu_{k+1}(a_1^{k+1}), \quad a_1^k \in A^k. \tag{1}$$

A process is considered to be defined by its joint distributions, that is, the particular space on which the functions X_n are defined is not important; all that really matters in probability theory is the distribution of the process. Thus one is free to choose the underlying space (X, Σ, μ) on which the X_n are defined in any convenient manner, as long as the the joint distributions are left unchanged. An important instance of this idea is the Kolmogorov model for a process, which represents it as the sequence of coordinate functions on the space of infinite sequences drawn from the alphabet A, equipped with a Borel measure constructed from the consistency conditions (1).

The rigorous construction of the Kolmogorov representation is carried out as follows. Let A^∞ denote the set of all infinite sequences

$$x = \{x_i\}, \quad x_i \in A, \quad 1 \leq i < \infty.$$

The *cylinder set determined by* a_m^n, denoted by $[a_m^n]$, is the subset of A^∞ defined by

$$[a_m^n] = \{x\colon\ x_i = a_i,\ m \leq i \leq n\}.$$

Let $\mathcal{C}_n$ be the collection of cylinder sets defined by sequences that belong to A^n, let $\mathcal{C} = \cup_n \mathcal{C}_n$, and let $\mathcal{R}_n = \mathcal{R}_n(\mathcal{C}_n)$ denote the ring generated by $\mathcal{C}_n$. The sequence $\{\mathcal{R}_n\}$ is increasing, and its union $\mathcal{R} = \mathcal{R}(\mathcal{C})$ is the ring generated by all the cylinder sets. Two important properties are summarized in the following lemma.

Lemma I.1.1

(a) *Each set in $\mathcal{R}_n$ is a finite disjoint union of cylinder sets from $\mathcal{C}_n$.*

(b) *If $\{B_n\} \subset \mathcal{R}$ is a decreasing sequence of sets with empty intersection then there is an N such that $B_n = \emptyset$, $n \geq N$.*

Proof. The proof of (a) is left as an exercise. Part (b) is an application of the finite intersection property for sequences of compact sets, since the space A^∞ is compact in the product topology and each set in $\mathcal{R}$ is closed, from (a). Thus the lemma is established.

Let Σ be the σ-algebra generated by the cylinder sets $\mathcal{C}$. The collection Σ can also be defined as the the σ-algebra generated by $\mathcal{R}$, or by the compact sets. The members of Σ are commonly called the *Borel* sets of A^∞.

For each $n \geq 1$ the coordinate function $\widehat{X}_n\colon A^\infty \mapsto A$ is defined by $\widehat{X}_n(x) = x_n$, $x \in A^\infty$. The Kolmogorov representation theorem states that every process with alphabet A can be thought of as the coordinate function process $\{\widehat{X}_n\}$, together with a Borel measure on A^∞.

Theorem I.1.2 (Kolmogorov representation theorem.)

If $\{\mu_k\}$ is a sequence of measures for which the consistency conditions (1) hold, then there is a unique Borel probability measure μ on A^∞ such that, $\mu([a_1^k]) = \mu_k(a_1^k)$, for each k and each a_1^k.

In other words, if $\{X_n\}$ is a process with finite alphabet A, there is a unique Borel measure μ on A^∞ for which the sequence of coordinate functions $\{\widehat{X}_n\}$ has the same distribution as $\{X_n\}$.

Proof. The process $\{X_n\}$ defines a set function μ on the collection $\mathcal{C}$ of cylinder sets by the formula

$$\mu([a_1^n]) = \text{Prob}(X_i = a_i,\ 1 \leq i \leq n). \tag{2}$$

The consistency conditions (1) together with Lemma I.1.1(a), imply that μ extends to a finitely additive set function on the ring $\mathcal{R} = \mathcal{R}(\mathcal{C})$ generated by the cylinder sets. The finite intersection property, Lemma I.1.1(b), implies that μ can be extended to a unique countably additive measure on the σ-algebra Σ generated by $\mathcal{R}$. Equation (2) translates into the statement that $\{\widehat{X}_n\}$ and $\{X_n\}$ have the same joint distribution. This proves Theorem I.1.2. □

The sequence of coordinate functions $\{\widehat{X}_n\}$ on the probability space (A^∞, Σ, μ) will be called the *Kolmogorov* representation of the process $\{X_n\}$. The measure μ will be called the *Kolmogorov measure of the process*, or the *Kolmogorov measure of the sequence* $\{\mu_k\}$. Process and measure language will often be used interchangeably; for example, "let μ be a process," means "let μ be the Kolmogorov measure for some process $\{X_n\}$."

As noted earlier, a process is considered to be defined by its joint distributions, that is, the particular space on which the functions X_n are defined is not important. The Kolmogorov model is simply one particular way to define a space and a sequence of functions with the given distributions. Another useful model is the *complete measure model*, that is, one in which subsets of sets of measure 0 are measurable. The Kolmogorov measure μ on A^∞ extends to a complete measure $\bar{\mu}$ on the completion $\bar{\Sigma}$ of the Borel sets Σ relative to μ. Furthermore, completion has no effect on joint distributions, for $\bar{\mu}([a_1^k]) = \mu_k(a_1^k)$ for all $k \geq 1$ and all a_1^k, and uniqueness is preserved, that is, different processes have Kolmogorov measures with different completions.

Many ideas are easier to express and many results are easier to establish in the framework of complete measures; thus, whenever it is convenient to do so the complete Kolmogorov model will be used in this book, though often without explicitly saying so. In particular, "Kolmogorov measure" is taken to refer to either the measure defined by Theorem I.1.2, or to its completion, whichever is appropriate to the context.

A process is *stationary*, if the joint distributions do not depend on the choice of time origin, that is,

$$\text{Prob}(X_i = a_i,\ m \leq i \leq n) = \text{Prob}(X_{i+1} = a_i,\ m \leq i \leq n), \tag{3}$$

for all m, n, and a_m^n. Stationary finite-alphabet processes serve as models in many settings of interest, including physics, data transmission and storage, and statistics. The phrase "stationary process" will mean a stationary, discrete-time, finite-alphabet process, unless stated otherwise.

The statement that a process $\{X_n\}$ with finite alphabet A is stationary translates into the statement that the Kolmogorov measure μ is invariant under the shift transformation T. The (left) shift $T\colon A^\infty \mapsto A^\infty$ is defined by

$$(Tx)_n = x_{n+1}, \quad x \in A^\infty,\ n \geq 1,$$

which is expressed in symbolic form by

$$x = (x_1, x_2, \ldots) \implies Tx = (x_2, x_3, \ldots).$$

The shift transformation is continuous relative to the product topology on A^∞, and defines a set transformation T^{-1} by the formula,

$$T^{-1}B = \{x\colon Tx \in B\}, \quad B \subseteq A^\infty.$$

Note that

$$T^{-1}[a_m^n] = [b_{m+1}^{n+1}], \quad b_{i+1} = a_i, \; m \le i \le n,$$

so that the set transformation T^{-1} maps a cylinder set onto a cylinder set. It follows that $T^{-1}B$ is a Borel set for each Borel set B, which means that T is a Borel measurable mapping. The condition that the process be stationary is just the condition that $\mu([a_m^n]) = \mu([b_{m+1}^{n+1}])$, that is, $\mu(B) = \mu(T^{-1}B)$ for each cylinder set B; which is, in turn, equivalent to the condition that $\mu(B) = \mu(T^{-1}B)$ for each Borel set B. The condition,

$$\mu(B) = \mu(T^{-1}B), \quad B \in \Sigma,$$

is usually summarized by saying that T *preserves the measure* μ or, alternatively, that μ *is* T*-invariant.*

In summary, the shift T is Borel measurable. It preserves a Borel probability measure μ if and only if the sequence of coordinate functions $\{\widehat{X}_n\}$ is a stationary process on the probability space (A^∞, Σ, μ).

The intuitive idea of stationarity is that the mechanism generating the random variables doesn't change in time. It is often convenient to think of a process has having run for a long time before the start of observation, so that only nontransient effects, that is, those that do not depend on the time origin, are seen. Such a model is well interpreted by another model, called *the doubly-infinite sequence model or two-sided model.*

Let Z denote the set of integers and let A^Z denote the set of doubly-infinite A-valued sequences, that is, the set of all sequences $x = \{x_n\}$, where each $x_n \in A$ and $n \in Z$ ranges from $-\infty$ to ∞. The shift $T\colon A^Z \mapsto A^Z$ is defined by $(Tx)_n = x_{n+1}$, $n \in Z$. The concept of cylinder set extends immediately to this new setting. Furthermore, if $\{\mu_k\}$ is a set of measures for which the consistency conditions (1) and the stationarity conditions (3) hold then there is a unique T-invariant Borel measure μ on A^Z such that $\mu([a_1^k]) = \mu_k(a_1^k)$, for each $k \ge 1$ and each a_1^k. The projection of μ onto A^∞ is, of course, the Kolmogorov measure on A^∞ for the process defined by $\{\mu_k\}$.

In summary, for stationary processes there are two standard representations, one as the Kolmogorov measure μ on A^∞, the other as the shift-invariant extension to the space A^Z. The latter model will be called the *two-sided Kolmogorov model of a stationary process.* Note that the shift T for the two-sided model is invertible, that is, T is one-to-one and onto and T^{-1} is measure preserving. (The two-sided shift is, in fact, a homeomorphism on the compact space A^Z.)

From a physical point of view it does not matter in the stationary case whether the one-sided or two-sided model is used. Proofs are sometimes simpler if the two-sided model is used, for invertibility of the shift on A^Z makes things easier; such results will then apply to the one-sided model in the stationary case. In some cases where such a simplification is possible, only the proof for the invertible case will be given. In the stationary case, "Kolmogorov measure" is taken to refer to either the one-sided measure or its completion, or to the two-sided measure or its completion, as appropriate to the context.

Remark I.1.3

Much of the success of modern probability theory is based on use of the Kolmogorov model, for asymptotic properties can often be easily formulated in terms of subsets of the infinite product space A^∞. In addition, the model provides a concrete uniform setting for all processes; different processes with the same alphabet A are distinguished by having different Kolmogorov measures on A^∞. When completeness is needed the complete model can be used, and, in the stationary case, when invertibility is useful the two-sided model on A^Z can be used. As will be shown in later sections, even more can be gained in the stationary case by retaining the abstract concept of a stationary process as a sequence of random functions with a specified joint distribution on an arbitrary probability space.

Remark I.1.4

While this book is primarily concerned with finite-alphabet processes, it is sometimes desirable to consider countable-alphabet processes, or even continuous-alphabet processes. For example, return-time processes associated with finite-alphabet processes are, except in trivial cases, countable-alphabet processes, and functions of uniformly distributed i.i.d. processes will be useful in Chapter 4. The Kolmogorov theorem extends easily to the countable case, as well as to the continuous-alphabet i.i.d. cases that will be considered.

I.1.a Examples.

Some standard examples of finite-alphabet processes will now be presented.

Example I.1.5 (Independent processes.)

The simplest example of a stationary finite-alphabet process is an independent, identically distributed (i.i.d.) process. A sequence of random variables $\{X_n\}$ is independent if

$$\text{Prob}\big(X_n = a_n | X_1^{n-1} = a_1^{n-1}\big) = \text{Prob}(X_n = a_n),$$

holds for all $n \geq 1$ and all $a_1^n \in A^n$. It is identically distributed if $\text{Prob}(X_n = a) = \text{Prob}(X_1 = a)$ holds for all $n \geq 1$ and all $a \in A$. An independent process is stationary if and only if it is identically distributed.

The i.i.d. condition holds if and only if the product formula

$$\mu_k(a_1^k) = \prod_{i=1}^{k} \mu_1(a_i),$$

holds. Such a measure is called a *product* measure defined by μ_1. Thus a process is i.i.d. if and only if its Kolmogorov measure is the product measure defined by some distribution on A.

Example I.1.6 (Markov chains.)

The simplest examples of finite-alphabet dependent processes are the Markov processes. A sequence of random variables, $\{X_n\}$, is a Markov chain if

$$\text{Prob}\big(X_n = a_n | X_1^{n-1} = a_1^{n-1}\big) = \text{Prob}(X_n = a_n | X_{n-1} = a_{n-1}) \tag{4}$$

holds for all $n \geq 2$ and all $a_1^n \in A^n$. A Markov chain is said to be homogeneous or to have stationary transitions if $\text{Prob}(X_n = b | X_{n-1} = a)$ does not depend on n, in which case the $|A| \times |A|$ matrix M defined by

$$M_{ab} = M_{b|a} = \text{Prob}(X_n = b | X_{n-1} = a)$$

is called the *transition matrix* of the chain. The start distribution is the (row) vector $\mu_1 = \{\mu_1(s)\}$, where

$$\mu_1(a) = \text{Prob}(X_1 = a),\ a \in A.$$

The start distribution is called a *stationary distribution* of a homogeneous chain if the matrix product $\mu_1 M$ is equal to the row vector μ_1. If this holds, then a direct calculation shows that, indeed, the process is stationary.

The condition that a Markov chain be a stationary process is that it have stationary transitions *and* that its start distribution is a stationary distribution. Unless stated otherwise, "stationary Markov process" will mean a Markov chain with stationary transitions for which $\mu_1 M = \mu_1$ and, for which $\mu_1(a) > 0$ for all $a \in A$. The positivity condition rules out transient states and is included so as to avoid trivialities.

Example I.1.7 (Multistep Markov chains.)

A generalization of the Markov property allows dependence on k steps in the past. A process is called *k-step Markov* if

$$\text{Prob}\big(X_n = a_n | X_1^{n-1} = a_1^{n-1}\big) = \text{Prob}\big(X_n = a_n | X_{n-k}^{n-1} = a_{n-k}^{n-1}\big) \tag{5}$$

holds for all $n > k$ and for all a_1^n. The two conditions for stationarity are, first, that transitions be stationary, that is,

$$\mu(a_n | a_{n-k}^{n-1}) = \text{Prob}\big(X_n = a_n | X_{n-k}^{n-1} = a_{n-k}^{n-1}\big)$$

does not depend on n as long as $n > k$, and, second, that $\mu_k M^{(k)} = \mu_k$, where $\mu_k(a_1^k) = \text{Prob}(X_1^k = a_1^k)$, $a_1^k \in A^k$, and $M^{(k)}$ is the $|B| \times |B|$ matrix defined by

$$M^{(k)}_{a_1^k, b_1^k} = \begin{cases} \text{Prob}\big(X_{k+1} = b_k | X_1^k = a_1^k\big), & \text{if } b_1^{k-1} = a_2^k \\ 0 & \text{otherwise,} \end{cases}$$

and $B = \{a_1^k\colon\ \mu_k(a_1^k) > 0\}$. Note that probabilities for a k-th order stationary chain can be calculated by thinking of it as a first-order stationary chain with alphabet B.

Example I.1.8 (Finite-state processes.)

Let A and S be finite sets. An A-valued process $\{Y_n\}$ is said to be a *finite-state process with state space S*, if

$$\text{Prob}\big(s_n = s, Y_n = b | s_1^{n-1}, Y_1^{n-1}\big) = \text{Prob}(s_n = s, Y_n = b | s_{n-1}), \tag{6}$$

for all $n > 1$. In other words, $\{Y_n\}$ is a finite-state process if there is a finite-alphabet process $\{s_n\}$ such that the pair process $\{X_n = (s_n, Y_n)\}$ is a Markov chain.

Note that the finite-state process $\{Y_n\}$ is stationary if the chain $\{X_n\}$ is stationary, but $\{Y_n\}$ is not, in general, a Markov chain, see Exercise 2.

The Markov chain $\{X_n\}$ is often referred to as the underlying Markov process or hidden Markov process associated with $\{Y_n\}$. In ergodic theory, any finite-alphabet process is called a finite-state process, and what is here called a finite-state process is called a function (or lumping) of a Markov chain. In information theory, what is here called a finite-state process is sometimes called a Markov source, following the terminology used in [15]. Here "Markov" will be reserved for processes that satisfy the Markov property (4) or its more general form (5), and "finite-state process" will be reserved for processes satisfying property (6).

Example I.1.9 (Stationary coding.)

The concept of finite-state process can be generalized to allow dependence on the past and future, an idea most easily expressed in terms of the two-sided representation of a stationary process. Suppose A and B are finite sets. A Borel measurable mapping $F\colon A^Z \mapsto B^Z$ is a *stationary coder* if

$$F(T_A x) = T_B F(x), \ \forall x \in A^Z,$$

where T_A and T_B denotes the shifts on the respective two-sided sequence spaces A^Z and B^Z. The coder F transports a Borel measure μ on A^Z into the measure $\nu = \mu \circ F^{-1}$, defined for Borel subsets C of B^Z by

$$\nu(C) = \mu(F^{-1}(C)).$$

The encoded process ν is said to be a *stationary coding of μ, with coder F.* Note that a stationary coding of a stationary process is automatically a stationary process.

Define the function $f\colon A^Z \mapsto B$ by the formula $f(x) = F(x)_0$, that is, $f(x)$ is just the 0-th coordinate of $y = F(x)$. The function f is called the *time-zero coder* associated with F. The stationary coder F and its time-zero coder f are connected by the formula

$$F(x)_n = f(T_A^n x), \ x \in A^Z, \ n \in Z,$$

that is, the image $y = F(x)$ is the sequence defined by $y_n = f(T_A^n x)$. Thus the stationary coding idea can be expressed in terms of the sequence-to-sequence coder F, sometimes called the full coder, or in terms of the sequence-to-symbol coder f, otherwise known as the per-symbol or time-zero coder. The word "coder" or "encoder" may refer to either F or f.

A case of particular interest occurs when the time-zero coder f depends on only a finite number of coordinates of the input sequence x, that is, when there is a nonnegative integer w such that

$$x_{-w}^{w} = \tilde{x}_{-w}^{w} \implies f(x) = f(\tilde{x}).$$

In this case it is common practice to write $f(x) = f(x_{-w}^{w})$ and say that f (or its associated full coder F) is a *finite coder* with *window half-width* w. The encoded sequence $y = F(x)$ is defined by the formula

$$y_n = f(x_{n-w}^{n+w}), \ n \in Z,$$

that is, the value y_n depends only on x_n and the values of its w past and future neighbors. In the special case when $w = 0$ the encoding is said to be *instantaneous*, and the encoded process $\{Y_n\}$ defined by $Y_n = f(X_n)$ is called an *instantaneous function* or simply a *function* of the $\{X_n\}$ process.

Finite coding is sometimes called *sliding-block or sliding-window coding*, for one can think of the coding as being done by sliding a window of width $2w + 1$ along x. At time n the window will contain

$$x_{n-w}, x_{n-w+1}, \ldots, x_n, \ldots, x_{n+w}$$

which determines the value y_n, through the function f. To determine y_{n+1}, the window is shifted to the right (that is, the contents of the window are shifted to the left) eliminating x_{n-w} and bringing in x_{n+w+1}.

The key property of finite codes is that the inverse image of a cylinder set of length n is a finite union of cylinder sets of length $n + 2w$, where w is the window half-width of the code. Such a finite coder f defines a mapping, also denoted by f, from A_{m-w}^{n+w} to B_m^n by the formula

$$f(x_{m-w}^{n+w}) = y_m^n, \text{ where } y_i = f(x_{i-w}^{i+w}),\ m \le i \le n.$$

If $F: A^Z \mapsto B^Z$ is the full coder defined by f, then

$$F^{-1}([y_m^n]) = \bigcup_{f(x_{m-w}^{n+w})=y_m^n} [x_{m-w}^{n+w}],$$

so that, in particular, $F^{-1}([y_m^n])$ is measurable with respect to the σ-algebra $\Sigma(X_{m-w}^{n+w})$ generated by the random variables $X_{m-w}, X_{m-w+1}, \ldots, X_{n+w}$. Furthermore, for any measure μ, cylinder set probabilities for the encoded measure $\nu = \mu \circ F^{-1}$ are given by the formula

$$\nu(y_m^n) = \mu(F^{-1}[y_m^n]) = \sum_{f(x_{m-w}^{n+w})=y_m^n} \mu(x_{m-w}^{n+w}). \tag{7}$$

A process $\{Y_n\}$ is a *finite coding* of a process $\{X_n\}$ if it is a stationary coding of $\{X_n\}$ with finite width time-zero coder. Note that a finite-state process is merely a finite coding of a Markov chain in which the window half-width is 0, that is, a function of a Markov chain.

A stationary coding of an i.i.d. process is called a *B-process*. More generally, a finite alphabet process is called a B-process if it is a stationary coding of some i.i.d. process with finite or *infinite* alphabet. Much more will be said about B-processes in later parts of this book, especially in Chapter 4, where several characterizations will be discussed.

Example I.1.10 (Block codings.)

Another type of coding, called block coding, destroys stationarity. It is frequently used in practice, however, and is often easier to analyze than is stationary coding. An *N-block code* is a function $C_N: A^N \mapsto B^N$, where A and B are finite sets. Such a code can be used to map an A-valued process $\{X_n\}$ into a B-valued process $\{Y_n\}$ by applying C_N to consecutive nonoverlapping blocks of length N, that is,

$$Y_{jN+1}^{(j+1)N} = C_N\left(X_{jN+1}^{(j+1)N}\right), \quad j = 0, 1, 2, \ldots$$

The process $\{Y_n\}$ is called the *N-block coding of the process $\{X_n\}$ defined by the N-block code C_N*. The Kolmogorov measures, μ and ν, of the respective processes, $\{X_n\}$ and $\{Y_n\}$, are connected by the formula $\nu = \mu \circ F^{-1}$, where $y = F(x)$ is the (measurable) mapping defined by

$$y_{jN+1}^{(j+1)N} = C_N\left(x_{jN+1}^{(j+1)N}\right), \quad j = 0, 1, 2, \ldots$$

If the process $\{X_n\}$ is stationary, the N-block coding $\{Y_n\}$ is not, in general, stationary, although it is N-stationary, that is, it is invariant under the N-fold shift T^N, where $T = T_B$, the shift on B. There is a simple way to convert an N-stationary process, such as the encoded process $\{Y_n\}$, into a stationary process, $\{\tilde{Y}_n\}$. The process $\{\tilde{Y}_n\}$ is defined by selecting an integer $u \in [1, N]$ according to the uniform distribution and defining $\tilde{Y}_i = Y_{u+i-1}$, $i = 1, 2, \ldots$. This method is called "randomizing the start." The relation between the Kolmogorov measures, ν and $\widetilde{\nu}$, of the respective processes, $\{Y_n\}$ and $\{\widetilde{Y}_n\}$, is expressed by the formula

$$\widetilde{\nu}(C) = \frac{1}{N}\sum_{i=0}^{N-1} \nu(T^{-i}C),$$

that is, $\widetilde{\nu}$ is the average of ν, together with its first $N-1$ shifts. The method of randomizing the start clearly generalizes to convert any N-stationary process into a stationary process.

In summary, an N-block code C_N induces a block coding of a stationary process $\{X_n\}$ onto an N-stationary process $\{Y_n\}$, which can then be converted to a stationary process, $\{\widetilde{Y}_n\}$, by randomizing the start. In general, the final stationary process $\{\widetilde{Y}_n\}$ is not a stationary encoding of the original process $\{X_n\}$ and many useful properties of $\{X_n\}$ may get destroyed. For example, block coding introduces a periodic structure in the encoded process, $\{Y_n\}$, a structure which is inherited, with a random delay, by the process $\{\widetilde{Y}_n\}$. In Section I.8, a general procedure will be developed for converting a block code to a stationary code by inserting spacers between blocks so that the resulting process is a stationary coding of the original process, and as such inherits many of its properties.

Example I.1.11 (Concatenated-block processes.)

Processes can also be constructed by concatenating N-blocks drawn independently at random according a given distribution, with stationarity produced by randomizing the start. There are several equivalent ways to describe such a process.

In random variable terms, suppose $X_1^N = (X_1, \ldots, X_N)$ is a random vector with values in A^N. Let $\{Y_i\}$ be the A-valued process defined by the requirement that the blocks $Y_{(j-1)N+1}^{jN}$ be independent and have the distribution of X_1^N. The process $\{\widetilde{Y}_i\}$ obtained from the N-stationary process $\{Y_i\}$ by randomizing the start is called the *concatenated-block process defined by the random vector* X_1^N. The concatenated-block process $\{\widetilde{Y}_i\}$ is characterized by the following two conditions.

(i) $\{\widetilde{Y}_i\}$ is stationary.

(ii) There is random variable U uniformly distributed on $\{1, 2, \ldots, N\}$ such that, conditioned on $U = u$, the sequence $\{\widetilde{Y}_{u+(j-1)N+1}^{u+jN}\colon j = 1, 2, \ldots\}$ is i.i.d. with distribution μ.

An alternative formulation of the same idea in terms of measures is obtained by starting with a probability measure μ on A^N, forming the product measure μ^* on $(A^N)^\infty$, transporting this to an N-stationary measure $\mu^* \circ \phi^{-1}$ on A^∞ via the mapping

$$\phi(w(1), w(2), \ldots) = x = w(1)w(2)\cdots$$

where

$$x_{(j-1)N+1}^{jN} = w(j), \quad j = 1, 2, \ldots,$$

then averaging to obtain the measure $\widetilde{\mu}$ defined by the formula

$$\widetilde{\mu}(B) = \frac{1}{N}\sum_{i=0}^{N-1} \mu^*(\phi^{-1}(T^{-i}B)), \ B \in \Sigma.$$

The measure $\widetilde{\mu}$ is called the *concatenated-block process defined by the measure* μ *on* A^N. The process defined by $\widetilde{\mu}$ has the properties (i) and (ii), hence this measure terminology is consistent with the random variable terminology.

A third formulation, which is quite useful, represents a concatenated-block process as a finite-state process, that is, a function of a Markov chain. To describe this representation fix a measure μ on A^N and let $\{Y_m\}$ be the (stationary) Markov chain with alphabet $S = A^N \times \{1, 2, \ldots, N\}$, defined by the following transition and starting rules.

(a) If $i < N$, (a_1^N, i) can only go to $(a_1^N, i+1)$.

(b) (a_1^N, N) goes to $(b_1^N, 1)$ with probability $\mu(b_1^N)$, for each $b_1^N \in A^N$.

(c) The process is started by selecting (a_1^N, i) with probability $\mu(a_1^N)/N$.

The process $\{\widetilde{Y}_m\}$ defined by setting $\widetilde{Y}_m = a_j$, if $Y_m = (a_1^N, j)$, is the same as the concatenated-block process based on the block distribution μ. (See Exercise 7.)

Concatenated-block processes will play an important role in many later discussions in this book.

Example I.1.12 (Blocked processes.)

Associated with a stationary process $\{X_n\}$ and a positive integer N are two different stationary processes, $\{Z_n\}$ and $\{W_n\}$, defined, for $n \geq 1$, by

$$Z_n = (X_{(n-1)N+1}, X_{(n-1)N+2}, \ldots, X_{nN});$$

$$W_n = (X_n, X_{n+1}, \ldots, X_{n+N-1}).$$

The Z process is called the *nonoverlapping* N-block process determined by $\{X_n\}$, while the W process is called the *overlapping* N-block process determined by $\{X_n\}$. Each is stationary if $\{X_n\}$ is stationary.

If the process $\{X_n\}$ is i.i.d. then its nonoverlapping blockings are i.i.d., but overlapping clearly destroys independence. The overlapping N-blocking of an i.i.d. or Markov process is Markov, however. In fact, if $\{X_n\}$ is Markov with transition matrix M_{ab}, then the overlapping N-block process is Markov with alphabet $B = \{a_1^N : \mu(a_1^N) > 0\}$ and transition matrix defined by

$$\widehat{M}_{a_1^N, b_1^N} = \begin{cases} M_{a_N b_N}, & \text{if } b_1^{N-1} = a_2^N \\ 0 & \text{otherwise.} \end{cases}$$

Note also that if $\{X_n\}$ is Markov with transition matrix M_{ab}, then the nonoverlapping N-block process is Markov with alphabet B and transition matrix defined by

$$\widetilde{M}_{a_1^N, b_1^N} = M_{a_N b_1} \prod_{i=1}^{N-1} M_{b_i b_{i+1}}$$

A related process of interest $\{Y_n\}$, called the *N-th term process*, selects every N-th term from the $\{X_n\}$ process, that is, $Y_n = X_{(n-1)N+1}$, $n \in Z$. If $\{X_n\}$ is Markov with transition matrix M, then $\{Y_n\}$ will be Markov with transition matrix M^N, the N-th power of M.

I.1.b Probability tools.

Two elementary results from probability theory will be frequently used, the Markov inequality and the Borel-Cantelli principle.

Lemma I.1.13 (The Markov inequality.)

Let f be a nonnegative, integrable function on a probability space (X, Σ, μ). If $\int f\, d\mu \leq \epsilon\delta$ then $f(x) \leq \epsilon$, except for a set of measure at most δ.

Lemma I.1.14 (The Borel-Cantelli principle.)

If $\{C_n\}$ is a sequence of measurable sets in a probability space (X, Σ, μ) such that $\sum \mu(C_n) < \infty$ then for almost every x there is an $N = N(x)$ such that $x \notin C_n$, $n \geq N$.

In general, a property P is said to be measurable if the set of all x for which $P(x)$ is true is a measurable set. If $\{P_n\}$ is a sequence of measurable properties then

(a) $P_n(x)$ holds *eventually almost surely*, if for almost every x there is an $N = N(x)$ such that $P_n(x)$ is true for $n \geq N$.

(b) $P_n(x)$ holds *infinitely often, almost surely*, if for almost every x there is an increasing sequence $\{n_i\}$ of integers, which may depend on x, such that $P_{n_i}(x)$ is true for $i = 1, 2, \ldots$.

For example, the Borel-Cantelli principle is often expressed by saying that if $\sum \mu(C_n) < \infty$ then $x \notin C_n$, eventually almost surely.

Almost-sure convergence is often established using the following generalization of the Borel-Cantelli principle.

Lemma I.1.15 (The iterated Borel-Cantelli principle.)

Suppose $\{G_n\}$ and $\{B_n\}$ are two sequences of measurable sets such that $x \in G_n$, eventually almost surely, and $x \notin B_n \cap G_n$, eventually almost surely. Then $x \notin B_n$, eventually almost surely.

The proof of this is left to the exercises. In many applications, the fact that $x \in B_n \cap G_n$, eventually almost surely, is established by showing that $\sum \mu(B_n \cap G_n) < \infty$, in which case the iterated Borel-Cantelli principle is, indeed, just a generalized Borel-Cantelli principle.

Frequent use will be made of various equivalent forms of almost sure convergence, summarized as follows.

Lemma I.1.16

The following are equivalent for measurable functions on a probability space.

(a) *$f_n \to f$, almost surely.*

(b) *$|f_n(x) - f(x)| < \epsilon$, eventually almost surely, for every $\epsilon > 0$.*

(c) *Given $\epsilon > 0$, there is an N and a set G of measure at least $1 - \epsilon$, such that $|f_n(x) - f(x)| < \epsilon$, $x \in G$, $n \geq N$.*

As several of the preceding examples suggest, a process is often specified as a function of some other process. Probabilities for the new process can then be calculated by using the inverse image to transfer back to the old process. Sometimes it is useful to go in the opposite direction, e. g., first see what is happening on a set of probability 1 in the old process then transfer to the new process. A complication arises, namely, that Borel images of Borel sets may not be Borel sets. For nice spaces, such as product spaces, this is not a serious problem, for such images are always measurable with respect to the completion of the image measure, [5]. This fact is summarized by the following lemma.

Lemma I.1.17 (The Borel mapping lemma.)
Let F be a Borel function from A^∞ into B^∞, let μ be a Borel measure on A^∞, let $\nu = \mu \circ F^{-1}$ and let $\bar{\nu}$ be the completion of ν. If X is a Borel set such that $\mu(X) = 1$, then $F(X)$ is measurable with respect to the completion $\bar{\nu}$ of ν and $\bar{\nu}(F(X)) = 1$.

A similar result holds with either A^∞ or B^∞ replaced, respectively, by A^Z or B^Z.

Use will also be made of two almost trivial facts about the connection between cardinality and probability, namely, that lower bounds on probability give upper bounds on cardinality, and upper bounds on cardinality "almost" imply lower bounds on probability. For ease of later reference these are stated here as the following lemma.

Lemma I.1.18 (Cardinality bounds.)
Let μ be a probability measure on the finite set A, let $B \subset A$, and let α be a positive number.

(a) *If $a \in B \Rightarrow \mu(a) \geq \alpha$, then $|B| \leq 1/\alpha$.*

(b) *For $b \in B$, $\mu(b) \geq \alpha/|B|$, except for a subset of B of measure at most α.*

Deeper results from probability theory, such as the martingale theorem, the central limit theorem, the law of the iterated logarithm, and the renewal theorem, will not play a major role in this book, though they may be used in various examples, and sometimes the martingale theorem is used to simplify an argument.

I.1.c Exercises.

1. Prove Lemma I.1.18.

2. Give an example of a function of a finite-alphabet Markov chain that is not Markov of any order. (Include a proof that your example is not Markov of any order.)

3. If $\mu([a_1^n] \cap [a_{n+m+1}^{n+m+k}]) = \mu([a_1^n])\mu([a_{n+m+1}^{n+m+k}])$, for all n and k, and all a_1^n and a_{n+m+1}^{n+m+k}, then a stationary process is said to be m-dependent. Show that a finite coding of an i.i.d. process is m-dependent for some m. (How is m related to window half-width?)

4. Let $\{U_n\}$ be i.i.d. with each U_n uniformly distributed on $[0, 1]$. Define $X_n = 1$, if $U_n \geq U_{n-1}$, otherwise $X_n = 0$. Show that $\{X_n\}$ is 1-dependent, and is not a finite coding of a finite-alphabet i.i.d. process. (Hint: show that such a coding would have the property that there is a number $c > 0$ such that $\mu(x_1^n) \geq c^n$, if $\mu(x_1^n) \neq 0$. Then show that the probability of n consecutive 0's is $1/(n+1)!$.)

5. Show that the process constructed in the preceding exercise is not Markov of any order.

6. Establish the Kolmogorov representation theorem for countable alphabet processes.

7. Show that the finite-state representation of a concatenated-block process in Example I.1.11 satisfies the two conditions (i) and (ii) of that example.

8. A measure μ on A^n defines a measure $\mu^{(N)}$ on A^N, where $N = Kn+r$, $0 \leq n < r$, by the formula

$$\mu^{(N)}(x_1^N) = \left(\prod_{k=0}^{K-1} \mu(x_{kn+1}^{kn+n})\right) \mu(x_{Kn+1}^{Kn+r}).$$

The measures $\{\mu^{(N)},\ N = 1, 2, \ldots\}$ satisfy the Kolmogorov consistency conditions, hence have a common extension μ^* to A^∞. Show that the concatenated-block process $\widetilde{\mu}$ defined by μ is an average of shifts of μ^*, that is, $\widetilde{\mu}(B) = (1/n)\sum_{i=1}^{n} \mu^*(T^{-i}B)$, for each Borel subset $B \subseteq A^\infty$.

9. Prove the iterated Borel-Cantelli principle, Lemma I.1.15.

Section I.2 The ergodic theory model.

The Kolmogorov measure for a stationary process is preserved by the shift T on the sequence space. This suggests the possibility of using ergodic theory ideas in the study of stationary processes. Ergodic theory is concerned with the orbits $x, Tx, T^2x, \ldots$ of a transformation $T: X \mapsto X$ on some given space X. In many cases of interest there is a natural probability measure preserved by T, relative to which information about orbit structure can be expressed in probability language. Finite measurements on the space X, which correspond to finite partitions of X, then give rise to stationary processes. This model, called the *transformation/partition model for a stationary process,* is the subject of this section and is the basis for much of the remainder of this book.

The Kolmogorov model for a stationary process implicitly contains the transformation/partition model. The shift T on the sequence space, $X = A^\infty$, (or on the space A^Z), is the transformation and the partition is $\mathcal{P} = \{P_a\colon\ a \in A\}$, where for each $a \in A$,

$$P_a = \{x\colon\ x_1 = a\}.$$

The partition $\mathcal{P}$ is called the *Kolmogorov partition* associated with the process.

The sequence of random variables and the joint distributions are expressed in terms of the shift and Kolmogorov partition, as follows. First, associate with the partition $\mathcal{P}$ the random variable $X_{\mathcal{P}}$ defined by $X_{\mathcal{P}}(x) = a$ if a is the label of member of $\mathcal{P}$ to which x belongs, that is, $x \in P_a$. The coordinate functions, $\{X_n\}$, are given by the formula

$$X_n(x) = X_{\mathcal{P}}(T^{n-1}x),\ n \geq 1. \tag{1}$$

The process can therefore be described as follows: Pick a point $x \in X$, that is, an infinite sequence, at random according to the Kolmogorov measure, and, for each n, let $X_n = X_n(x)$ be the label of the member of $\mathcal{P}$ to which $T^{n-1}x$ belongs. Since to say that

$T^{n-1}x \in P_a$ is the same as saying that $x \in T^{-n+1}P_a$, cylinder sets and joint distributions may be expressed by the respective formulas,

$$[a_m^n] = \cap_{i=m}^n T^{-i+1} P_{a_i},$$

and

$$\mu_k(a_1^k) = \mu([a_1^k]) = \mu\left(\cap_{i=1}^k T^{-i+1} P_{a_i}\right). \tag{2}$$

In summary, the coordinate functions, the cylinder sets in the Kolmogorov representation, and the joint distributions can all be expressed directly in terms of the Kolmogorov partition and the shift transformation.

The concept of stationary process is formulated in terms of the abstract concepts of measure-preserving transformation and partition, as follows. Let (X, Σ, μ) be a probability space. A mapping $T: X \mapsto X$ is said to be *measurable* if $T^{-1}B \in \Sigma$, for all $B \in \Sigma$, and *measure preserving* if it is measurable and if

$$\mu(T^{-1}B) = \mu(B), \quad B \in \Sigma,$$

A *partition*

$$\mathcal{P} = \{P_a \colon a \in A\}$$

of X is a finite, disjoint collection of measurable sets, indexed by a finite set A, whose union has measure 1, that is, $X - \cup_a P_a$ is a null set. (In some situations, countable partitions, that is, partitions into countably many sets, are useful.)

Associated with the partition $\mathcal{P} = \{P_a \colon a \in A\}$ is the random variable $X_{\mathcal{P}}$ defined by $X_{\mathcal{P}}(x) = a$ if $x \in P_a$. The random variable $X_{\mathcal{P}}$ and the measure-preserving transformation T together define a process by the formula

$$X_n(x) = X_{\mathcal{P}}(T^{n-1}x), \ n \geq 1. \tag{3}$$

The k-th order distribution μ_k of the process $\{X_n\}$ is given by the formula

$$\mu_k(a_1^k) = \mu\left(\cap_{i=1}^k T^{-i+1} P_{a_i}\right), \tag{4}$$

the direct analogue of the sequence space formula, (2).

The process $\{X_n \colon n \geq 1\}$ defined by (3), or equivalently, by (4), is called the *process defined by the transformation T and partition $\mathcal{P}$*, or, more simply, the *$(T, \mathcal{P})$-process*. The sequence $\{x_n = X_n(x) \colon n \geq 1\}$ defined for a point $x \in X$ by the formula $T^{n-1}x \in P_{x_n}$, is called the *$(T, \mathcal{P})$-name* of x.

The $(T, \mathcal{P})$-process may also be described as follows. Pick $x \in X$ at random according to the μ-distribution and let $X_1(x)$ be the label of the set in $\mathcal{P}$ to which x belongs. Then apply T to x to obtain Tx and let $X_2(x)$ be the label of the set in $\mathcal{P}$ to which Tx belongs. Continuing in this manner, the values,

$$X_1(x), X_2(x), X_3(x), \ldots, X_n(x), \ldots$$

tell to which set of the partition the corresponding member of the random orbit

$$x, Tx, T^2x, \ldots, T^{n-1}x, \ldots$$

belongs. Of course, in the Kolmogorov representation a random point x is a sequence in A^∞ or A^Z, and the $(T, \mathcal{P})$-name of x is the same as x, (or the forward part of x in the two-sided case.)

The $(T, \mathcal{P})$-process concept is, in essence, just an abstract form of the stationary coding concept, in that a partition $\mathcal{P}$ of (X, Σ, μ) gives rise to a measurable function $F: X \mapsto B^\infty$ which carries μ onto the Kolmogorov measure ν of the $(T, \mathcal{P})$-process, such that $F(Tx) = T_A F(x)$, where T_B is the shift on B^∞. The mapping F extends to a stationary coding to B^Z in the case when $X = A^Z$ and T is the shift. Conversely, a stationary coding $F: A^Z \mapsto B^Z$ carrying μ onto ν determines a partition $\mathcal{P} = \{P_b: b \in B\}$ of A^Z such that ν is the Kolmogorov measure of the $(T_A, \mathcal{P})$-process. The partition $\mathcal{P}$ is defined by $P_b = \{x: f(x) = b\}$. See Exercise 2 and Exercise 3.

In summary, a stationary process can be thought of as a shift-invariant measure on a sequence space, or, equivalently, as a measure-preserving transformation T and partition $\mathcal{P}$ of an arbitrary probability space. The ergodic theory point of view starts with the transformation/partition concept, while modern probability theory starts with the sequence space concept.

I.2.a Ergodic processes.

It is natural to study the orbits of a transformation by looking at its action on invariant sets, since once an orbit enters an invariant set it never leaves it. In particular, the natural object of study becomes the restriction of the transformation to sets that cannot be split into nontrivial invariant sets. This leads to the concept of ergodic transformation.

A measurable set B is said to be T-invariant if $TB \subseteq B$. The space X is T-decomposable if it can be expressed as the disjoint union $X = X_1 \cup X_2$ of two measurable invariant sets, each of positive measure. The condition that $TX_i \subseteq X_i$, $i = 1, 2$ translates into the statement that $T^{-1}X_i = X_i$, $i = 1, 2$, hence to say that the space is indecomposable is to say that if $T^{-1}B = B$ then $\mu(B)$ is 0 or 1. It is standard practice to use the word "ergodic" to mean that the space is indecomposable.

A measure-preserving transformation T is said to be *ergodic* if

$$T^{-1}B = B \implies \mu(B) = 0 \text{ or } \mu(B) = 1. \tag{5}$$

The following lemma contains several equivalent formulations of the ergodicity condition. The notation, $C \Delta D = (C - D) \cup (D - C)$, denotes symmetric difference, and U_T denotes the operator on functions defined by $(U_T f)(x) = f(Tx)$, where the domain can be taken to be any one of the L^p-spaces or the space of measurable functions.

Lemma I.2.1 (Ergodicity equivalents.)

The following are equivalent for a measure-preserving transformation T on a probability space.

(a) T is ergodic.

(b) $T^{-1}B \subseteq B, \implies \mu(B) = 0$ or $\mu(B) = 1$.

(c) $T^{-1}B \supseteq B, \implies \mu(B) = 0$ or $\mu(B) = 1$.

(d) $\mu(T^{-1}B \Delta B) = 0 \implies \mu(B) = 0$ or $\mu(B) = 1$.

(e) $U_T f = f$, a.e., implies that f is constant, a.e.

Proof. The equivalence of the first two follows from the fact that if $T^{-1}B \subseteq B$ and $C = \bigcap_{n\geq 0} T^{-n}B$, then $T^{-1}C = C$ and $\mu(C) = \mu(B)$. The proofs of the other equivalences are left to the reader. □

Remark I.2.2

In the particular case when T is invertible, that is, when T is one-to-one and for each measurable set C the set TC is measurable and has the same measure as C, the conditions for ergodicity can be expressed in terms of the action of T, rather than T^{-1}. In particular, an invertible T is ergodic if and only if any T-invariant set has measure 0 or 1. Also, note that if T is invertible then U_T is a unitary operator on L^2.

A stationary process is *ergodic,* if the shift in the Kolmogorov representation is ergodic relative to the Kolmogorov measure. As will be shown in Section I.4, to say that a stationary process is ergodic is equivalent to saying that measures of cylinder sets can be determined by counting limiting relative frequencies along a sample path x, for almost every x. Furthermore, a shift-invariant measure shift-invariant measures. Thus the concept of ergodic process, which is natural from the transformation point of view, is equivalent to an important probability concept.

I.2.b Examples of ergodic processes.

Examples of ergodic processes include i.i.d. processes, irreducible Markov chains and functions thereof, stationary codings of ergodic processes, concatenated-block processes, and some, but not all, processes obtained from a block coding of an ergodic process by randomizing the start. These and other examples will now be discussed.

Example I.2.3 (The Baker's transformation.)

A simple geometric example provides a transformation and partition for which the resulting process is the familiar coin-tossing process. Let $X = [0, 1) \times [0, 1)$ denote the unit square and define a transformation T by

$$T(s,t) = \begin{cases} (2s, t/2) & \text{if } s < 1/2 \\ (2s-1, (t+1)/2) & \text{it } s \geq 1/2. \end{cases}$$

The transformation T is called the Baker's transformation since its action can be described as follows. (See Figure I.2.4.)

1. Cut the unit square into two columns of equal width.
2. Squeeze each column down to height 1/2 and stretch it to width 1.
3. Place the right rectangle on top of the left to obtain a square.

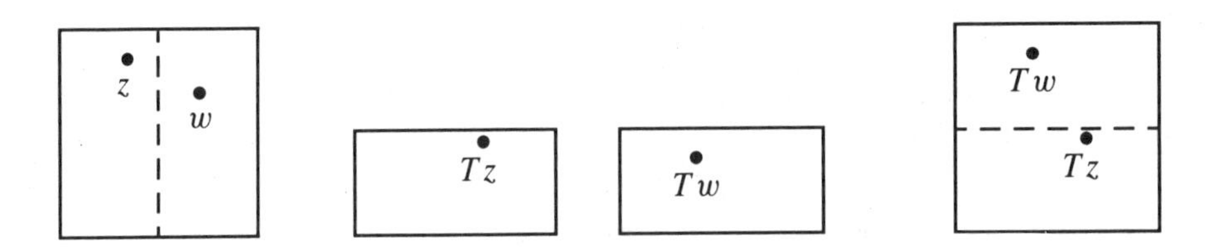

Figure I.2.4 The Baker's transformation.

The Baker's transformation T preserves Lebesgue measure in the square, for dyadic subsquares (which generate the Borel field) are mapped into rectangles of the same area. To obtain the coin-tossing process define the two-set partition $\mathcal{P} = \{P_0, P_1\}$ by setting

$$P_0 = \{(s,t)\colon\ s < 1/2\},\quad P_1 = \{(s,t)\colon\ s \geq 1/2\}. \tag{6}$$

To assist in showing that the $(T, \mathcal{P})$-process is, indeed, the binary, symmetric, i.i.d. process, some useful partition notation and terminology will be developed.

Let $\mathcal{P} = \{P_a\colon\ a \in A\}$ be a partition of the probability space (X, Σ, μ). The *distribution* of $\mathcal{P}$ is the probability distribution $\{\mu(P_a),\ a \in A\}$. The *join*, $\mathcal{P} \vee \mathcal{Q}$, of two partitions $\mathcal{P}$ and $\mathcal{Q}$ is their common refinement, that is, the partition

$$\mathcal{P} \vee \mathcal{Q} = \{P_a \cap Q_b\colon\ P_a \in \mathcal{P},\ Q_b \in \mathcal{Q}\}.$$

The join $\vee_1^k \mathcal{P}^{(i)}$ of a finite sequence $\mathcal{P}^{(1)}, \mathcal{P}^{(2)}, \ldots, \mathcal{P}^{(k)}$ of partitions is their common refinement; defined inductively by $\vee_1^k \mathcal{P}^{(i)} = (\vee_1^{k-1} \mathcal{P}^{(i)}) \vee \mathcal{P}^{(k)}$.

Two partitions $\mathcal{P}$ and $\mathcal{Q}$ are *independent* if the distribution of the restriction of $\mathcal{P}$ to each $Q_b \in \mathcal{Q}$ does not depend on b, that is, if

$$\mu(P_a \cap Q_b) = \mu(P_a)\mu(Q_b), \forall a, b.$$

The sequence of partitions $\{\mathcal{P}^{(i)}\colon\ i \geq 1\}$ is said to be *independent* if $\mathcal{P}^{(k+1)}$ and $\vee_1^k \mathcal{P}^{(i)}$ are independent for each $k \geq 1$.

For the Baker's transformation and partition (6), Figure I.2.5 illustrates the partitions $\mathcal{P}, T\mathcal{P}, T^2\mathcal{P}, T^{-1}\mathcal{P}$, along with the join of these partitions. Note that $T^2\mathcal{P}$ partitions each set of $T^{-1}\mathcal{P} \vee \mathcal{P} \vee T\mathcal{P}$ into exactly the same proportions as it partitions the entire space X. This is precisely the meaning of independence. In particular, it can be shown that the partition $T^n\mathcal{P}$ is independent of $\vee_0^{n-1} T^i\mathcal{P}$, for each $n \geq 1$, so that $\{T^i\mathcal{P}\}$ is an independent sequence. This, together with the fact that each of the two sets P_0, P_1 has measure 1/2, implies, of course, that the $(T, \mathcal{P})$-process is just the coin-tossing process.

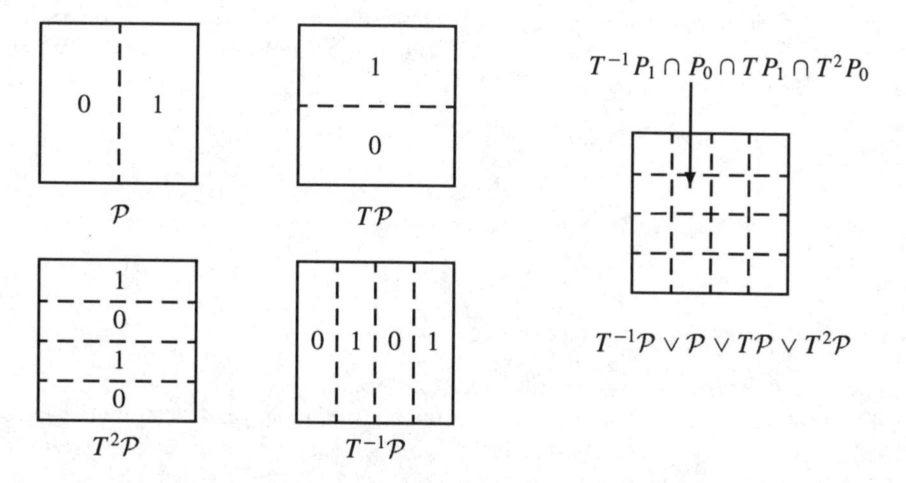

Figure I.2.5 The join of $\mathcal{P}$, $T\mathcal{P}$, $T^2\mathcal{P}$, and $T^{-1}\mathcal{P}$.

In general, the i.i.d. process μ defined by the first-order distribution $\mu_1(a)$, $a \in A$, is given by a generalized Baker's transformation, described as follows. (See Figure I.2.6.)

1. Cut the unit square into $|A|$ columns, labeled by the letters of A, such that the column labeled a has width $\mu_1(a)$.

2. For each $a \in A$ squeeze the column labeled a down and stretch it to obtain a rectangle of height $\mu_1(a)$ and width 1.

3. Stack the rectangles to obtain a square.

The corresponding Baker's transformation T preserves Lebesgue measure. The partition $\mathcal{P}$ into columns defined in part (a) of the definition of T then produces the desired i.i.d. process.

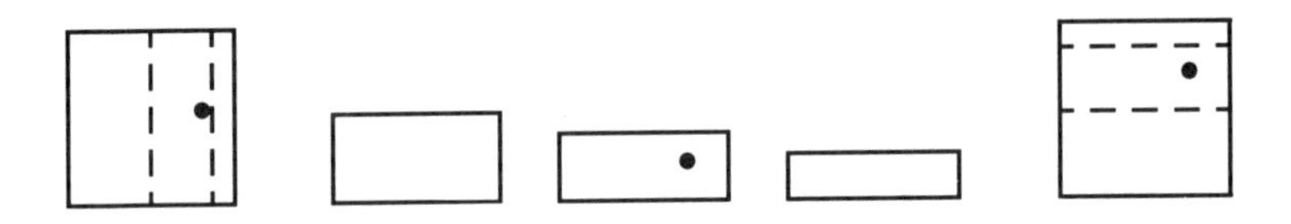

Figure I.2.6 The generalized Baker's transformation.

Example I.2.7 (Ergodicity for i.i.d. processes.)
To show that a process is ergodic it is sometimes easier to verify a stronger property, called mixing. A transformation is *mixing* if

$$\lim_{n\to\infty} \mu(T^{-n}C \cap D) = \mu(C)\mu(D), \quad C, D \in \Sigma. \tag{7}$$

A stationary process is mixing if the shift is mixing for its Kolmogorov measure. Mixing clearly implies ergodicity, for if $T^{-1}C = C$ then $T^{-n}C \cap D = C \cap D$, for all sets D and positive integers n, so that the mixing property, (7), implies that $\mu(C \cap D) = \mu(C)\mu(D)$. Since this holds for all sets D, the choice $D = C$ gives $\mu(C) = \mu(C)^2$ and hence $\mu(C)$ is 0 or 1.

Suppose μ is a product measure on A^∞. To show that the mixing condition (7) holds for the shift first note that it is enough to establish the condition for any two sets in a generating algebra, hence it is enough to show that it holds for any two cylinder sets. But this is easy, for if $C = [a_1^n]$ and $N > 0$ then

$$T^{-N}C = [b_{N+1}^{N+n}], \quad b_{N+i} = a_i, \; 1 \le i \le n.$$

Thus, if $D = [b_1^m]$ and $N > m$ then $T^{-N}C$ and D depend on values of x_i for indices i in disjoint sets of integers. Since the measure is product measure this means that

$$\mu(T^{-N}C \cap D) = \mu(T^{-N}C)\mu(D) = \mu(C)\mu(D).$$

Thus i.i.d. measures satisfy the mixing condition.

Example I.2.8 (Ergodicity for Markov chains.)
The location of the zero entries, if any, in the transition matrix M determine whether a Markov process is ergodic. The results are summarized here; the reader may refer to standard probability books or to the extensive discussion in [29] for details.

The stochastic matrix M is said to be *irreducible*, if for any pair i, j there is a sequence $i_0, i_1, \ldots, i_n$ with $i_0 = i$ and $i_n = j$ such that

$$M_{i_m i_{m+1}} > 0, \quad m = 0, 1, \ldots, n-1.$$

This is just the assertion that for any pair i, j of states, if the chain is at state i at some time then there is a positive probability that it will be in state j at some later time.

If M is irreducible there is a unique probability vector π such that $\pi M = \pi$. Furthermore, each entry of π is positive and

$$\lim_{N\to\infty} \frac{1}{N}\sum_{n=1}^{N} M^n = P, \tag{8}$$

where each row of the $k \times k$ limit matrix P is equal to π. In fact, the limit of the averages of powers can be shown to exist for an arbitrary finite stochastic matrix M. In the irreducible case, there is only one probability vector π such that $\pi M = \pi$ and the limit matrix P has it all its rows equal to π. The condition $\pi M = \pi$ shows that the Markov chain with start distribution $\mu_1 = \pi$ and transition matrix M is stationary.

An irreducible Markov chain is ergodic. To prove this it is enough to show that

$$\lim_{N\to\infty} \frac{1}{N}\sum_{j=1}^{N} \mu(T^{-j}C \cap D) = \mu(C)\mu(D), \tag{9}$$

for any cylinder sets C and D, hence for any measurable sets C and D. This condition, which is weaker than the mixing condition (7), implies ergodicity, for if $T^{-1}C = C$ and $D = C$, then the limit formula (9) gives $\mu(C) = \mu(C)^2$, so that C must have measure 0 or 1.

To establish (9) for an irreducible chain, let $D = [d_1^n]$ and $C = [c_1^m]$ and write $T^{-k-n}C = [b_{n+k+1}^{n+k+m}]$, where $b_{n+k+i} = c_i$, $1 \le i \le m$. Thus when $k > 0$ the sets $T^{-k-n}C$ and D depend on different coordinates. Summing over all the ways to get from d_n to b_{n+k+1} yields the product

$$\mu([d_1^n] \cap [b_{n+k+1}^{n+k+m}]) = UVW,$$

where

$$\begin{aligned} U &= \pi(d_1)\prod_{i=1}^{n-1} M_{d_i d_{i+1}} = \mu([d_1^n]), \\ V &= \sum_{d_{n+1}^{n+k}} \left(\prod_{i=n}^{n+k-1} M_{d_i d_{i+1}}\right) M_{d_{n+k} b_{n+k+1}}, \\ W &= \prod_{i=n+k+1}^{n+k+m-1} M_{b_i b_{i+1}}. \end{aligned}$$

Note that V, which is the probability of transition from d_n to $b_{n+k+1} = c_1$ in k steps, is equal to $M_{d_n c_1}^k = [M^k]_{d_n c_1}$, the $d_n c_1$ term in the k-th power of M, and hence

$$\mu([d_1^n] \cap [b_{n+k+1}^{n+k+m}]) = \mu([d_1^n]) M_{d_n c_1}^k \prod_{i=1}^{m-1} M_{c_i c_{i+1}}. \tag{10}$$

The sequence $\{M_{d_n c_1}^k\}$ converges in the sense of Cesaro to $\pi(c_1)$, by (8), which establishes (9), since $\mu(C) = \pi(c_1)\prod_{i=1}^{m-1} M_{c_i c_{i+1}}$. This proves that irreducible Markov chains are ergodic.

The converse is also true, at least with the additional assumption being used in this book that every state has positive probability. Indeed, if every state has positive probability and if $P_j = \{x\colon x_1 = j\}$, then the set

$$B = T^{-1}P_j \cup T^{-2}P_j \cup \ldots$$

has positive probability and satisfies $T^{-1}B \supseteq B$. Thus if the chain is ergodic then $\mu(B) = 1$, so that, $\mu(B \cap P_i) > 0$, for any state i. But if $\mu(B \cap P_i) > 0$, then $\mu(T^{-n}P_j \cap P_i) > 0$, for some n, which means that transition from i to j in n steps occurs with positive probability, and hence the chain is irreducible. In summary,

Proposition I.2.9
A stationary Markov chain is ergodic if and only if its transition matrix is irreducible.

If some power of M is positive (that is, has all positive entries) then M is certainly irreducible. In this case, the Cesaro limit theorem (8) can be strengthened to $\lim_{N\to\infty} M^N = P$. The argument used to prove (9) then shows that the shift T must be mixing. The converse is also true, again, assuming that all states have positive probability.

Proposition I.2.10
A stationary Markov chain is mixing if and only if some power of its transition matrix has all positive entries.

Remark I.2.11
In some probability books, for example, Feller's book, [11], "ergodic" for Markov chains is equivalent to the condition that some power of the transition matrix be positive. To be consistent with the general concept of ergodic process as used in this book, "ergodic" for Markov chains will mean merely irreducible with "mixing Markov" reserved for the additional property that some power of the transition matrix has all positive entries.

Example I.2.12 (Codings and ergodicity.)
Stationary coding preserves the ergodicity property. This follows easily from the definitions, for suppose $F\colon A^Z \mapsto B^Z$ is a stationary encoder, suppose μ is the Kolmogorov measure of an ergodic A-valued process, and suppose $\nu = \mu \circ F^{-1}$ is the Kolmogorov measure of the encoded B-valued process. If C is a shift-invariant subset of B^Z then $F^{-1}C$ is a shift-invariant subset of A^Z so that $\mu(F^{-1}C)$ is 0 or 1. Since $\nu(C) = \mu(F^{-1}C)$ it follows that $\nu(C)$ is 0 or 1, and hence that ν is the Kolmogorov measure of an ergodic process. It is not important that the domain of F be the sequence space A^Z; any probability space will do. Thus, if T is ergodic the $(T, \mathcal{P})$-process is ergodic for any finite partition $\mathcal{P}$. It should be noted, however, that the $(T, \mathcal{P})$-process can be ergodic even though T is not. (See Exercise 5, below.)

A simple extension of the above argument shows that stationary coding also preserves the mixing property, see Exercise 19.

A finite-state process is a stationary coding with window width 0 of a Markov chain. Thus, if the chain is ergodic then the finite-state process will also be ergodic. Likewise, a finite-state process for which the underlying Markov chain is mixing must itself be mixing.

A concatenated-block process is always ergodic, since it can be represented as a stationary coding of an irreducible Markov chain; see Example I.1.11. The underlying

Markov chain is not generally mixing, however, for it has a periodic structure due to the blocking. Since this periodic structure is inherited, up to a shift, concatenated-block processes are not mixing except in special cases.

A transformation T on a probability space (X, Σ, μ) is said to be *totally ergodic* if every power T^N is ergodic. If T is the shift on a sequence space then T is totally ergodic if and only if for each N, the nonoverlapping N-block process defined by

$$Z_n = (X_{(n-1)N+1}, X_{(n-1)N+2}, \ldots, X_{nN}),$$

is ergodic. (See Example I.1.12.) Since the condition that $F(T_A x) = T_B F(x)$, for all x, implies that $F(T_A^N x) = T_B^N F(x)$, for all x *and all* N, it follows that a stationary coding of a totally ergodic process must be totally ergodic.

As noted in Section I.1, N-block codings destroy stationarity, but a stationary process can be constructed from the encoded process by randomizing the start, see Example I.1.10. The final randomized-start process may not be ergodic, however, even if the original process was ergodic. For example, let μ give measure 1/2 to each of the two sequences $1010\ldots$ and $0101\ldots$, that is, μ is the stationary Markov measure with transition matrix M and start distribution π given, respectively, by

$$M = \begin{bmatrix} 0 & 1 \\ 1 & 0 \end{bmatrix}, \quad \pi = \begin{bmatrix} \frac{1}{2} & \frac{1}{2} \end{bmatrix}.$$

Let ν be the encoding of μ defined by the 2-block code $C(01) = 00$, $C(10) = 11$, so that ν is concentrated on the two sequences $000\ldots$ and $111\ldots$. The measure $\tilde{\nu}$ obtained by randomizing the start is, in this case, the same as ν, hence $\tilde{\nu}$ is not an ergodic process.

A condition insuring ergodicity of the process obtained by N-block coding and randomizing the start is that the original process be ergodic relative to the N-shift T^N. The proof of this is left to the reader. In particular, applying a block code to a totally ergodic process and randomizing the start produces an ergodic process.

Example I.2.13 (Rotation processes.)

Let α be a fixed real number and let T be defined on $X = [0, 1)$ by the formula

$$Tx = x \oplus \alpha,$$

where $\oplus$ indicates addition modulo 1. The mapping T is called *translation* by α. It is also called *rotation*, since X can be thought of as the unit circle by identifying x with the angle $2\pi x$, so that translation by α becomes rotation by $2\pi\alpha$. A subinterval of the circle corresponds to a subinterval or the complement of a subinterval in X, hence the word "interval" can be used for subsets of X that are connected or whose complements are connected. The transformation T is one-to-one and maps intervals onto intervals of the same length, hence preserves Lebesgue measure μ. (The measure-preserving property is often established by proving, as in this case, that it holds on a family of sets that generates the σ-algebra.)

Any partition $\mathcal{P}$ of the interval gives rise to a process; such processes are called *translation processes*, (or *rotation processes* if the circle representation is used.) As an example, let $\mathcal{P}$ consist of the two intervals $P_0 = [0, 1/2)$, $P_1 = [1/2, 1)$, which correspond to the upper and lower halves of the circle in the circle representation. The $(T, \mathcal{P})$-process $\{X_n\}$ is then described as follows. Pick a point x at random in the unit interval according to the uniform (Lebesgue) measure. The value $X_n(x) = X_{\mathcal{P}}(T^{n-1}x)$ is then 0 or 1, depending on whether $T^{n-1}x = x \oplus (n-1)\alpha$ belongs to P_0 or P_1.

The following proposition is basic.

Proposition I.2.14
T is ergodic if and only if α is irrational.

Proof. It is left to the reader to show that T is not ergodic if α is rational. Assume that α is irrational. Two proofs of ergodicity will be given. The first proof is based on the following classical result of Kronecker.

Proposition I.2.15 (Kronecker's theorem.)
If α is irrational the forward orbit $\{T^n x\colon\ n \geq 1\}$ is dense in X, for each x.

Proof. To establish Kronecker's theorem let F be the closure of the forward orbit, $\{T^n x\colon\ n \geq 1\}$, and suppose that F is not dense in X. Then there is an interval I of positive length which is maximal with respect to the property of not meeting F. Furthermore, for each positive integer n, $T^{-n}I$ cannot be the same as I, since α is irrational, and hence the maximality of I implies that $T^{-n}I \cap I = \emptyset$. It follows that $\{T^{-n}I\}$ is a disjoint sequence and

$$1 = \mu([0, 1)) \geq \sum_{n=1}^{\infty} \mu(T^{-n}I) = \infty,$$

which is a contradiction. This proves Kronecker's theorem. □

Proof of Proposition I.2.14 continued.
To proceed with the proof that T is ergodic if α is irrational, suppose A is an invariant set of positive measure. Given $\epsilon > 0$ choose an interval I such that

$$\mu(I) < \epsilon \text{ and } \mu(A \cap I) \geq (1 - \epsilon)\mu(I).$$

Kronecker's theorem, applied to the end points of I, produces integers $n_1 < n_2 < \ldots < n_k$ such that $\{T^{n_i}I\}$ is a disjoint sequence and $\sum_{i=1}^{k} \mu(T^{n_i}I) \geq 1 - 2\epsilon$. The assumption that $\mu(A \cap I) \geq (1 - \epsilon)$ gives

$$\begin{aligned} \mu(A) &\geq \sum_{i=1}^{k} \mu(A \cap T^{n_i}I) \\ &\geq (1 - \epsilon)\sum_{i=1}^{k} \mu(T^{n_i}I) \geq (1 - \epsilon)(1 - 2\epsilon). \end{aligned}$$

Thus $\mu(A) = 1$, which completes the proof of Proposition I.2.14. □

The preceding proof is typical of proofs of ergodicity for many transformations defined on the unit interval. One first shows that orbits are dense, then that small disjoint intervals can be placed around each point in the orbit. Such a technique does not work in all cases, but does establish ergodicity for many transformations of interest. A generalization of this idea will be used in Section I.10 to establish ergodicity of a large class of transformations.

An alternate proof of ergodicity can be obtained using Fourier series. Suppose f is square integrable and has Fourier series $\sum a_n e^{2\pi i n x}$. The Fourier series of $g(x) = f(Tx)$ is

$$\sum a_n e^{2\pi i n(x+\alpha)} = \sum a_n e^{2\pi i n\alpha} e^{2\pi i n x}.$$

If $g(x) = f(x)$, a. e., then $a_n = a_n e^{2\pi i n\alpha}$ holds for each integer n. If α is irrational and $a_n = a_n e^{2\pi i n\alpha}$ then $e^{2\pi i n\alpha} \neq 0$ unless $n = 0$, which means that $a_n = 0$, unless n=0, which, in turn, implies that the function f is constant, a. e. In other words, in the irrational case, the rotation T has no invariant functions and hence is ergodic.

In general, translation of a compact group will be ergodic with respect to Haar measure if orbits are dense. For example, consider the torus $\mathcal{T}$, that is, the product of two intervals, $\mathcal{T} = [0, 1) \times [0, 1)$, with addition mod 1, (or, alternatively, the product of two circles.) The proof of the following is left to the reader.

Proposition I.2.16
The following are equivalent for the mapping $T: (x, y) \mapsto (x \oplus \alpha, y \oplus \beta)$.

(i) $\{T^n(x, y)\}$ *is dense for any pair* (x, y).

(ii) T *is ergodic.*

(iii) α *and* β *are rationally independent.*

I.2.c The return-time picture.

A general return-time concept has been developed in ergodic theory, which, along with a suggestive picture, provides an alternative and often simpler view of two standard probability models, the renewal processes and the regenerative processes.

Let T be a measure-preserving transformation on the probability space (X, Σ, μ) and let B be a measurable subset of X of positive measure. If $x \in B$, let $n(x)$ be the least positive integer n such that $T^n x \in B$, that is, $n(x)$ is the time of first return to B.

Theorem I.2.17 (The Poincare recurrence theorem.)
Return to B *is almost certain, that is,* $n(x) < \infty$, *almost surely.*

Proof. To prove this, define the sets $B_n = \{x \in B: n(x) = n\}$, for $n \geq 1$, and define $B_\infty = \{x \in B: T^n x \notin B, \forall n \geq 1\}$. These sets are clearly disjoint and have union B, and, furthermore, are measurable, for $B_1 = B \cap T^{-1}B$ and

$$B_n = \cap_{j=1}^{n-1} T^{-j}(X - B) \cap B \cap T^{-n}B, \ n \geq 2.$$

Furthermore, if $x \in B_\infty$ then $T^n x \notin B_\infty$ for all $n \geq 1$, from which it follows that the sequence of sets $\{T^{-i}B_\infty\}$ is disjoint. Since they all have the same measure as B_∞ and $\mu(X) = 1$, it follows that $\mu(B_\infty) = 0$. This proves Theorem I.2.17. □

To simplify further discussion it will be assumed in the remainder of this subsection that T is invertible. (Some extensions to the noninvertible case are outlined in the exercises.) For each n, the sets

$$B_n, TB_n, \ldots, T^{n-1}B_n \tag{11}$$

are disjoint and have the same measure. Furthermore, only the first set B_n in this sequence meets B, while applying T to the last set $T^{n-1}B_n$ returns it to B, that is, $T^n B_n \subseteq B$.

The return-time picture is obtained by representing the sets (11) as intervals, one above the other. Furthermore, by reassembling within each level it can be assumed that T just moves points directly upwards one level. (See Figure I.2.18). Points in the top level of each column are moved by T to the base B in some manner unspecified by the picture.

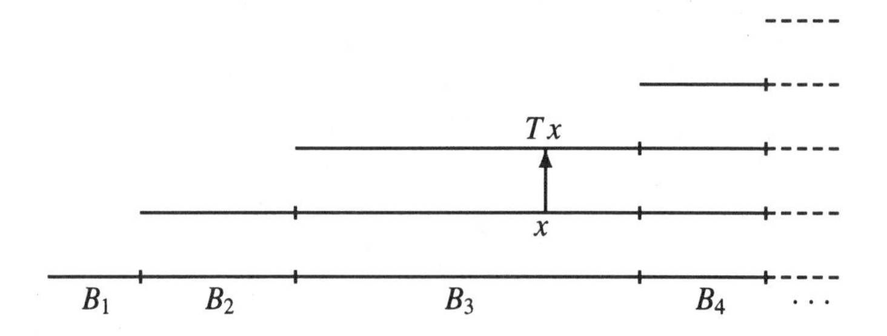

Figure I.2.18 Return-time picture.

Note that the set $\cup_{i=1}^{n} T^{i-1} B_n$ is just the union of the column whose base is B_n, so that the picture is a representation of the set

$$\cup_{j=0}^{\infty} T^j B = \cup_{n=1}^{\infty} \cup_{i=1}^{n} T^{i-1} B_n. \tag{12}$$

This set is T-invariant and has positive measure. If T is ergodic then it must have measure 1, in which case, the picture represents the whole space, modulo a null set, of course. The picture suggests the following terminology. The ordered set

$$B_n, TB_n, \ldots, T^{n-1}B_n$$

is called the *column* $\mathcal{C} = \mathcal{C}_n$ *with base* B_n, *width* $w(\mathcal{C}) = \mu(B_n)$, *height* $h(\mathcal{C}) = n$, *levels* $L_i = T^{i-1}B_n$, $1 \leq i \leq n$, *and top* $T^{n-1}B_n$. Note that the measure of column $\mathcal{C}$ is just its width times its height, that is, $\mu(\mathcal{C}) = h(\mathcal{C})w(\mathcal{C})$.

Various quantities of interest can be easily expressed in terms of the return-time picture. For example, the return-time distribution is given by

$$\text{Prob}\left(n(x) = n \mid x \in B\right) = \frac{w(\mathcal{C}_n)}{\sum_m w(\mathcal{C}_m)},$$

and the expected return time is given by

$$\begin{aligned} E(n(x)|x \in B) &= \sum_n h(\mathcal{C}_n)\text{Prob}\left(n(x) = n \mid x \in B\right) \\ &= \frac{\sum_n h(\mathcal{C}_n)w(\mathcal{C}_n)}{\sum_m w(\mathcal{C}_m)}. \end{aligned} \tag{13}$$

In the ergodic case the latter takes the form

$$E(n(x)|x \in B) = \frac{1}{\sum_m w(\mathcal{C}_m)} = \frac{1}{\mu(B)}, \tag{14}$$

since $\sum_n h(\mathcal{C}_n)w(\mathcal{C}_n)$ is the measure of the set (12), which equals 1, when T is ergodic. This formula is due to Kac, [23].

The transformation $\widehat{T} = T_B$ defined on B by the formula $\widehat{T}x = T^{n(x)}x$ is called the transformation *induced by* T *on the subset* B. The basic theorem about induced transformations is due to Kakutani.

Theorem I.2.19 (The induced-transformation theorem.)
If $\mu(B) > 0$ the induced transformation preserves the conditional measure $\mu(\cdot|B)$ and is ergodic if T is ergodic.

Proof. If $C \subseteq B$, then $x \in \widehat{T}^{-1}C$ if and only if there is an $n \geq 1$ such that $x \in B_n$ and $T^n x \in C$. This translates into the equation

$$\widehat{T}^{-1}C = \bigcup_{n=1}^{\infty}(B_n \cap T^{-n}C),$$

which shows that $\widehat{T}^{-1}C$ is measurable for any measurable $C \subseteq B$, and also that

$$\mu(\widehat{T}^{-1}C) = \sum_{n=1}^{\infty} \mu(B_n \cap T^{-n}C),$$

since $B_n \cap B_m = \emptyset$, if $m \neq n$, with $m, n \geq 1$. More is true, namely,

$$T^n B_n \cap T^m B_m = \emptyset,\ m, n \geq 1,\ m \neq n, \tag{15}$$

by the definition of "first return" and the assumption that T is invertible. But this implies

$$\sum_{n=1}^{\infty} \mu(T^n B_n) = \sum_{n=1}^{\infty} \mu(B_n) = \mu(B),$$

which, together with (15) yields $\sum \mu(B_n \cap T^{-n}C) = \sum \mu(T^n B_n \cap C) = \mu(C)$. Thus the induced transformation $\widehat{T}$ preserves the conditional measure $\mu(\cdot|B)$.

The induced transformation is ergodic if the original transformation is ergodic. The picture in Figure I.2.18 shows why this is so. If $\widehat{T}^{-1}C = C$, then each $C \cap B_n$ can be pushed upwards along its column to obtain the set

$$D = \bigcup_{n=1}^{\infty}\bigcup_{i=0}^{n-1} T^i(C \cap B_n).$$

But $\mu(TD \Delta D) = 0$ and T is invertible, so the measure of D must be 0 or 1, which, in turn, implies that $\mu(C|B)$ is 0 or 1, since $\mu((D \cap B)\Delta C) = 0$.

This completes the proof of the induced-transformation theorem. □

I.2.c.1 Processes associated with the return-time picture.

Several processes of interest are associated with the induced transformation and the return-time picture. It will be assumed throughout this discussion that T is an invertible, ergodic transformation on the probability space (X, Σ, μ), partitioned into a finite partition $\mathcal{P} = \{P_a\colon a \in A\}$; that B is a set of positive measure; that $n(x) = \min\{n\colon T^n x \in B\}$, $x \in B$, is the return-time function; and that $\widehat{T}$ is the transformation induced on B by T.

Two simple processes are connected to returns to B. The first of these is $\{R_n\}$, the $(T, \mathcal{B})$-process defined by the partition $\mathcal{B} = \{B, X - B\}$, with B labeled by 1 and $X - B$ labeled by 0, in other words, the binary process defined by

$$R_n(x) = \chi_B(T^{n-1}x), \quad n \geq 1,$$

where χ_B denotes the indicator function of B. The $(T, \mathcal{B})$-process is called the *generalized renewal process* defined by T and B. This terminology comes from the classical definition of a (stationary) renewal process as a binary process in which the times between occurrences of 1's are independent and identically distributed, with a finite expectation, the only difference being that now these times are not required to be i.i.d., but only stationary with finite expected return-time.

The process $\{R_n\}$ is a stationary coding of the $(T, \mathcal{P})$-process with time-zero encoder χ_B, hence it is ergodic. Any ergodic, binary process which is not the all 0 process is, in fact, the generalized renewal process defined by some transformation T and set B; see Exercise 20.

The second process connected to returns to B is $\{\widehat{R}_j\}$, the $(\widehat{T}, \widehat{\mathcal{B}})$-process defined by the (countable) partition

$$\widehat{\mathcal{B}} = \{B_1, B_2, \ldots\},$$

of B, where B_n is the set of points in B whose first return to B occurs at time n. The process $\{\widehat{R}_j\}$ is called the *return-time process* defined by T and B. It takes its values in the positive integers, and has finite expectation given by (14). Also, it is ergodic, since $\widehat{T}$ is ergodic. Later, it will be shown that any ergodic positive-integer valued process with finite expectation is the return-time process for some transformation and partition; see Theorem I.2.24.

The generalized renewal process $\{R_n\}$ and the return-time process $\{\widehat{R}_j\}$ are connected, for conditioned on starting in B, the times between successive occurrences of 1's in $\{R_n\}$ are distributed according to the return-time process. In other words, if $R_1 = 1$ and the sequence $\{S_j\}$ of successive returns to 1 is defined inductively by setting $S_0 = 1$ and

$$S_j = \min\{n > S_{j-1}\colon\ R_n = 1\}, \quad j \geq 1,$$

then the sequence of random variables defined by

$$\widehat{R}_j = S_j - S_{j-1}, \quad j \geq 1 \tag{16}$$

has the distribution of the return-time process. Thus, the return-time process is a function of the generalized renewal process.

Except for a random shift, the generalized renewal process $\{R_n\}$ is a function of the return-time process, for given a random sample path $\widehat{R} = \{\widehat{R}_m\}$ for the return-time process, a random sample path $R = \{R_n\}$ for the generalized renewal process can be expressed as a concatenation

$$R = \tilde{w}(1)w(2)w(3)\cdots \tag{17}$$

of blocks, where, for $m > 1$, $w(m) = 0^{\widehat{R}_m - 1}1$, that is, a block of 0's of length $\widehat{R}_m$-1, followed by a 1. The initial block $\tilde{w}(1)$ is a tail of the block $w(1) = 0^{\widehat{R}_1 - 1}1$. The only real problem is how to choose the start position in $w(1)$ in such a way that the process is stationary, in other words how to determine the waiting time τ until the first 1 occurs.

The start position problem is solved by the return-time picture and the definition of the $(T, \mathcal{P})$-process. The $(T, \mathcal{P})$-process is defined by selecting $x \in X$ at random according to the μ-measure, then setting $X_n(x) = a$ if and only if $T^{n-1}x \in P_a$. In the ergodic case, the return-time picture is a representation of X, hence selecting x at random is the same as selecting a random point in the return-time picture. But this, in turn, is the same as selecting a column C at random according to column measure $\mu(C)$, then selecting j at random according to the uniform distribution on $\{1, 2, \ldots, h(\mathcal{C})\}$, then selecting x at random in the j-th level of $\mathcal{C}$.

In summary,

Theorem I.2.20

The generalized renewal process $\{R_n\}$ and the return-time process $\{\widehat{R}_m\}$ are connected via the successive-returns formula, (16), and the concatenation representation, (17) with initial block $\tilde{w}(1) = 0^{\tau-1}1$, where $\tau = \widehat{R}_1 - j + 1$ and j is uniformly distributed on $[1, \widehat{R}_1]$.

The distribution of τ can be found by noting that

$$\tau(x) = i \iff x \in \cup_{n=i}^{\infty} T^{n-i} B_n,$$

from which it follows that

$$\text{Prob}\,(\tau = i) = \sum_{n=i}^{\infty} \mu(T^{n-i} B_n) = \sum_{n \geq i} w(\mathcal{C}_n).$$

The latter sum, however, is equal to $\text{Prob}(x \in B \text{ and } n(x) \geq i)$, which gives the alternative form

$$\text{Prob}\,(\tau = i) = \frac{1}{E(n(x)|x \in B)} \text{Prob}\,(n(x) \geq i | x \in B)\,, \tag{18}$$

since $E(n(x)|x \in B) = 1/\mu(B)$ holds for ergodic processes, by (13). The formula (18) was derived in [11, Section XIII.5] for renewal processes by using generating functions. As the preceding argument shows, it is a very general and quite simple result about ergodic binary processes with finite expected return time.

The return-time process keeps track of returns to the base, but may lose information about what is happening outside B. Another process, called the induced process, carries along such information. Let A^* be the set all finite sequences drawn from A and define the countable partition $\widehat{\mathcal{P}} = \{P_w \colon w \in A^*\}$ of B, where, for $w = a_1^k$,

$$P_w = P_{a_1^k} = \{x \in B_k \colon T^{i-1}x \in P_{a_i}, 1 \leq i \leq k\}.$$

The $(\widehat{T}, \widehat{\mathcal{P}})$-process is called the *process induced by the $(T, \mathcal{P})$-process on the set B.*

The relation between the (T, P)-process $\{X_n\}$ and its induced $(\widehat{T}, \widehat{\mathcal{P}})$-process $\{\widehat{X}_m\}$ parallels the relationship (17) between the generalized renewal process and the return-time process. Thus, given a random sample path $\widehat{X} = \{\widehat{W}_m\}$ for the induced process, a random sample path $x = \{X_n\}$ for the (T, P)-process can be expressed as a concatenation

$$x = \tilde{w}(1)w(2)w(3)\cdots, \tag{19}$$

into blocks, where, for $m > 1$, $w(m) = \widehat{X}_m$, and the initial block $\tilde{w}(1)$ is a tail of the block $w(1) = \widehat{X}_1$. Again, the only real problem is how to choose the start position in $w(1)$ in such a way that the process is stationary, in other words how to determine the length of the tail $\tilde{w}(1)$. A generalization of the return-time picture provides the answer.

The generalized version of the return-time picture is obtained by further partitioning the columns of the return-time picture, Figure I.2.18, into subcolumns according to the conditional distribution of $(T, \mathcal{P})$-names. Thus the column with base B_k is partitioned into subcolumns $\mathcal{C}_w$ labeled by k-length names w, so that the subcolumn corresponding to $w = a_1^k$ has width

$$\mu(\cap_{i=1}^k T^{-i+1} P_{a_i} \cap B_n),$$

the ith-level being labeled by a_i. Furthermore, by reassembling within each level of a subcolumn, it can again be assumed that T moves points directly upwards one level. Points in the top level of each subcolumn are moved by T to the base B in some manner unspecified by the picture. For example, in Figure I.2.21, the fact that the third subcolumn over B_3, which has levels labeled $1, 1, 0$, is twice the width of the second subcolumn, indicates that given that a return occurs in three steps, it is twice as likely to visit $1, 1, 0$ as it is to visit $1, 0, 0$ in its next three steps.

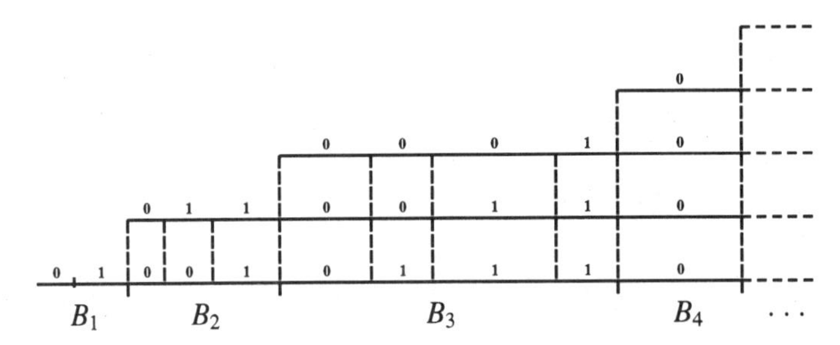

Figure I.2.21 The generalized return-time picture.

The start position problem is solved by the generalized return-time picture and the definition of the $(T, \mathcal{P})$-process. The $(T, \mathcal{P})$-process is defined by selecting $x \in X$ at random according to the μ-measure, then setting $X_n(x) = a$ if and only if $T^{n-1}x \in P_a$. In the ergodic case, the generalized return-time picture is a representation of X, hence selecting x at random is the same as selecting a random point in the return-time picture. But this, in turn, is the same as selecting a column $\mathcal{C}_w$ at random according to column measure $\mu(\mathcal{C}_w)$, then selecting j at random according to the uniform distribution on $\{1, 2, \ldots, h(w)\}$, where $h(w) = h(\mathcal{C}_w)$, then selecting x at random in the j-th level of $\mathcal{C}_w$. This proves the following theorem.

Theorem I.2.22

The process $\{X_n\}$ has the stationary concatenation representation (19) in terms of the induced process $\{\widehat{X}_m\}$, if and only if the initial block is $\tilde{w}(1) = w(1)_j^{h(w(1))}$, where j is uniformly distributed on $[1, h(w(1))]$.

A special case of the induced process occurs when B is one of the sets of $\mathcal{P}$, say $B = B_b$. In this case, the induced process outputs the blocks that occur between successive occurrences of the symbol b. In general case, however, knowledge of where the blocking occurs may require knowledge of the entire past and future, or may even be fully hidden.

I.2.c.2 The tower construction.

The return-time construction has an inverse construction, called the tower construction, described as follows. Let T be a measure-preserving mapping on the probability space (X, Σ, μ), and let f be a measurable mapping from X into the positive integers $\mathcal{N} = \{1, 2, \ldots\}$, of finite expected value. Let $\widetilde{X}$ be the subset of the product space $X \times \mathcal{N}$ defined by

$$\widetilde{X} = \{(x, i): 1 \leq i \leq f(x)\}.$$

A useful picture is obtained by defining $B_n = \{x: f(x) = n\}$, and for each $n \geq 1$, stacking the sets

$$B_n \times \{1\}, \; B_n \times \{2\}, \; \ldots, \; B_n \times \{n\} \tag{20}$$

as a column in order of increasing i, as shown in Figure I.2.23.

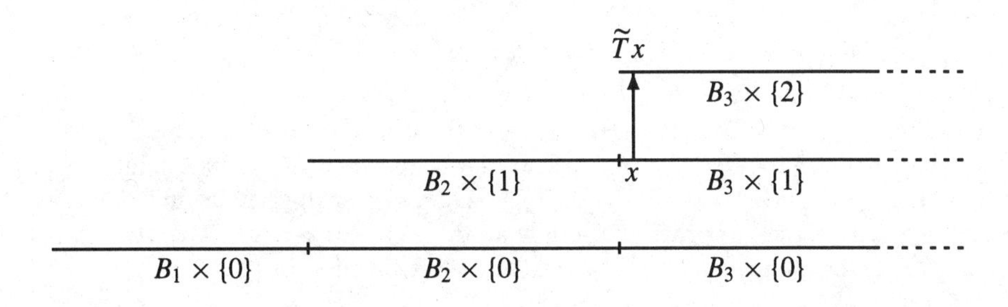

Figure I.2.23 The tower transformation.

The measure μ is extended to a measure $\widetilde{\mu}$ on $\widetilde{X}$, by thinking of each set in (20) as a copy of B_n with measure $\mu(B_n)$, then normalizing to get a probability measure, the normalizing factor being

$$\sum_{n=1}^{\infty} n\mu(B_n) = E(f),$$

the expected value of f. In other words, $\widetilde{B} \subseteq \widetilde{X}$ is defined to be measurable if and only if

$$\{x: (x, i) \in \widetilde{B}\} \cap B_n \in \Sigma, \quad n \geq 1, \; i \leq n,$$

with its measure given by

$$\widetilde{\mu}(\widetilde{B}) = \frac{1}{E(f)} \sum_{n=1}^{\infty} \sum_{i=1}^{n} \mu\Big(\{x: (x, i) \in \widetilde{B}\} \cap B_n\}\Big).$$

A transformation $\widetilde{T}$ is obtained by mapping points upwards, with points in the top of each column sent back to the base according to T. The formal definition is

$$\widetilde{T}(x, i) = \begin{cases} (x, i+1) & i < f(x) \\ (Tx, 0) & i = f(x). \end{cases}$$

The transformation $\widetilde{T}$ is easily shown to preserve the measure $\widetilde{\mu}$ and to be invertible (and ergodic) if T is invertible (and ergodic). The transformation $\widetilde{T}$ is called the *tower extension of T by the (height) function f.*

The tower extension $\widetilde{T}$ induces the original transformation T on the base $B = X \times \{0\}$. This and other ways in which inducing and tower extensions are inverse operations are explored in the exercises.

As an application of the tower construction it will be shown that return-time processes are really just stationary positive-integer valued processes with finite expected value.

Theorem I.2.24

An ergodic, positive-integer valued process $\{\widehat{R}_j\}$ with $E(\widehat{R}_1) < \infty$ is the return-time process for some transformation and set. In particular, $\{\widehat{R}_j\}$ defines a stationary, binary process via the concatenation-representation (17), with the distribution of the waiting time τ until the first 1 occurs given by (18).

Proof. Let μ be the two-sided Kolmogorov measure on the space $\mathcal{N}^Z$ of doubly-infinite positive-integer valued sequences, and note that the shift T on $\mathcal{N}^Z$ is invertible and ergodic and the function $f(x) = x_0$ has finite expectation. Let $\widetilde{T}$ be the tower extension of T by the function f and let $\widetilde{B} = X \times \{1\}$. A point $(x, 1)$ returns to $\widetilde{B} = X \times \{1\}$ at the point $(Tx, 1)$ at time $f(x)$, that is, the return-time process defined by $\widetilde{T}$ and $\widetilde{B}$ has the same distribution as $\{\widehat{R}_j\}$. This proves the theorem. □

A stationary process $\{X_n\}$ is *regenerative* if there is a renewal process $\{R_n\}$ such that the future joint process $\{(X_n, R_n): n > t\}$ is independent of the past joint process $\{(X_m, R_m): n \leq t\}$, given that $R_t = 1$. The induced transformation and the tower construction can be used to prove the following theorem, see Exercise 24. (Recall that if $Y_n = f(X_n)$, for all n, then $\{Y_n\}$ is an instantaneous function of $\{X_n\}$.)

Theorem I.2.25

An ergodic process is regenerative if and only if it is an instantaneous function of an induced process whose return-time process is i.i.d.

I.2.d Exercises.

1. Complete the proof of Lemma I.2.1.

2. Let T be an invertible measure-preserving transformation and $\mathcal{P} = \{P_a: a \in A\}$ a finite partition of the probability space (X, Σ, μ). Show that there is a measurable mapping $F: X \mapsto A^Z$ which carries μ onto the (two-sided) Kolmogorov measure ν of the $(T, \mathcal{P})$-process, such that $F(Tx) = T_A F(x)$, where T_A is the shift on A^Z. (Thus, a partition gives rise to a stationary coding.)

3. Suppose $F: A^Z \mapsto B^Z$ is a stationary coding which carries μ onto ν. Determine a partition $\mathcal{P}$ of A^Z such that ν is the Kolmogorov measure of the $(T_A, \mathcal{P})$-process. (Thus, a stationary coding gives rise to a partition.)

4. Let T be a measure-preserving transformation. A nonnegative measurable function f is said to be T-*subinvariant*, if $f(Tx) \leq f(x)$. Show that a T-subinvariant function is almost-surely invariant. (Hint: consider $g(x) = \min\{f(x), N\}$.)

5. Show that the $(T, \mathcal{P})$-process can be ergodic even if T is not. (Hint: start with a suitable nonergodic Markov chain and lump states to produce an ergodic chain, then use Exercise 2.)

6. Show that the $(T, \mathcal{P})$-process can be mixing even if T is only ergodic.

7. Let $Z_n = (X_{(n-1)N+1}, X_{(n-1)N+2}, \ldots, X_{nN})$ be the nonoverlapping N-block process, let $W_n = (X_n, X_{n+1}, \ldots, X_{n+N-1})$ be the overlapping N-block process, and let $Y_n = X_{(n-1)N+1}$ be the N-th term process defined by the $(T, \mathcal{P})$-process $\{X_n\}$.

   (a) Show that $\{Z_n\}$ is the $(T^N, \vee_1^N T^{i-1}\mathcal{P})$-process.

(b) Show that $\{W_n\}$ is the $(T, \vee_1^N T^{i-1}\mathcal{P})$-process.

(c) Show that $\{Y_n\}$ is the $(T^N, \mathcal{P})$-process.

8. Let T be the shift on $X = \{-1, 1\}^Z$, and let μ be the product measure on X defined by $\mu(-1) = \mu(1) = 1/2$. Let $Y = X \times X$, with the product measure $\nu = \mu \times \mu$, and define $S(x, y) = (Tx, T^{x_0}y)$.

   (a) Show that S preserves ν.

   (b) Let $\mathcal{P}$ be the time-zero partition of X and let $\mathcal{Q} = \mathcal{P} \times \mathcal{P}$. Give a formula expressing the $(S, \mathcal{Q})$-name of (x, y) in terms of the coordinates of x and y. (The $(S, \mathcal{Q})$-process is called "random walk with random scenery.")

9. Generalize the preceding exercise by showing that if T is a measure-preserving transformation on a probability space (X, μ) and $\{T_x\colon x \in X\}$ is a family of measure-preserving transformations on a probability space (Y, ν), such that the mapping $(x, y) \mapsto (Tx, T_x y)$ is measurable, then $S(x, y) = (Tx, T_x y)$ is a measure-preserving mapping on the product space $(X \times Y, \mu \times \nu)$. S is called a skew product. If $T_x = R$, for every x, then S is called the direct product of T and R.

10. Show that a concatenated-block process is ergodic.

11. Show that the overlapping n-blocking of an ergodic process is ergodic.

12. Show that the nonoverlapping n-blocking of an ergodic process may fail to be ergodic.

13. Insertion of random spacers between blocks can be used to change a concatenated-block process into a mixing process. Fix a measure μ on A^N, a symbol $a \in A^N$, and number $p \in (0, 1)$. Let $\{Z_m\}$ be the stationary Markov chain with alphabet $S = A^N \times \{0, 1, 2, \ldots, N\}$, defined by the following transition rules.

   (i) If $i < N$, (a_1^N, i) can only go to $(a_1^N, i+1)$.

   (ii) (a_1^N, N) goes to $(b_1^N, 0)$ with probability $p\mu(b_1^N)$, for each $b_1^N \in A^N$.

   (iii) (a_1^N, N) goes to $(b_1^N, 1)$ with probability $(1-p)\mu(b_1^N)$, for each $b_1^N \in A^N$.

   (a) Show that there is indeed a unique stationary Markov chain with these properties. (Hint: the transition matrix is irreducible.)

   (b) Show that the process $\{\widetilde{Y}_m\}$ defined by setting $\widetilde{Y}_m = a$ if $Y_m = (a_1^N, 0)$ and $Y_m = a_j$, if $Y_m = (a_1^N, j)$, $j > 0$, is a mixing finite-state process.

14. Show that if T is mixing then it is totally ergodic.

15. Show that an ergodic rotation is not mixing, but is totally ergodic.

16. Show that if T is mixing then so is the direct product $T \times T\colon X \times X \mapsto X \times X$, where $X \times X$ is given the measure $\mu \times \mu$.

17. Show directly, that is, without using Proposition 6, that if α is irrational then the direct product $(x, y) \mapsto (x+\alpha, y+\alpha)$ is not ergodic.

18. Prove that even if T is not invertible, the induced transformation is measurable, preserves the conditional measure, and is ergodic if T is ergodic. (Hint: obtain a picture like Figure I.2.18 for T^{-1}, then use this to guide a proof.)

19. Prove that a stationary coding of a mixing process is mixing. (Hint: use the formula $F^{-1}(C \cap T_B^{-i} D) = F^{-1}C \cap T_A^{-i} F^{-1} D$.)

20. Show that if $\{X_n\}$ is a binary, ergodic process which is not identically 0, then it is the generalized renewal process for some transformation T and set B. (Hint: let T be the shift in the two-sided Kolmogorov representation and take $B = \{x\colon x_0 = 1\}$.)

21. Prove: a tower $\widetilde{T}$ over T induces on the base a transformation isomorphic to T.

22. Show that the tower $\widetilde{T}$ defined by the induced transformation $\widehat{T}$ and return-time function $n(x)$ is isomorphic to T.

23. Let T be an invertible measure-preserving transformation and $\mathcal{P} = \{P_a\colon a \in A\}$ a finite partition of the probability space (X, Σ, μ). Let $\widetilde{X}$ be the tower over X defined by f, and let $S = \widetilde{T}$ be the tower transformation. Show how to extend $\mathcal{P}$ to a partition $\mathcal{Q}$ of $\widetilde{X}$, such that the $(T, \mathcal{P})$-process is an instantaneous coding of the induced $(\widehat{S}, \widehat{\mathcal{Q}})$-process.

24. Prove Theorem I.2.25.

25. Define $\mathcal{P} = \{P_a\colon a \in A\}$ and $\mathcal{Q} = \{Q_b\colon b \in B\}$ to be ϵ-independent if $\sum_{a,b} |\mu(P_a \cap Q_b) - \mu(P_a)\mu(Q_b)| \le \epsilon$.

    (a) Show that $\mathcal{P}$ and $\mathcal{Q}$ are independent if and only if they are ϵ-independent for each $\epsilon > 0$.

    (b) Show that if $\mathcal{P}$ and $\mathcal{Q}$ are ϵ-independent then $\sum_a |\mu(P_a|Q_b) - \mu(P_a)| \le \sqrt{\epsilon}$, except for a set of Q_b's of total measure at most $\sqrt{\epsilon}$.

    (c) Show that if $\sum_a |\mu(P_a|Q_b) - \mu(P_a)| \le \epsilon$, except for a set of Q_b's of total measure at most ϵ, then $\mathcal{P}$ and $\mathcal{Q}$ are 3ϵ-independent.

Section I.3 The ergodic theorem.

The ergodic theorem extends the strong law of large numbers from i.i.d. processes to the general class of stationary processes.

Theorem I.3.1 (The ergodic theorem.)
If T is a measure-preserving transformation on a probability space (X, Σ, μ) and if f is integrable then the average $(1/n)\sum_{i=1}^{n} f(T^{i-1}x)$ converges almost surely and in L^1-norm to a T-invariant function $f^*(x)$.

The ergodic theorem in the almost-sure form presented here is due to G. D. Birkhoff and is often called *Birkhoff's ergodic theorem* or the *individual ergodic theorem.*

The L^1-convergence implies that $\int f^*\, d\mu = \int f\, d\mu$, since

$$\int \frac{1}{n}\sum_{i=1}^{n} f(T^{i-1}x)\, d\mu = \frac{1}{n}\sum_{i=1}^{n}\int f(T^{i-1}x)\, d\mu = \int f\, d\mu,$$

by the measure-preserving property of T. Thus if T is ergodic then the limit function f^*, since it is T-invariant, must be almost-surely equal to the constant value $\int f\, d\mu$. In particular, if f is taken to be 0,1-valued and T to be ergodic the following form of the ergodic theorem is obtained.

Theorem I.3.2 (Ergodic theorem: binary ergodic form.)
If $\{X_n\}$ is a binary ergodic process then the average $(X_1+X_2+\ldots+X_n)/n$ converges almost surely to the constant value $E(X_1) = \mu(1)$.

This binary version of the ergodic theorem is sufficient for many results in this book, but the more general version will also be needed and is quite useful in many situations not treated in this book.

The proof of the ergodic theorem will be based on a rather simple combinatorial result discussed in the next subsection. The combinatorial idea is not a merely step on the way to the ergodic theorem, however, for it is an important tool in its own right and will be used frequently in later parts of this book.

I.3.a Packings from coverings.

A general technique for extracting "almost packings" of integer intervals from certain kinds of "coverings" of the natural numbers will be described in this subsection.

In this discussion intervals are subsets of the natural numbers $\mathcal{N} = \{1, 2, \ldots\}$, of the form $[n, m] = \{j \in \mathcal{N}\colon n \leq j \leq m\}$. A *strong cover* $\mathcal{C}$ of $\mathcal{N}$ is defined by an integer-valued function $n \mapsto m(n)$ for which $m(n) \geq n$, and consists of all intervals of the form $[n, m(n)]$, $n \in \mathcal{N}$. (The word "strong" is used to indicate that every natural number is required to be the left endpoint of a member of the cover.)

A strong cover $\mathcal{C}$ has a packing property, namely, there is a subcover $\mathcal{C}'$ whose members are *disjoint*. This is a trivial observation; just set $\mathcal{C}' = \{[n_i, m(n_i)]\}$, where $n_1 = 1$ and $n_{i+1} = 1 + m(n_i)$, $i \geq 1$. The finite problem has a different character, for, unless the function $m(n)$ is severely restricted, it may not be possible, even asymptotically, to pack an initial segment $[1, K]$ by disjoint subcollections of a given strong cover $\mathcal{C}$ of the natural numbers. If it is only required, however, there be a disjoint subcollection that fills most of $[1, K]$, then a positive and useful result is possible.

To motivate the positive result, suppose all the intervals of $\mathcal{C}$ have the same length, say L, and apply a sequential greedy algorithm, that is, start from the left and select successive disjoint intervals, stopping when within L of the end. This produces the disjoint collection

$$\mathcal{C}' = \{[iL+1, (i+1)L]\colon\ 0 \le i \le (K-L)/L\}$$

which covers all but at most the final $L-1$ members of $[1, K]$. In particular, if $\delta > 0$ is given and $K > L/\delta$ then all but at most a δ-fraction is covered. The desired positive result is just an extension of this idea to the case when most of the intervals have length bounded by some $L < \delta K$.

Let $\mathcal{C}$ be a strong cover of the natural numbers $\mathcal{N}$. The interval $[1, K]$ is said to be (L, δ)*-strongly-covered by* $\mathcal{C}$ if

$$\frac{|\{n \in [1, K]\colon\ m(n) - n + 1 > L\}|}{K} \le \delta.$$

A collection of subintervals $\mathcal{C}'$ of the interval $[1, K]$ is called a $(1-\delta)$*-packing* of $[1, K]$ if the intervals in $\mathcal{C}'$ are disjoint and their union has cardinality at least $(1-\delta)K$.

Lemma I.3.3 (The packing lemma.)

Let $\mathcal{C}$ be a strong cover of $\mathcal{N}$ and let $\delta > 0$ be given. If $K > L/\delta$ and if $[1, K]$ is (L, δ)-strongly-covered by $\mathcal{C}$, then there is a subcollection $\mathcal{C}' \subset \mathcal{C}$ which is a $(1-2\delta)$-packing of $[1, K]$.

Proof. By hypothesis $K > L/\delta$ and

$$|\{n \in [1, K]\colon\ m(n) - n + 1 > L\}| \le \delta K. \tag{1}$$

The construction of the $(1-2\delta)$-packing will proceed sequentially from left to right, selecting the first interval of length no more than L that is disjoint from the previous selections, stopping when within L of the end of $[1, K]$. To carry this out rigorously set $m(0) = n_0 = 0$ and, by induction, define

$$n_i = \min\{j \in [1 + m(n_{i-1}), K - L]\colon\ m(j) - j + 1 \le L\}.$$

The construction stops after $I = I(\mathcal{C}, K)$ steps if $m(n_I) > K - L$ or there is no $j \in [1 + m(n_I), K - L]$ for which $m(j) - j + 1 \le L$.

The claim is that $\mathcal{C}' = \{[n_i, m(n_i)]\colon\ 1 \le i \le I\}$ is a $(1-2\delta)$-packing of $[1, K]$. The intervals are disjoint, by construction, and are contained in $[1, K]$, since $m(n_i) - n_i + 1 \le L$, and hence $m(n_I) \le K$. Thus it is only necessary to show that the union $\mathcal{U}$ of the $[n_i, m(n_i)]$ has length at least $(1-2\delta)K$. The interval $(K-L, K]$ has length at most $L - 1 \le \delta K$, so that

$$|(K-L, K] - \mathcal{U}]| \le \delta K.$$

For the interval $[1, K-L]$, the definition of the n_i implies the following fact.

$$\text{If } j \in [1, K-L] - \mathcal{U} \text{ then } m(j) - j + 1 > L. \tag{2}$$

The (L, δ)-strong-cover assumption, (1), thus guarantees that

$$|[1, K-L] - \mathcal{U}| \le \delta K.$$

This completes the proof of the packing lemma. □

Remark I.3.4
In most applications of the packing lemma all that matters is the result, but in the proof of the general ergodic theorem use will be made of the explicit construction given in the proof of Lemma I.3.3, including the description (2) of those indices that are neither within L of the end of the interval nor are contained in sets of the packing. Some extensions of the packing lemma will be mentioned in the exercises.

I.3.b The binary, ergodic process proof.

A proof of the ergodic theorem for the binary ergodic process case will be given first, as it illustrates the ideas in simplest form. In this case μ is invariant and ergodic with respect to the shift T on $\{0, 1\}^\infty$, and the goal is to show that

$$\lim_{n\to\infty} \frac{1}{n}\sum_{i=1}^{n} x_i = \mu(1),\ a.s, \tag{3}$$

where $\mu(1) = \mu\{x\colon x_1 = 1\} = E(X_1)$.

Suppose (3) is false. Then either the limit superior of the averages is too large on a set of positive measure or the limit inferior is too small on a set of positive measure. Without loss of generality the first of these possibilities can be assumed, and hence there is an $\epsilon > 0$ such that the set

$$B = \left\{x : \limsup_{n\to\infty} \frac{1}{n}\sum_{i=1}^{n} x_i > \mu(1) + \epsilon\right\}$$

has positive measure. Since

$$\limsup_{n\to\infty} \frac{x_1 + \ldots + x_n}{n} = \limsup_{n\to\infty} \frac{x_2 + \ldots + x_{n+1}}{n},$$

the set B is T-invariant and therefore $\mu(B) = 1$. (Here is where the ergodicity assumption is used.)

Now suppose $x \in B$. Since B is T-invariant, for each integer n there will be a first integer $m(n) \geq n$ such that

$$\frac{x_n + x_{n+1} + \ldots + x_{m(n)}}{m(n) - n + 1} > \mu(1) + \epsilon.$$

Thus the collection $\mathcal{C}(x) = \{[n, m(n)]\colon n \in \mathcal{N}\}$ is a (random) strong cover of the natural numbers $\mathcal{N}$. Furthermore, each interval in $\mathcal{C}(x)$ has the property that the average of the x_i over that interval is too big by a fixed amount. These intervals overlap, however, so it is not easy to see what happens to the average over a fixed, large interval.

The packing lemma can be used to reduce the problem to a nonoverlapping interval problem, for, combined with a simple observation about almost-surely finite variables, it will imply that if K is large enough then with high probability, most of the terms x_i, $i \leq K$ are contained in *disjoint* intervals over which the average is too big. But, if the average over most such disjoint blocks is too big, then the average over the entire interval $[1, K]$ must also be too big by a somewhat smaller fixed amount. Since this occurs with high probability it implies that the expected value of the average over the entire set A^K must be too big, which is a contradiction.

A bit of preparation is needed before the packing lemma can be applied. First, since the random variable $m(1)$ is almost-surely finite, it is bounded except for a set of small probability. Thus given $\delta > 0$ there is a number L such that if $D = \{x\colon m(1) > L\}$, then $\mu(D) < \delta^2$. Second, note that the function

$$g_K(x) = \frac{1}{K}\sum_{i=1}^{K} \chi_D(T^{i-1}x),$$

where χ denotes the indicator function, has integral $\mu(D) < \delta^2$, since T is measure preserving. The Markov inequality implies that for each K, the set $G_K = \{x\colon g_K(x) \leq \delta\}$ has measure at least $1-\delta$.

The definitions of D and G_K imply that if $x \in G_K$ then $\mathcal{C}(x)$ is an (L,δ)-strong-covering of $[1, K]$. Thus the packing lemma implies that if $K \geq L/\delta$ and $x \in G_K$ then there is a subcollection

$$\mathcal{C}'(x) = \{[n_i, m(n_i)]\colon i \leq I(x)\} \subset \mathcal{C}(x)$$

which is a $(1-2\delta)$-packing of $[1, K]$. Since the x_i are nonnegative, by assumption, and since the intervals in $\mathcal{C}'(x)$ are disjoint, the sum over the intervals in $\mathcal{C}'(x)$ lower bounds the sum over the entire interval, that is,

$$\sum_{j=1}^{K} x_j \geq \sum_{i=1}^{I(x)} \sum_{j=n_i}^{m(n_i)} x_j \geq (1-2\delta)K\,[\mu(1)+\epsilon]. \tag{4}$$

Note that while the collection $\mathcal{C}(x)$ and subcollection $\mathcal{C}'(x)$ both depend on x, the lower bound is independent of x, as long as $x \in G_K$. Thus, taking expected values yields

$$\begin{aligned} E\left(\sum_{j=1}^{K} x_j\right) &\geq (1-2\delta)K\,[\mu(1)+\epsilon]\,\mu(G_K) \\ &\geq (1-2\delta)(1-\delta)K\,[\mu(1)+\epsilon], \end{aligned}$$

so that

$$\mu(1) = E\left(\frac{1}{K}\sum_{j=1}^{K} x_j\right) \geq (1-2\delta)(1-\delta)\,[\mu(1)+\epsilon],$$

which cannot be true for all δ. This completes the proof of (3), and thereby the proof of Theorem I.3.2, the binary, ergodic process form of the ergodic theorem. □

I.3.c The proof in the general case.

The preceding argument will now be extended to obtain the general ergodic theorem. The following lemma generalizes the essence of what was actually proved; it will be used to obtain the general theorem.

Lemma I.3.5

Let T be a measure-preserving transformation on (X, Σ, μ), let $f \in L^1(X, \Sigma, \mu)$, let α be an arbitrary real number, and define the set

$$B = \{x\colon \limsup_{n\to\infty} \frac{1}{n}\sum_{i=1}^{n} f(T^{i-1}x) > \alpha\}.$$

Then $\int_B f(x)\,d\mu(x) \geq \alpha\mu(B)$.

Proof. Note that in the special case where the process is ergodic, where f is the indicator function of a set C and $\alpha = \mu(C) + \epsilon$, this lemma is essentially what was just proved. The same argument can be used to show that Lemma I.3.5 is true in the ergodic case for bounded functions. Only a bit more is needed to handle the general case, where f is allowed to be unbounded, though integrable.

The lemma is clearly true if $\mu(B) = 0$, so it can be assumed that $\mu(B) > 0$. The set B is T-invariant and the restriction of T to it preserves the conditional measure $\mu(\cdot\,|B)$. Thus, it is enough to prove

$$\int_B f(x)\,d\mu(x|B) \geq \alpha.$$

For $x \in B$ and $n \in \mathcal{N}$ there is a least integer $m(n) \geq n$ such that

$$\frac{\sum_{i=n}^{m(n)} f(T^{i-1}x)}{m(n) - n + 1} > \alpha. \tag{5}$$

Since B is T-invariant, the collection $\mathcal{C}(x) = \{[n, m(n)]\colon n \in \mathcal{N}\}$ is a (random) strong cover of the natural numbers $\mathcal{N}$. As before, given $\delta > 0$ there is an L such that if

$$D = \{x \in B\colon m(1) > L\}$$

then $\mu(D|B) < \delta^2$ and for every K the set

$$G_K = \{x \in B\colon \frac{1}{K}\sum_{i=1}^{K} \chi_D(T^{i-1}x) \leq \delta\}$$

has conditional measure at least $1 - \delta$. Furthermore, if $K \geq L/\delta$ then the packing lemma can be applied and the argument used to prove (4) yields a lower bound of the form

$$\begin{aligned}\sum_{j=1}^{K} f(T^{j-1}x) &\geq R(K,x) + \sum_{i=1}^{I(x)}\sum_{j=n_i}^{m(n_i)} f(T^{i-1}x)\\ &\geq R(K,x) + (1-2\delta)K\alpha,\end{aligned} \tag{6}$$

where

$$R(K,x) = \sum_{j\in[1,K]-\cup[n_i,m(n_i)]} f(T^{j-1}x)$$

is the sum of $f(T^{j-1}x)$ over the indices that do not belong to the $[n_i, m(n_i)]$ intervals. (Keep in mind that the collection $\{[n_i, m(n_i)]\}$ depends on x.)

In the bounded case, say $|f(x)| \leq M$, the sum $R(K,x)$ is bounded from below by $-2MK\delta$ and the earlier argument produces the desired conclusion. In the unbounded case, a bit more care is needed to control the effect of $R(K,x)$ as well as the effect of integrating over G_K in place of X. An integration, together with the lower bound (6), does give

$$\begin{aligned}\int_B f(x)\,d\mu(x|B) &= \int_B \frac{1}{K}\sum_{j=1}^{K} f(T^{j-1}x)\,d\mu(x|B)\\ &\geq (1-\delta)\alpha\mu(G_K|B) + I_1 + I_2 + I_3,\end{aligned} \tag{7}$$

where

$$\begin{aligned} I_1 &= \frac{1}{K}\int_{B-G_K}\sum_{j=1}^{K} f(T^{j-1}x)\,d\mu(x|B), \\ I_2 &= \frac{1}{K}\int_{G_K}\sum_{j\in[1,K-L]-\cup[n_i,m(n_i)]} f(T^{j-1}x)\,d\mu(x|B), \\ I_3 &= \frac{1}{K}\int_{G_K}\sum_{j\in(K-L,K]-\cup[n_i,m(n_i)]} f(T^{j-1}x)\,d\mu(x|B). \end{aligned}$$

The measure-preserving property of T gives

$$\int_{B-G_K} f(T^j x)\,d\mu(x|B) = \int_{T^{-j}(B-G_K)} f(x)\,d\mu(x|B).$$

Since $f\in L^1$ and since all the sets $T^{-j}(B-G_K)$ have the same measure, which is upper bounded by δ, all of the terms in I_1, and hence I_1 itself, will be small if δ is small enough.

The integral I_3 is also easy to bound for it satisfies

$$|I_3| \le \frac{1}{K}\int_B \sum_{j\in(K-L,K]} |f|(T^{j-1}x)\,d\mu(x|B),$$

and hence the measure-preserving property of T yields

$$|I_3| \le \frac{L}{K}\int_B |f|(x)\,d\mu(x|B),$$

which is small if K is large enough, for any fixed L.

To bound I_2, recall from the proof of the packing lemma, see (2), that if $j\in[1,K-L]-\cup[n_i,m(n_i)]$ then $m(j)-j+1>L$. In the current setting this translates into the statement that if $j\in[1,K-L]-\cup[n_i,m(n_i)]$ then $T^{j-1}x\in D$. Thus

$$|I_2| \le \frac{1}{K}\int_B \sum_{j=1}^{K} \chi_D(T^{j-1}x)|f(T^{j-1}x)|\,d\mu(x|B),$$

and the measure-preserving property of T yields the bound

$$|I_2| \le \int_D |f(x)|\,d\mu(x|B).$$

which is small if L is large enough.

In summary, in the inequality

$$\int_B f(x)\,d\mu(x|B) \ge (1-\delta)\alpha\mu(G_K|B) + I_1 + I_2 + I_3,$$

see (7), the final three terms can all be made arbitrarily small, so that passage to the appropriate limit shows that indeed $\int_B f(x)\,d\mu(x|B)\ge\alpha$, which completes the proof of Lemma I.3.5. □

Proof of the Birkhoff ergodic theorem: general case.

To prove almost-sure convergence first note that Lemma I.3.5 remains true for any T-invariant subset of B. Thus if

$$C = \left\{x: \liminf_{n\to\infty} \frac{1}{n}\sum_{i=1}^{n} f(T^i x) < \alpha < \beta < \limsup_{n\to\infty} \frac{1}{n}\sum_{i=1}^{n} f(T^i x)\right\}$$

then

$$\beta\mu(C) \le \int_C f\, d\mu \le \alpha\mu(C),$$

which can only be true if $\mu(C) = 0$. This proves almost sure convergence.

To complete the discussion of the ergodic theorem L^1-convergence must be established. To do this define the operator U_T on L^1 by the formula,

$$U_T f(x) = f(Tx),\ x \in X,\ f \in L^1.$$

The operator U_T is linear and is an isometry because T is measure preserving, that is, $\|U_T f\| = \|f\|$, where $\|f\| = \int |f|\, d\mu$ is the L^1-norm. Thus the averaging operator

$$A_n f = \frac{1}{n}\sum_{i=1}^{n} U_T^{i-1} f$$

is linear and is a contraction, that is, $\|A_n f\| \le \|f\|$, $f \in L^1$. Moreover, if f is a bounded function, say $|f(x)| \le M$, almost surely, then $|A_n f(x)| \le M$, almost surely. Since almost-sure convergence holds, the dominated convergence theorem implies L^1-convergence of $A_n f$ for bounded functions. But if $f, g \in L^1$ then for all m, n,

$$\begin{aligned}\|A_n f - A_m f\| &\le \|A_n(f-g) - A_m(f-g)\| + \|A_n g - A_m g\| \\ &\le 2\|f-g\| + \|A_n g - A_m g\|\end{aligned}$$

The L_1-convergence of $A_n f$ follows immediately from this inequality, for g can be taken to be a bounded approximation to f. Since, as was just shown, $A_n g$ converges in L^1-norm, and since the average $A_n f$ converges almost surely to a limit function f^*, it follows that $f^* \in L^1$ and that $A_n f$ converges in L^1-norm to f^*. This completes the proof of the Birkhoff ergodic theorem. □

Remark I.3.6

Birkhoff's proof was based on a proof of Lemma I.3.5. Most subsequent proofs of the ergodic theorem reduce it to a stronger result, called the maximal ergodic theorem, whose statement is obtained by replacing lim sup by sup in Lemma I.3.5. There are several different proofs of the maximal theorem, including the short, elegant proof due to Garsia that appears in many books on ergodic theory. The proof given here is less elegant, but it is much closer in spirit to the general partitioning, covering, and packing ideas which have been used recently in ergodic and information theory and which will be used frequently in this book. The packing lemma is a simpler version of a packing idea used to establish entropy theorems for amenable groups in [51]; its application to prove the ergodic theorem as given here was developed by this author in 1980 and published much later in [68]. Other proofs that were stimulated by [51] are given in [25, 27]. Proofs of the maximal theorem and Kingman's subadditive ergodic theorem which use ideas from the packing lemma will be sketched in the exercises, as well as von Neumann's elegant functional analytic proof of L^2-convergence. The reader is referred to Krengel's book for historical commentary and numerous other ergodic theorems, [32].

I.3.d Extensions of the packing lemma.

The given proof of the ergodic theorem was based on the purely combinatorial packing lemma. Several later results will be proved using extensions of the packing lemma. Now that the ergodic theorem is available, however, the packing lemma and its relatives can be reformulated in a more convenient way as lemmas about stopping times.

A *(generalized) stopping time* is a measurable function τ defined on the probability space (X, Σ, μ) with values in the extended natural numbers $\mathcal{N} \cup \{\infty\}$. Note that the concept of stopping time used here does not require that X be a sequence space, nor that the set $\{x\colon \tau(x) = n\}$ be measurable with respect to the first n coordinates, which is a common requirement in probability theory. The stopping time idea was used implicitly in the proof of the ergodic theorem, for the first time m that $\sum_{i=1}^{m} f(T^{i-1}x) > m\alpha$ is a stopping time, see (5).

Suppose μ is a stationary measure on A^∞. A stopping time τ is μ-almost-surely finite if

$$\mu(\{x\colon \tau(x) = \infty\}) = 0.$$

An almost-surely finite stopping time τ has the property that for almost every x, the stopping time for each shift $T^{n-1}x$ is finite. Thus almost-sure finiteness guarantees that, for almost every x, the function $\tau(T^{n-1}x)$ is finite for every n. The interval $[n, \tau(T^{n-1}x) + n - 1]$ is called the *stopping interval for x, starting at n.* In particular, for an almost-surely finite stopping time, the collection

$$\mathcal{C} = \mathcal{C}(x, \tau) = \{[n, \tau(T^{n-1}x) + n - 1]\colon n \in \mathcal{N}\}$$

of stopping intervals is, almost surely, a (random) strong cover of the natural numbers $\mathcal{N}$.

Frequent use will be made of the following almost-sure stopping-time packing result for ergodic processes.

Lemma I.3.7 (The ergodic stopping-time packing lemma.)
If τ is an almost-surely finite stopping time for an ergodic process μ then for any $\delta > 0$ and almost every x there is an $N = N(\delta, x)$ such that if $n \geq N$ then $[1, n]$ is $(1 - \delta)$-packed by intervals from $\mathcal{C}(x, \tau)$.

Proof. Since $\tau(x)$ is almost-surely finite, it is bounded except for a set of small measure, hence, given $\delta > 0$, there is an integer L such that

$$\mu(\{x\colon \tau(x) > L\}) < \delta/4.$$

Let D be the set of all x for which $\tau(x) > L$. By the ergodic theorem,

$$\lim_{n\to\infty} \frac{1}{n} \sum_{i=1}^{n} \chi_D(T^{i-1}x)\, d\mu(x) = \mu(D) < \delta/4,$$

almost surely, so that if

$$G_n = \left\{x : \frac{1}{n} \sum_{i=1}^{n} \chi_D(T^{i-1}x) < \delta/4\right\}$$

then $x \in G_n$, eventually almost surely.

Suppose $x \in G_n$ and $n \geq 4L/\delta$. The definition of G_n implies that $\tau(T^{-(i-1)}x) \leq L$ for at least $(1-\delta/4)n$ indices i in the interval $[1, n]$ and hence for at least $(1-\delta/2)n$ indices i in the interval $[1, n-L]$. But this means that $[1, n]$ is $(L, 1-\delta/2)$-strongly-covered by $\mathcal{C}(x, \tau)$. The packing lemma produces a $(L, 1-\delta/2)$-packing of $[1, n]$ by a subset of $\mathcal{C}(x, \tau)$. This proves the lemma. □

Two variations on these general ideas will also be useful later, a partial-packing result that holds in the case when the stopping time is assumed to be finite only on a set of positive measure, and a two-packings variation. These are stated as the next two lemmas, with the proofs left as exercises.

For a stopping time τ which is not almost-surely infinite, put $F(\tau) = \{x\colon \tau(x) < \infty\}$, and define $\mathcal{C}(x, \tau)$ to be the set of all intervals $[n, \tau(T^{n-1}x) + n - 1]$, for which $T^{n-1}x \in F(\tau)$.

Lemma I.3.8 (The partial-packing lemma.)
Let μ be an ergodic measure on A^∞, let τ be a stopping time such that $\mu(F(\tau)) > \gamma > 0$. For almost every x, there is an $N = N(x, \tau)$ such that if $n \geq N$, then $[1, n]$ has a γ-packing $\mathcal{C}' \subset \mathcal{C}(x, \tau)$

Lemma I.3.9 (The two-packings lemma.)
Let $\{[u_i, v_i]\colon i \in I\}$ be a disjoint collection of subintervals of $[1, K]$ of total length αK, and let $\{[s_j, t_j]\colon j \in J\}$ be another disjoint collection of subintervals of $[1, K]$ of total length βK. Suppose each $[u_i, v_i]$ has length m, and each $[s_j, t_j]$ has length at least M, where $M > m/\delta$. Let I^* be the set of indices $i \in I$ such that $[u_i, v_i]$ meets at least one of the $[s_j, t_j]$. Then the disjoint collection $\{[s_j, t_j]\colon j \in J\} \cup \{[u_i, v_i]\colon i \in I^*\}$ has total length at least $(\alpha + \beta - 2\delta)K$.

I.3.e Exercises.

1. A $(1-\delta)$-packing $\mathcal{C}'$ of $[1, n]$ is said to be *separated* if there is at least one integer between each interval in $\mathcal{C}'$. Let μ be an ergodic measure on A^∞, and let τ be an almost-surely finite stopping time which is almost-surely bounded below by a positive integer $M > 1/\delta$. Show that for almost every x, there is an integer $N = N(\delta, x)$ such that if $n \geq N$ then $[1, n]$ has a separated $(1-2\delta)$-packing $\mathcal{C}' \subset \mathcal{C}(x, \tau)$. (Hint: define $\tilde{\tau}(x) = \tau(x) + 1$ and apply the ergodic stopping-time lemma to $\tilde{\tau}$.)

2. Show that a separated $(1-\delta)$-packing of $[1, n]$ is determined by its complement. Show that this need not be true if some intervals are not separated by at least one integer.

3. Prove the partial-packing lemma.

4. Formulate the partial-packing lemma as a strictly combinatorial lemma about the natural numbers.

5. Prove the two-packings lemma. (Hint: use the fact that the cardinality of J is at most K/M to estimate the total length of those $[u_i, v_i]$ that meet the boundary of some $[s_j, t_j]$.)

6. Show that if τ is an almost-surely finite stopping time for a stationary process μ and $G_n(\tau, \delta)$ is the set of all x such that $[1, n]$ is $(1-\delta)$-packed by intervals from $\mathcal{C}(x, \tau)$, then $\mu(G_n(\tau, \delta)) \to 1$, as $n \to \infty$.

7. Show that if μ is ergodic and if f is a nonnegative measurable, but not integrable function, then the averages $(1/N)\sum_1^N f(T^{i-1}x)$ converge almost surely to ∞.

8. The mean ergodic theorem of von Neumann asserts that $(1/n)\sum_{i=1}^n U_T^{i-1} f$ converges in L^2-norm, for any L^2 function f. Assume T is invertible so that U_T is unitary.

   (a) Show that the theorem is true for $f \in \mathcal{F} = \{f \in L^2\colon U_T f = f\}$ and for $f \in \mathcal{M} = (I - U_T)L^2$.

   (b) Show that $\mathcal{F}+\mathcal{M}$ is dense in L^2. (Hint: show that its orthogonal complement is 0.)

   (c) Deduce von Neumann's theorem from the two preceding results.

   (d) Deduce that L^1-convergence holds, if $f \in L^1$.

   (e) Extend the preceding arguments to the case when it is only assumed that T is measure preserving.

9. Assume T is measure preserving and f is integrable, and for each n, let U_n be the set of x's for which $\sum_{i=0}^{n-1} f(T^i x) \geq 0$. Fix N and let $E = \cup_{n \leq N} U_n$. One form of the maximal ergodic theorem asserts that $\int_E f(x)\, d\mu(x) \geq 0$. The ideas of the following proof are due to R.Jones, [33].

   (a) For bounded f, show that

   $$\sum_{i=0}^{\tau(x)-1} f(T^i x)\chi_E(T^i x) \geq 0, \text{ for some } \tau(x) \leq N.$$

   (Hint: let $\tau(x) = n,\ x \in U_n,\ \tau(x) = 1,\ x \notin E$.)

   (b) For bounded f and $L > N$, show that

   $$\sum_{i=0}^{L} f(T^i x)\chi_E(T^i x) \geq -(N-1)\|f\|_\infty.$$

   (Hint: sum from 0 to $\tau(x) - 1$, then from $\tau(x)$ to $\tau(T^{\tau(x)}) - 1$, continuing until within N of L.)

   (c) Show that the theorem holds for bounded functions.

   (d) Show that the theorem holds for integrable functions.

   (e) For $B = \{x\colon \sup_n \frac{1}{n}\sum_{i=0}^{n-1} f(T^i x) > \alpha\}$, show that $\int_B f\, d\mu \geq \alpha\mu(B)$.

10. Assume T is measure preserving and $\{g_n\}$ is a sequence of integrable functions such that $g_{n+m}(x) \leq g_n(x) + g_m(T^n x)$. Kingman's subadditive ergodic theorem asserts that $g_n(x)/n$ converges almost surely to an invariant function $g(x) \geq -\infty$. The ideas of the following proof are due to M.Steele, [80].

    (a) Prove the theorem under the additional assumption that $g_n \leq 0$. (Hint: first show that $g(x) = \liminf g_n(x)/n$ is an invariant function, then apply packing.)

(b) Prove the theorem by reducing it to the case when $g_n \leq 0$. (Hint: let $g'_m(x) = g_m(x) - \sum_1^{m-1} g_1(T^i x)$.)

(c) Show that if $\alpha = \inf_n \int g_n/n \, d\mu < \infty$, then convergence is also in L_1-norm, and $\alpha = \int g \, d\mu$.

(d) Show that the same conclusions hold if $g_n \leq 0$ and $g_{n+g+m}(x) \leq g_n(x_1^n) + g_m(T^{n+g}x) + \gamma_g$, where $\gamma_g \to 0$ as $g \to \infty$.

Section I.4 Frequencies of finite blocks.

An important consequence of the ergodic theorem for stationary finite alphabet processes is that relative frequencies of overlapping k-blocks in n-blocks converge almost surely as $n \to \infty$. In the ergodic case, limiting relative frequencies converge almost surely to the corresponding probabilities, a property characterizing ergodic processes. In the nonergodic case, the existence of limiting frequencies allows a stationary process to be represented as an average of ergodic processes. These and related ideas will be discussed in this section.

I.4.a Frequencies for ergodic processes.

The *frequency* of the block a_1^k in the sequence x_1^n is defined for $n \geq k$ by the formula

$$f(a_1^k|x_1^n) = \sum_{i=1}^{n-k+1} \chi_{[a_1^k]}(T^{i-1}x),$$

where $\chi_{[a_1^k]}$ denotes the indicator function of the cylinder set $[a_1^k]$. The frequency can also be expressed in the alternate form

$$f(a_1^k|x_1^n) = |\{i \in [1, n-k+1] \colon x_i^{i+k-1} = a_1^k\}|,$$

where $|\cdot|$ denotes cardinality, that is, $f(a_1^k|x_1^n)$ is obtained by sliding a window of length k along x_1^n and counting the number of times a_1^k is seen in the window.

The *relative frequency* is defined by dividing the frequency by the maximum possible number of occurrences, $n-k+1$, to obtain

$$p_k(a_1^k|x_1^n) = \frac{f(a_1^k|x_1^n)}{n-k+1} = \frac{|\{i \in [1, n-k+1] \colon x_i^{i+k-1} = a_1^k\}|}{n-k+1}.$$

If x_1^n and $k \leq n$ are fixed the relative frequency defines a measure $p_k(\cdot\,|x_1^n)$ on A^k, called the *empirical distribution of overlapping k-blocks*, or the *(overlapping) k-type* of x_1^n. When k is understood, the subscript k on $p_k(a_1^k|x_1^n)$ may be omitted.

The *limiting (relative) frequency* of a_1^k in the infinite sequence x is defined by

$$p(a_1^k|x) = \lim_{n\to\infty} p(a_1^k|x_1^n), \tag{1}$$

provided, of course, that this limit exists. A sequence x is said to be *(frequency) typical* for the process μ if each block a_1^k appears in x with limiting relative frequency equal to $\mu(a_1^k)$. Thus, x is typical for μ if for each k the empirical distribution of each k-block converges to its theoretical probability $\mu(a_1^k)$.

Let $\mathcal{T}(\mu)$ denote the set of sequences that are frequency typical for μ. A basic result for ergodic processes is that almost every sequence is typical. Thus the entire structure of an ergodic process can be almost surely recovered merely by observing limiting relative frequencies along a single sample path.

Theorem I.4.1 (The typical-sequence theorem.)
If μ is ergodic then $\mu(\mathcal{T}(\mu)) = 1$, that is, for almost every x,

$$\lim_{n\to\infty} p(a_1^k|x_1^n) = \mu(a_1^k),$$

for all k and all a_1^k.

Proof. In the ergodic case,

$$p(a_1^k|x_1^n) = \frac{1}{n-k+1}\sum_{i=1}^{n-k+1} \chi_{[a_1^k]}(T^{i-1}x),$$

which, by the ergodic theorem, converges almost surely to $\int \chi_{[a_1^k]}\,d\mu = \mu(a_1^k)$, for fixed a_1^k. In other words, there is a set $B(a_1^k)$ of measure 0, such that $p(a_1^k|x_1^n) \to \mu(a_1^k)$, for $x \notin B(a_1^k)$, as $n \to \infty$. Since there are only a countable number of possible a_1^k, the set

$$B = \bigcup_{k=1}^{\infty} \bigcup_{a_1^k \in A^k} B(a_1^k)$$

has measure 0 and convergence holds for all k and all a_1^k, for all $x \in A^\infty - B$, a set of measure 1. This proves the theorem. □

The converse of the preceding theorem is also true.

Theorem I.4.2 (The typical-sequence converse.)
Suppose μ is a stationary measure such that for each k and for each block a_1^k, the limiting relative frequencies $p(a_1^k|x)$ exist and are constant in x, μ-almost surely. Then μ is an ergodic measure.

Proof. If the limiting relative frequency $p(a_1^k|x)$ exists and is a constant c, with probability 1, then this constant c must equal $\mu(a_1^k)$, since the ergodic theorem includes L^1-convergence of the averages to a function with the same mean. Thus the hypotheses imply that the formula

$$\lim_{n\to\infty} \frac{1}{n}\sum_{i=1}^{n} \chi_B(T^{i-1}x) = \mu(B),$$

holds for almost all x for any cylinder set B. Multiplying each side of this equation by the indicator function $\chi_C(x)$, where C is an arbitrary measurable set, produces

$$\lim_{n\to\infty} \frac{1}{n}\sum_{i=1}^{n} \chi_B(T^{i-1}x)\chi_C(x) = \mu(B)\chi_C(x),$$

and an integration then yields

$$\lim_{n\to\infty} \frac{1}{n}\sum_{i=1}^{n} \mu(T^{-i}B\cap C) = \mu(B)\mu(C). \tag{2}$$

The latter holds for any cylinder set B and measurable set C. The usual approximation argument then shows that formula (2) holds for any measurable sets B and C. But if $T^{-1}B = B$ then the formula gives $\mu(B) = \mu(B)^2$, so that $\mu(B)$ must be 0 or 1. This proves Theorem I.4.2. □

Note that as part of the preceding proof the following characterizations of ergodicity were established.

Theorem I.4.3
The following are equivalent.

(i) *T is ergodic.*

(ii) $\lim_n \frac{1}{n}\sum_1^n \chi_E(T^{i-1}x) = \mu(E)$, *a.e., for each E in a generating algebra.*

(iii) $\lim_n \frac{1}{n}\sum_1^n \mu(T^{-i}A \cap B) = \mu(A)\mu(B)$, *for each A and B in a generating algebra.*

Some finite forms of the typical sequence theorems will be discussed next. Sequences of length n can be partitioned into two classes, G_n and B_n so that those in G_n will have "good" k-block frequencies and the "bad" set B_n has low probability when n is large. The precise definition of the "good" set, which depends on k and a measure of error, is

$$G_n(k, \epsilon) = \{x_1^n\colon\ |p(a_1^k|x_1^n) - \mu(a_1^k)| < \epsilon,\ a_1^k \in A^k\}. \tag{3}$$

The "bad" set is the complement, $B_n(k, \epsilon) = A^n - G_n(k, \epsilon)$. The members of the "good" set, $G_n(k, \epsilon)$, are often called the *frequency-(k, ϵ)-typical sequences*, or when k and ϵ are understood, just the *typical sequences*.

Theorem I.4.4 (The "good set" form of ergodicity.)

(a) *If μ is ergodic then $x_1^n \in G_n(k, \epsilon)$, eventually almost surely, for every k and $\epsilon > 0$.*

(b) *If $\lim_n \mu_n(G_n(k, \epsilon)) = 1$, for every integer $k > 0$ and every $\epsilon > 0$, then μ is ergodic.*

Proof. Part (a) is just a finite version of the typical-sequence theorem, Theorem I.4.1. The condition that $\mu_n(G_n(k, \epsilon))$ converge to 1 for any $\epsilon > 0$ is just the condition that $p(a_1^k|x_1^n)$ converges in probability to $\mu(a_1^k)$. Since $p(a_1^k|x_1^n)$ is bounded and converges almost surely to some limit, the limit must be almost surely constant, and hence (b) follows from Theorem I.4.2. □

It is also useful to have the following finite-sequence form of the typical-sequence characterization of ergodic processes.

Theorem I.4.5 (The finite form of ergodicity.)
A measure μ is ergodic if and only if for each block a_1^k and $\epsilon > 0$ there is an $N = N(a_1^k, \epsilon)$ such that if $n > N$ then there is a collection $C_n \subset A^n$ such that

(i) $\mu(C_n) > 1 - \epsilon$.

(ii) *If $x_1^n, \tilde{x}_1^n \in C_n$ then* $|p(a_1^k|x_1^n) - p(a_1^k|\tilde{x}_1^n)| < \epsilon$.

Proof. If μ is ergodic, just set $C_n = G_n(k, \epsilon)$, the "good" set defined by (3), and use Theorem I.4.4. To establish the converse fix a_1^k and $\epsilon > 0$, then define

$$\widetilde{G}_n = \{x\colon\ |p(a_1^k|x_1^n) - p(a_1^k|x)| < \epsilon\},$$

and use convergence in probability to select $n > N(a_1^k, \epsilon)$ such that $\mu(\widetilde{G}_n) > 1 - \epsilon$. If C_n satisfies (i) and (ii) put

$$\widetilde{C}_n = \{x\colon\ x_1^n \in C_n\}$$

and note that $\mu(\widetilde{G}_n \cap \widetilde{C}_n) > 1 - 2\epsilon$. Thus,

$$|p(a_1^k|x) - p(a_1^k|\widetilde{x})| \le 3\epsilon,\quad x, \widetilde{x} \in \widetilde{G}_n \cap \widetilde{C}_n,$$

and hence $p(a_1^k|x)$ is constant almost surely, proving the theorem. □

I.4.b The ergodic theorem and covering properties.

The ergodic theorem is often used to derive covering properties of finite sample paths for ergodic processes. In these applications, a set $\mathcal{B}_n \subseteq A^n$ with desired properties is determined in some way, then the ergodic theorem is applied to show that, eventually almost surely as $M \to \infty$, there are approximately $\mu(\mathcal{B}_n)M$ indices $i \in [1, M-n]$ for which $x_i^{i+n-1} \in \mathcal{B}_n$. This simple consequence of the ergodic theorem can be stated as either a limit result or as an approximation result.

Theorem I.4.6 (The covering theorem.)
If μ is an ergodic process with alphabet A, and if $\mathcal{B}_n$ is a subset of A^n of positive measure, then

$$\lim_{M\to\infty} p_n(\mathcal{B}_n|x_1^M) = \mu(\mathcal{B}_n),\ a.s.$$

In other words, for any $\delta > 0$,

$$\Big||\{i \in [1, M-n]\colon\ x_i^{i+n-1} \in \mathcal{B}_n\}| - M\mu(\mathcal{B}_n)\Big| \le \delta M, \tag{4}$$

eventually almost surely.

Proof. Apply the ergodic theorem to the indicator function of $\mathcal{B}_n$. □

In many situations the set $\mathcal{B}_n$ is chosen to have large measure, in which case the conclusion (4) is expressed, informally, by saying that, eventually almost surely, x_1^M is "mostly strongly covered" by blocks from $\mathcal{B}_n$. A sequence x_1^M is said to be $(1-\delta)$-*strongly-covered by* $\mathcal{B}_n$ if

$$|\{i \in [1, M-n+1]\colon\ x_i^{i+n-1} \notin \mathcal{B}_n\}| \le \delta M,$$

that is, if $p_n(\mathcal{B}_n|x_1^M) \le \delta$.

Theorem I.4.7 (The almost-covering principle.)
If μ is an ergodic process with alphabet A, and if $\mathcal{B}_n \subseteq A^n$, satisfies $\mu(\mathcal{B}_n) > 1-\delta$, then x_1^M is eventually almost surely $(1-\delta)$-strongly-covered by $\mathcal{B}_n$.

The almost strong-covering idea and a related almost-packing idea are discussed in more detail in Section I.7. The following example illustrates one simple use of almost strong-covering. Note that the ergodic theorem is used twice in the example, first to select the set $\mathcal{B}_n$, and then to obtain almost strong-covering by $\mathcal{B}_n$. Such "doubling" is common in applications.

Example I.4.8

Let μ be an ergodic process and fix a symbol $a \in A$ of positive probability. Let $F(x_1^M)$ and $L(x_1^M)$ denote the first and last occurrences of $a \in x_1^M$, that is,

$$F(x_1^M) = \min\{i \in [1, M]: x_i = a\}$$
$$L(x_1^M) = \max\{i \in [1, M]: x_i = a\},$$

with the convention $F(x_1^M) = L(x_1^M) = 0$, if no a appears in x_1^M. The almost-covering idea will be used to show that

$$\lim_{M\to\infty} \frac{L(x_1^M) - F(x_1^M)}{M} = 1, \text{ almost surely.} \tag{5}$$

To prove this, let

$$\mathcal{B}_n = \{a_1^n: a_i = a, \text{ for some } i \in [1, n]\}.$$

Given $\delta > 0$, choose n so large that $\mu(\mathcal{B}_n) > 1 - \delta$. Such an n exists by the ergodic theorem. The almost-covering principle, Theorem I.4.7, implies that x_1^M is eventually almost surely $(1 - \delta)$-strongly-covered by $\mathcal{B}_n$. Note, however, that if x_1^M is $(1 - \delta)$-strongly-covered by $\mathcal{B}_n$, then at least one member of $\mathcal{B}_n$ must start within δM of the beginning of x_1^M and at least one member of $\mathcal{B}_n$ must start within δM of the end of x_1^M. But if this is so and $M > n/\delta$, then $F(x_1^M) \le 2\delta M$ and $L(x_1^M) \ge (1 - \delta)M$, so that $L(x_1^M) - F(x_1^M) \ge (1 - 3\delta)M$. Since δ is arbitrary, the desired result, (5), follows.

It follows easily from (5), together with the ergodic theorem, that the expected waiting time between occurrences of a along a sample path is, in the limit, almost surely equal to $1/\mu(a)$, see Exercise 2. This result also follows easily from the return-time picture discussed in Section I.2.c.

I.4.c The ergodic decomposition.

In the stationary nonergodic case, limiting relative frequencies of all orders exist almost surely, but the limit measure varies from sequence to sequence. The sequences that produce the same limit measure can be grouped into classes and the process measure can then be represented as an average of the limit measures, almost all of which will, in fact, be ergodic. These ideas will be made precise in this section.

The set of all x for which the limit

$$p(a_1^k|x) = \lim_{n\to\infty} p(a_1^k|x_1^n)$$

exists for all k and all a_1^k will be denoted by $\mathcal{L}$. Thus $\mathcal{L}$ is the set of all sequences for which all limiting relative frequencies exist for all possible blocks of all possible sizes. The set $\mathcal{L}$ is shift invariant and any shift-invariant measure has its support in $\mathcal{L}$. This support result, which is a simple consequence of the ergodic theorem, is stated as the following theorem.

Theorem I.4.9

If μ is a shift-invariant measure then $\mu(\mathcal{L}) = 1$.

Proof. Since

$$p(a_1^k|x_1^n) = \frac{1}{n-k+1} \sum_{i=1}^{n-k+1} \chi_{[a_1^k]}(T^{i-1}x),$$

the ergodic theorem implies that $p(a_1^k|x) = \lim_n p(a_1^k|x_1^n)$ exists almost surely, for each k and a_1^k. Since there are only a countable number of the a_1^k, the complement of $\mathcal{L}$ must have measure 0 with respect to μ, which establishes the theorem. □

Theorem I.4.9 can be strengthened to assert that a shift-invariant measure μ must actually be supported by the sequences for which the limiting frequencies not only exist but define ergodic measures. To make this assertion precise first note that the ergodic theorem gives the formula

$$\mu(a_1^k) = \int p(a_1^k|x)\, d\mu(x). \tag{6}$$

The frequencies $p(a_1^k|x)$ can be thought of as measures, for if $x \in \mathcal{L}$ then the formula

$$\mu_x(a_1^k) = p(a_1^k|x),\ a_1^k \in A^k$$

defines a measure μ_x on sequences of length k, for each k, which can be extended, by the Kolmogorov consistency theorem, to a Borel measure μ_x on A^∞. The measure μ_x will be called the *(limiting) empirical measure determined by x.*

Two sequences x and y are said to be *frequency-equivalent* if $\mu_x = \mu_y$. Let π denote the projection onto frequency equivalence classes. The projection π is measurable and transforms the measure μ onto a measure $\omega = \mu \circ \pi^{-1}$ on equivalence classes. Next define $\mu_{\pi(x)} = \mu_x$, which is well-defined on the frequency equivalence class of x. Formula (6) can then be expressed in the form

$$\mu(B) = \int \mu_{\pi(x)}(B)\, d\omega(\pi(x)),\ B \in \Sigma. \tag{7}$$

This formula indeed holds for any cylinder set, by (6), and extends by countable additivity to any Borel set B.

Let $\mathcal{E}$ denote the set of all sequences $x \in \mathcal{L}$ for which the measure μ_x is ergodic. Of course, each frequency-equivalence class is either entirely contained in $\mathcal{E}$ or disjoint from $\mathcal{E}$. Theorem I.4.9 may now be expressed in the following much stronger form.

Theorem I.4.10 (The ergodic decomposition theorem.)

If μ is shift-invariant then $\mu(\mathcal{E}) = 1$ and hence the representation (7) takes the form,

$$\mu(B) = \int_{\mathcal{E}} \mu_{\pi(x)}(B)\, d\omega(\pi(x)),\ B \in \Sigma.$$

In other words, a shift-invariant measure always has its support in the set of sequences whose empirical measures are ergodic. The ergodic measures, $\mu_{\pi(x)}$, $x \in \mathcal{E}$, are called the *ergodic components* of μ and the formula represents μ as an average of its ergodic components. An excellent discussion of ergodic components and their interpretation in communications theory can be found in [19].

The usual measure theory argument, together with the finite form of ergodicity, Theorem I.4.5, shows that to prove the ergodic decomposition theorem it is enough to prove the following lemma.

Lemma I.4.11

Given $\epsilon > 0$ and a_1^k there is a set X_1 with $\mu(X_1) \geq 1 - \epsilon$, such that for any $x \in X_1$ there is an N such that if $n \geq N$ there is a collection $C_n \subset A^n$, with $\mu_x(C_n) \geq 1 - \epsilon$, such that the frequency of occurrence of a_1^k in any two members of C_n differs by no more than ϵ.

Proof. Fix $\epsilon > 0$, fix a block a_1^k, and fix a sequence $y \in \mathcal{L}$, the set of sequences with limiting frequencies of all orders. Let $\mathcal{L}_1 = \mathcal{L}_1(y)$ denote the set of all $x \in \mathcal{L}$ such that

$$|p(a_1^k|x) - p(a_1^k|y)| \leq \epsilon/4. \tag{8}$$

Note, in particular, that the relative frequency of occurrence of a_1^k in any two members of $\mathcal{L}_1$ differ by no more than $\epsilon/2$.

For each $x \in \mathcal{L}_1$, the sequence $\{p(a_1^k|x_1^n)\}$ converges to $p(a_1^k|x)$, as $n \to \infty$, hence there is an integer N and a set $\mathcal{L}_2 \subset \mathcal{L}_1$ of measure at least $(1 - \epsilon^2)\mu(\mathcal{L}_1)$ such that

$$|p(a_1^k|x_1^n) - p(a_1^k|x)| \leq \epsilon/4, \ x \in \mathcal{L}_2, \ n \geq N. \tag{9}$$

Fix $n \geq N$ and put $C_n = \{x_1^n \colon x \in \mathcal{L}_2\}$. The conditions (9) and (8) guarantee that the relative frequency of occurrence of a_1^k in any two members $x_1^n, \bar{x}_1^n$ of C_n differ by no more ϵ.

The set $\mathcal{L}_1$ is invariant since $p(a_1^k|x) = p(a_1^k|Tx)$, for any $x \in \mathcal{L}$. Thus the conditional measure $\mu(\cdot\ |\mathcal{L}_1)$ is shift-invariant and

$$\mu(x_1^n|\mathcal{L}_1) = \frac{1}{\mu(\mathcal{L}_1)} \int_{\mathcal{L}_1} p(x_1^n|z)\, d\mu(z), \ x_1^n \in A^n.$$

In particular,

$$\mu(C_n|\mathcal{L}_1) = \frac{1}{\mu(\mathcal{L}_1)} \int_{\mathcal{L}_1} p(C_n|z)\, d\mu(z) \geq 1 - \epsilon^2$$

so the Markov inequality yields a set $\mathcal{L}_3 \subset \mathcal{L}_2$ such that $\mu(\mathcal{L}_3) \geq (1 - \epsilon)\mu(\mathcal{L}_1)$ and for which

$$\sum_{x_1^n \in C_n} p(x_1^n|z) \geq 1 - \epsilon, \ z \in \mathcal{L}_3,$$

which translates into the statement that $\mu_z(C_n) \geq 1 - \epsilon, \ z \in \mathcal{L}_3$.

The unit interval $[0, 1]$ is bounded so $\mathcal{L}$ can be covered by a finite number of the sets $\mathcal{L}_1(y)$ and hence the lemma is proved. This completes the proof of the ergodic decomposition theorem, Theorem I.4.10. □

Remark I.4.12

The ergodic decomposition can also be viewed as an application of the Choquet-Bishop-deLeeuw theorem of functional analysis, [57]. Indeed, the set $\mathcal{P}_s$ of shift-invariant probability measures is compact in the weak*-topology (obtained by thinking of continuous functions as linear functionals on the space of all measures via the mapping $\mu \mapsto \int f\, d\mu$.) The extreme points of $\mathcal{P}_s$ correspond exactly to the ergodic measures, see Exercise 1 of Section I.9.

I.4.d Exercises

1. Let $q_k(\cdot\,|x_1^n)$ be the empirical distribution of nonoverlapping k-blocks in x_1^n, that is,
$$q_k(a_1^k|x_1^n) = |\{j \in [0, t): x_{jk+1}^{jk+k} = a_1^k\}|/t,$$
where $n = tk + r, 0 \le r < k$.
   (a) Show that if T is totally ergodic then $q_k(a_1^k|x_1^n) \to \mu(x_1^k)$, almost surely.
   (b) Show that the preceding result is not true if T is not totally ergodic.

2. Combine (5) with the ergodic theorem to establish that for an ergodic process the expected waiting time between occurrences of a symbol $a \in A$ is just $1/\mu(a)$. (Hint: show that the average time between occurrences of a in x_1^n is close to $1/p_1(a|x_1^n)$.)

3. Let $\{X_n: n \ge 1\}$ be a stationary process. The reversed process $\{Y_n: n \ge 1\}$ is the $(T^{-1}, \mathcal{P})$-process, where $\mathcal{P}$ is the Kolmogorov partition in the two-sided representation of $\{X_n: n \ge 1\}$. Show that $\{X_n\}$ is ergodic if and only its reversed process is ergodic. (Hint: combine Exercise 2 with Theorem I.4.5.)

4. Use the central limit theorem to show that the "random walk with random scenery" process is ergodic. (Hint: with high probability (x_m^{m+k}, z_m^{m+k}) and (x_1^k, z_1^k) are looking at different parts of y. See Exercise 8 in Section I.2 for a definition of this process.)

5. Let $\mathcal{P}$ be the time-0 partition for an ergodic Markov chain with period $d = 3$.
   (a) Determine the ergodic components of the $(T^3, \mathcal{P})$-process.
   (b) Determine the ergodic components of the $(T^3, \mathcal{P} \vee T\mathcal{P} \vee T^2\mathcal{P})$-process.

6. Show that an ergodic finite-state process is a function of an ergodic Markov chain. (It is an open question whether a mixing finite-state process is a function of a mixing Markov chain.)

7. Let T be an ergodic rotation of the circle and $\mathcal{P}$ the two-set partition of the unit circle into upper and lower halves. Describe the ergodic components of the $(T \times T, \mathcal{P} \times \mathcal{P})$-process.

8. This exercise illustrates a direct method for transporting a sample-path theorem from an ergodic process to a sample-path theorem for a (nonstationary) function of the process.

   Fix $a \in A$ and let A_a^∞ be the set of all $x \in [a]$ such that $T^n x \in [a]$, for infinitely many n. Define the mapping $x \mapsto R(x) = \{R_1(x), R_2(x), \ldots\}$ by
$$\begin{aligned} R_1(x) &= \min\{m > 1: x_m = a\} - 1 \\ R_{n+1}(x) &= \min\{m > \sum_{j=1}^n R_j(x): x_m = a\} - \sum_{j=1}^n R_j(x). \end{aligned}$$
   (This is just the return-time process associated with the set $B = [a]$.) A measure μ on A_a^∞ transports to the measure $\nu = \mu \circ R^{-1}$ on $\mathcal{N}^\infty$. Assume μ is the conditional measure on $[a]$ defined by an ergodic process. Show that

(a) The mapping $x \mapsto R(X)$ is Borel.

(b) $R(T^{R_1(x)}x) = SR(x)$, where S is the shift on $\mathcal{N}^\infty$.

(c) If $\mathcal{T}$ is the set of the set of frequency-typical sequences for μ, then $\nu(R(\mathcal{T})) = 1$. (Hint: use the Borel mapping lemma, Lemma I.1.17.)

(d) $(1/n)\sum_1^n R_i \to 1/\mu(a)$, ν-almost surely. (Hint: use the preceding result.)

Section I.5 The entropy theorem.

For stationary processes, the measure $\mu(x_1^n)$ is nonincreasing in n, and, except for some interesting cases, has limit 0; see Exercise 4c. The entropy theorem asserts that, for ergodic processes, the decrease is almost surely exponential in n, with a constant exponential rate called the *entropy* or *entropy-rate*, of the process, and denoted by $h = h(\mu)$.

Theorem I.5.1 (The entropy theorem.)

Let μ be an ergodic measure for the shift T on the space A^∞, where A is finite. There is a nonnegative number $h = h(\mu)$ such that

$$\lim_{n\to\infty} \frac{1}{n} \log \frac{1}{\mu(x_1^n)} = h, \text{ almost surely.}$$

In the theorem and henceforth, log means the base 2 logarithm, and the natural logarithm will be denoted by ln.

The proof of the entropy theorem will use the packing lemma, some counting arguments, and the concept of the entropy of a finite distribution. The *entropy* of a probability distribution $\pi(a)$ on a finite set A is defined by

$$H(\pi) = -\sum_{a\in A} \pi(a) \log \pi(a).$$

Let

$$h_n(x) = \frac{1}{n} \log \frac{1}{\mu(x_1^n)} = -\frac{1}{n} \log \mu(x_1^n),$$

so that if μ_n is the measure on A^n defined by μ then $H(\mu_n) = nE(h_n(x))$. The next three lemmas contain the facts about the entropy function and its connection to counting that will be needed to prove the entropy theorem.

Lemma I.5.2

The entropy function $H(\pi)$ is concave in π and attains its maximum value $\log|A|$ *only for the uniform distribution,* $\pi(a) \equiv 1/|A|$.

Proof. An elementary calculus exercise. □

Lemma I.5.3

$E(h_n(x)) \le \log|A|$.

Proof. Since $H(\mu_n) = nE(h_n(x))$, the lemma follows from the preceding lemma. □

Lemma I.5.4 (The combinations bound.)
If $\binom{n}{k}$ *denotes the number of combinations of n objects taken k at a time and* $\delta < 1/2$ *then*

$$\sum_{k \le n\delta} \binom{n}{k} \le 2^{nH(\delta)},$$

where $H(\delta) = -\delta \log \delta - (1 - \delta) \log(1 - \delta)$.

Proof. The function $-q \log \delta - (1 - q) \log(1 - \delta)$ is increasing in the interval $0 \le q \le \delta$ and hence

$$2^{-nH(\delta)} \le \delta^k (1 - \delta)^{n-k},\ k \le n\delta.$$

Multiplying by $\binom{n}{k}$ and summing gives

$$2^{-nH(\delta)} \sum_{k \le n\delta} \binom{n}{k} \le \sum_{k \le n\delta} \binom{n}{k} \delta^k (1 - \delta)^{n-k} \le \sum_{k=0}^{n} \binom{n}{k} \delta^k (1 - \delta)^{n-k} = 1,$$

which proves the lemma. □

The function $H(\delta) = -\delta \log \delta - (1-\delta) \log(1-\delta)$ is called the *binary entropy function.* It is just the entropy of the binary distribution with $\mu(1) = \delta$. It can be shown, see [7], that the binary entropy function is the correct (asymptotic) exponent in the number of binary sequences of length n that have no more than δn ones, that is

$$\lim_{n \to \infty} \frac{1}{n} \log \sum_{k \le n\delta} \binom{n}{k} = H(\delta).$$

I.5.a The proof of the entropy theorem.

Define

$$h(x) = \liminf_{n \to \infty} h_n(x),$$

where $h_n(x) = -(1/n) \log \mu(x_1^n)$. The first task is to show that $h(x)$ is constant, almost surely. To establish this note that if $y = Tx$ then

$$[y_1^n] = [x_2^{n+1}] \supseteq [x_1^{n+1}],$$

so that $\mu(y_1^n) \ge \mu(x_1^{n+1})$ and hence

$$h(Tx) \le h(x),$$

that is, $h(x)$ is subinvariant. Thus $h(Tx) = h(x)$, almost surely, since subinvariant functions are almost surely invariant, see Exercise 4, Section I.2. Since T is assumed to be ergodic the invariant function $h(x)$ must indeed be constant, almost surely.

Define h to be the constant such that

$$h = \liminf_{n \to \infty} h_n(x), \quad \text{almost surely.}$$

The goal is to show that h must be equal to $\limsup_n h_n(x)$, almost surely. Towards this end, let ϵ be a positive number. The definition of limit inferior implies that for almost all x the inequality $h_n(x) \leq h + \epsilon$ holds for infinitely many values of n, a fact expressed in exponential form as

$$\mu(x_1^n) \geq 2^{-n(h+\epsilon)} \tag{1}$$

infinitely often, almost surely. The goal is to show that "infinitely often, almost surely" can be replaced by "eventually, almost surely," for a suitable multiple of ϵ.

Three ideas are used to complete the proof. The first idea is a *packing* idea. Eventually almost surely, most of a sample path is filled by disjoint blocks of varying lengths for each of which the inequality (1) holds. This is a simple application of the ergodic stopping-time packing lemma.

The second idea is a *counting* idea. The set of sample paths of length K that can be mostly filled by disjoint, long subblocks for which (1) holds cannot have cardinality exponentially much larger than $2^{K(h+\epsilon)}$, for large enough K. Indeed, if these subblocks are to be long, then there are not too many ways to specify their locations, and if they mostly fill, then once their locations are specified, there are not too many ways the parts outside the long blocks can be filled. Most important of all is that once locations for these subblocks are specified, then since a location of length n can be filled in at most $2^{n(h+\epsilon)}$ ways if (1) is to hold, there will be a total of at most $2^{K(h+\epsilon)}$ ways to fill all the locations.

The third idea is a *probability* idea. If a set of K-length sample paths has cardinality only a bit more than $2^{K(h+\epsilon)}$, then it is very unlikely that a sample path in the set has probability exponentially much smaller than $2^{-K(h+\epsilon)}$. This is just an application of the fact that upper bounds on cardinality "almost" imply lower bounds on probability, Lemma I.1.18(b).

To fill in the details of the first idea, let δ be a positive number and $M > 1/\delta$ an integer, both to be specified later. For each $K \geq M$, let $G_K(\delta, M)$ be the set of all X_1^K that are $(1-2\delta)$-packed by disjoint blocks of length at least M for which the inequality (1) holds. In other words, $x_1^K \in G_K(\delta, M)$ if and only if there is a collection

$$\mathcal{S} = \mathcal{S}(x_1^K) = \{[n_i, m_i]\}$$

of disjoint subintervals of $[1, K]$ with the following properties.

(a) $m_i - n_i + 1 \geq M, \quad [n_i, m_i] \in \mathcal{S}.$

(b) $\mu(x_{n_i}^{m_i}) \geq 2^{-(m_i-n_i+1)(h+\epsilon)}, \quad [n_i, m_i] \in \mathcal{S}.$

(c) $\sum_{[n_i,m_i]\in\mathcal{S}}(m_i - n_i + 1) \geq (1-2\delta)K.$

An application of the packing lemma produces the following result.

Lemma I.5.5

$x_1^K \in G_K(\delta, M)$, eventually almost surely.

Proof. Define $\tau(x)$ to be the first time $n \geq M$ such that $\mu([x_1^n]) \geq 2^{-n(h+\epsilon)}$. Since τ is measurable and (1) holds infinitely often, almost surely, τ is an almost surely finite stopping time. An application of the ergodic stopping-time lemma, Lemma I.3.7, then yields the lemma. □

The second idea, the counting idea, is expressed as the following lemma.

Lemma I.5.6

There is a $\delta > 0$ and an $M \geq 1/\delta$ such that $|G_K(\delta, M)| \leq 2^{K(h+2\epsilon)}$, for all sufficiently large K.

Proof. A collection $\mathcal{S} = \{[n_i, m_i]\}$ of disjoint subintervals of $[1, K]$, will be called a *skeleton* if it satisfies the requirement that $m_i - n_i + 1 \geq M$, for each i, and if it covers all but a 2δ-fraction of $[1, K]$, that is,

$$\sum_{[n_i, m_i] \in \mathcal{S}} (m_i - n_i + 1) \geq (1 - 2\delta)K. \tag{2}$$

A sequence x_1^K is said to be *compatible* with such a skeleton $\mathcal{S}$ if

$$\mu(x_{n_i}^{m_i}) \geq 2^{-(m_i - n_i + 1)(h+\epsilon)},$$

for each i. The bound of the lemma will be obtained by first upper bounding the number of possible skeletons, then upper bounding the number of sequences x_1^K that are compatible with a given skeleton. The product of these two numbers is an upper bound for the cardinality of $G_K(\delta, M)$ and a suitable choice of δ will then establish the lemma.

First note that the requirement that each member of a skeleton $\mathcal{S}$ have length at least M, means that $|\mathcal{S}| \leq K/M$, and hence there are at most K/M ways to choose the starting points of the intervals in $\mathcal{S}$. Thus the number of possible skeletons is upper bounded by

$$\sum_{k \leq K/M} \binom{K}{k} \leq 2^{KH(1/M)}, \tag{3}$$

where the upper bound is provided by Lemma I.5.4, with $H(\cdot)$ denoting the binary entropy function.

Fix a skeleton $\mathcal{S} = \{[n_i, m_i]\}$. The condition that x_1^K be compatible with $\mathcal{S}$ means that the compatibility condition

$$\mu(x_{n_i}^{m_i}) \geq 2^{-(m_i - n_i + 1)(h+\epsilon)}, \tag{4}$$

must hold for each $[n_i, m_i] \in \mathcal{S}$. For a given $[n_i, m_i]$ the number of ways $x_{n_i}^{m_i}$ can be chosen so that the compatibility condition (4) holds, is upper bounded by $2^{(m_i - n_i + 1)(h+\epsilon)}$, by the principle that lower bounds on probability imply upper bounds on cardinality, Lemma I.1.18(a). Thus, the number of ways x_j can be chosen so that $j \in \cup_i [n_i, m_i]$ and so that the compatibility conditions hold is upper bounded by

$$\prod_i 2^{(m_i - n_i + 1)(h+\epsilon)} \leq 2^{K(h+\epsilon)}$$

Outside the union of the $[n_i, m_i]$ there are no conditions on x_j. Since, however, there are fewer than $2\delta K$ such j these positions can be filled in at most $|A|^{2\delta K}$ ways. Thus, there are at most

$$|A|^{2\delta K} 2^{K(h+\epsilon)}$$

sequences compatible with a given skeleton $\mathcal{S} = \{[n_i, m_i]\}$. Combining this with the bound, (3), on the number of possible skeletons yields

$$|G_K(\delta, M)| \leq 2^{KH(1/M)} |A|^{2K\delta} 2^{K(h+\epsilon)}. \tag{5}$$

Since the binary entropy function $H(1/M)$ approaches 0 as $M \to \infty$ and since $|A|$ is finite, the numbers $\delta > 0$ and $M > 1/\delta$ can indeed be chosen so that $|G_K(\delta, M)| \leq 2^{K(h+2\epsilon)}$, for all sufficiently large K. This completes the proof of Lemma I.5.6. □

Fix $\delta > 0$ and $M \geq 1/\delta$ for which Lemma I.5.6 holds, put $G_K = G_K(\delta, M)$, and let B_K be the set of all x_1^K for which $\mu(x_1^K) < 2^{-K(h+3\epsilon)}$. Then

$$\mu(B_K \cap G_K) \leq |G_K| 2^{-K(h+3\epsilon)} \leq 2^{-K\epsilon},$$

holds for all sufficiently large K. Thus, $x_1^K \notin B_K \cap G_K$, eventually almost surely, by the Borel-Cantelli principle. Since $x_1^K \in G_K$, eventually almost surely, the iterated almost-sure principle, Lemma I.1.15, implies that $x_1^K \notin B_K$, eventually almost surely, that is,

$$\limsup_{K \to \infty} h_K(x) \leq h + 3\epsilon, \text{ a.s.}$$

In summary, for each $\epsilon > 0$,

$$h = \liminf_{K \to \infty} h_K(x) \leq \limsup_{K \to \infty} h_K(x) \leq h + 3\epsilon, \text{ a.s.},$$

which completes the proof of the entropy theorem, since ϵ is arbitrary. □

Remark I.5.7

The entropy theorem was first proved for Markov processes by Shannon, with convergence in probability established by McMillan and almost-sure convergence later obtained by Breiman, see [4] for references to these results. In information theory the entropy theorem is called the asymptotic equipartition property, or AEP. In ergodic theory it has been traditionally known as the Shannon-McMillan-Breiman theorem. The more descriptive name "entropy theorem" is used in this book. The proof given is due to Ornstein and Weiss, [51], and appeared as part of their extension of ergodic theory ideas to random fields and general amenable group actions. A slight variant of their proof, based on the separated packing idea discussed in Exercise 1, Section I.3, appeared in [68].

I.5.b Exercises.

1. Prove the entropy theorem for the i.i.d. case by using the product formula on $\mu(x_1^n)$, then taking the logarithm and using the strong law of large numbers. This yields the formula $h = -\sum_a \mu(a) \log \mu(a)$.

2. Use the idea suggested by the preceding exercise to prove the entropy theorem for ergodic Markov chains. What does it give for the value of h?

3. Suppose for each k, $\mathcal{T}_k$ is a subset of A^k of cardinality at most $2^{k\alpha}$. A sequence x_1^n is said to be $(K, \delta, \{\mathcal{T}\})$-packed if it can be expressed as the concatenation $x_1^n = w(1) \ldots w(t)$, such that the sum of the lengths of the $w(i)$ which belong to $\cup_{k=K}^{\infty} \mathcal{T}_k$ is at least $(1-\delta)n$. Let G_n be the set of all $(K, \delta, \{\mathcal{T}\})$-packed sequences x_1^n and let ϵ be positive number. Show that if K is large enough, if δ is small enough, and if n is large enough relative to K and δ, then $|G_n| \leq 2^{n(\alpha+\epsilon)}$.

4. Assume μ is ergodic and define $c(x) = \lim_{n\to\infty} \mu(x_1^n)$, $x \in A^\infty$.

   (a) Show that $c(x)$ is almost surely a constant c.

   (b) Show that if $c > 0$ then μ is concentrated on a finite set.

   (c) Show that if μ is mixing then $c(x) = 0$ for every x.

Section I.6 Entropy as expected value.

Entropy for ergodic processes, as defined by the entropy theorem, is given by the almost-sure limit

$$h = \lim_{n\to\infty} \frac{1}{n} \log \frac{1}{\mu(x_1^n)}.$$

Entropy can also be thought of as the limit of the expected value of the random quantity $-(1/n)\log \mu(x_1^n)$. The expected value formulation of entropy will be developed in this section.

I.6.a The entropy of a random variable.

Let X be a finite-valued random variable with distribution defined by $p(x) = \text{Prob}(X = x)$, $x \in A$. The *entropy* of X is defined as the expected value of the random variable $-\log p(X)$, that is,

$$H(X) = \sum_{x\in A} p(x) \log \frac{1}{p(x)} = -\sum_{x\in A} p(x) \log p(x).$$

The logarithm base is 2 and the conventions $0 \log 0 = 0$ and $\log 0 = -\infty$ are used. If p is the distribution of X, then $H(p)$ may be used in place of $H(X)$. For a pair (X, Y) of random variables with a joint distribution $p(x, y) = \text{Prob}(X = x, Y = y)$, the notation $H(X, Y) = -\sum_{x,y} p(x, y) \log p(x, y)$ will be used, a notation which extends to random vectors.

Most of the useful properties of entropy depend on the concavity of the logarithm function. One way to organize the concavity idea is expressed as follows.

Lemma I.6.1

If p and q are probability k-vectors then

$$-\sum p_i \log p_i \leq -\sum p_i \log q_i,$$

with equality if and only if $p = q$.

Proof. The natural logarithm is strictly concave so that, $\ln x \le x - 1$, with equality if and only if $x = 0$. Thus

$$\sum p_i \ln \frac{q_i}{p_i} \le \sum p_i \left(\frac{q_i}{p_i} - 1 \right) \le \sum q_i - \sum p_i = 0,$$

with equality if and only if $q_i = p_i$, $1 \le i \le k$. This proves the lemma, since $\log x = (\ln x)/(\ln 2)$. □

The proof only requires that q be a sub-probability vector, that is nonnegative with $\sum_i q_i \le 1$. The sum

$$D(p\|q) = \sum_i p_i \ln \frac{p_i}{q_i}$$

is called the (informational) *divergence*, or *cross-entropy*, and the preceding lemma is expressed in the following form.

Lemma I.6.2 (The divergence inequality.)
If p is a probability k-vector and q is a sub-probability k-vector then $D(p\|q) \ge 0$, with equality if and only if $p = q$.

A further generalization of the lemma, called the log-sum inequality, is included in the exercises.

The basic inequalities for entropy are summarized in the following theorem.

Theorem I.6.3 (Entropy inequalities.)

(a) Positivity. $H(X) \ge 0$, *with equality if and only if X is constant.*

(b) Boundedness. *If X has k values then $H(X) \le \log k$, with equality if and only if each $p(x) = 1/k$.*

(c) Subadditivity. $H(X, Y) \le H(X) + H(Y)$, *with equality if and only if X and Y are independent.*

Proof. Positivity is easy to prove, while boundedness is obtained from Lemma I.6.1 by setting $p_x = p(x)$, $q_x \equiv 1/k$. To establish subadditivity note that

$$H(X, Y) = -\sum_{x,y} p(x, y) \log p(x, y),$$

then replace $p(x, y)$ in the logarithm factor by $p(x)p(y)$, and use Lemma I.6.1 to obtain the inequality

$$H(X, Y) \le H(X) + H(Y),$$

with equality if and only if $p(x, y) \equiv p(x)p(y)$, that is, if and only if X and Y are independent. This completes the proof of the theorem. □

The concept of conditional entropy provides a convenient tool for organizing further results. If $p(x, y)$ is a given joint distribution, with corresponding conditional distribution $p(x|y) = p(x, y)/p(y)$, then

$$H(X|Y) = -\sum_{x,y} p(x, y) \log p(x|y) = -\sum_{x,y} p(x, y) \log \frac{p(x, y)}{p(y)}$$

is called the *conditional entropy* of X, given Y. (Note that this is a slight variation on standard probability language which would call $-\sum_x p(x|y)\log p(x|y)$ the conditional entropy. In information theory, however, the common practice is to take expected values with respect to the marginal, $p(y)$, as is done here.)

The key identity for conditional entropy is the following *addition law*.

$$H(X,Y) = H(Y) + H(X|Y). \tag{1}$$

This is easily proved using the additive property of the logarithm, $\log ab = \log a + \log b$.

The previous unconditional inequalities extend to conditional entropy as follows. (The proofs are left to the reader.)

Theorem I.6.4 (Conditional entropy inequalities.)

(a) Positivity. $H(X|Y) \geq 0$, *with equality if and only if* X *is a function of* Y.

(b) Boundedness. $H(X|Y) \leq H(X)$ *with equality if and only if* X *and* Y *are independent.*

(c) Subadditivity. $H((X,Y)|Z) \leq H(X|Z) + H(Y|Z)$, *with equality if and only if* X *and* Y *are conditionally independent given* Z.

A useful fact is that conditional entropy $H(X|Y)$ increases as more is known about the first variable and decreases as more is known about the second variable, that is, for any functions f and g,

Lemma I.6.5
$H(f(X)|Y) \leq H(X|Y) \leq H(X|g(Y))$.

The proof follows from the concavity of the logarithm function. This can be done directly (left to the reader), or using the partition formulation of entropy which is developed in the following paragraphs.

The entropy of a random variable X really depends only on the partition $\mathcal{P}_X = \{P_a\colon a \in A\}$ defined by $P_a = \{x\colon X(x) = a\}$, which is called the *partition defined by* X. The entropy $H(\mathcal{P})$ is defined as $H(X)$, where X is any random variable such that $\mathcal{P}_X = \mathcal{P}$. Note that the join $\mathcal{P}_X \vee \mathcal{P}_Y$ is just the partition defined by the vector (X,Y) so that $H(\mathcal{P}_X \vee \mathcal{P}_Y) = H(X,Y)$. The conditional entropy of $\mathcal{P}$ relative to $\mathcal{Q}$ is then defined by

$$H(\mathcal{P}|\mathcal{Q}) = H(\mathcal{P} \vee \mathcal{Q}) - H(\mathcal{Q}).$$

The partition point of view provides a useful geometric framework for interpretation of the inequalities in Lemma I.6.5, because the partition $\mathcal{P}_X$ is a refinement of the partition $\mathcal{P}_{f(X)}$, since each atom of $\mathcal{P}_{f(X)}$ is a union of atoms of $\mathcal{P}_X$. The inequalities in Lemma I.6.5 are expressed in partition form as follows.

Lemma I.6.6

(a) *If* $\mathcal{P}$ *refines* $\mathcal{Q}$ *then* $H(\mathcal{Q}|\mathcal{R}) \leq H(\mathcal{P}|\mathcal{R})$.

(b) *If* $\mathcal{R}$ *refines* $\mathcal{S}$ *then* $H(\mathcal{P}|\mathcal{S}) \geq H(\mathcal{P}|\mathcal{R})$.

Proof. The proof of (a) is accomplished by manipulating with entropy formulas, as follows.

$$\begin{aligned} H(\mathcal{P}|\mathcal{R}) &\overset{(i)}{=} H(\mathcal{P}\vee\mathcal{Q}|\mathcal{R}) \\ &\overset{(ii)}{=} H(\mathcal{Q}|\mathcal{R}) + H(\mathcal{P}|\mathcal{Q}\vee\mathcal{R}) \\ &\overset{(iii)}{\geq} H(\mathcal{Q}|\mathcal{R}). \end{aligned}$$

The equality (i) follows from the fact that $\mathcal{P}$ refines $\mathcal{Q}$, so that $\mathcal{P} = \mathcal{P}\vee\mathcal{Q}$; the equality (ii) is just the general addition law; and the inequality (iii) uses the fact that $H(\mathcal{P}|\mathcal{Q}\vee\mathcal{R}) \geq 0$.

Let $\mathcal{P}_C$ denote the partition of the set C defined by restricting the sets in $\mathcal{P}$ to C, where the conditional measure, $\mu(\cdot\ |C)$ is used on C. To prove (b) it is enough to consider the case when $\mathcal{R}$ is obtained from $\mathcal{S}$ by splitting one atom of $\mathcal{S}$ into two pieces, say

$$R_a = S_a,\ a \neq b; \quad S_b = R_{b_1} \cup R_{b_2}.$$

The quantity $H(\mathcal{P}|\mathcal{R})$ can be expressed as

$$\begin{aligned} &-\sum_t \sum_{a\neq b} \mu(P_t \cap R_a) \log \frac{\mu(P_t \cap R_a)}{\mu(R_a)} + \mu(S_b) H(\mathcal{P}_{S_b}|\mathcal{R}_{S_b}) \\ &\leq -\sum_t \sum_{a\neq b} \mu(P_t \cap R_a) \log \frac{\mu(P_t \cap R_a)}{\mu(R_a)} + \mu(S_b) H(\mathcal{P}_{S_b}). \end{aligned}$$

The latter is the same as $H(\mathcal{P}|\mathcal{S})$, which establishes inequality (b). □

I.6.b The entropy of a process.

The entropy of a process is defined by a suitable passage to the limit. The *n-th order entropy* of a sequence $\{X_1, X_2, \ldots\}$ of A-valued random variables is defined by

$$H(X_1^n) = \sum_{x_1^n \in A^n} p(x_1^n) \log \frac{1}{p(x_1^n)} = -\sum_{x_1^n \in A^n} p(x_1^n) \log p(x_1^n).$$

The *process entropy* is defined by passing to a limit, namely,

$$H(\{X_n\}) = \limsup_n \frac{1}{n} H((X_1^n) = \limsup_n \frac{1}{n} \sum_{x_1^n} p(x_1^n) \log \frac{1}{p(x_1^n)}. \tag{2}$$

In the stationary case the limit superior is a limit. This is a consequence of the following basic subadditivity property of nonnegative sequences.

Lemma I.6.7 (The subadditivity lemma.)

If $\{a_n\}$ is a sequence of nonnegative numbers which is subadditive, that is, $a_{n+m} \leq a_n + a_m$, then $\lim_n a_n/n$ exists and equals $\inf_n a_n/n$.

Proof. Let $a = \inf_n a_n/n$. Given $\epsilon > 0$ choose n so that $a_n < n(a+\epsilon)$. If $m \geq n$ write $m = np + r, 0 \leq r < n$. Subadditivity gives $a_{np} \leq pa_n$, so that if $b = \sup_{i\leq n} a_i$ then another use of subadditivity yields

$$a_m \leq a_{np} + a_r \leq pa_n + b.$$

Division by m, and the fact that $np/m \to 1$, as $m \to \infty$, gives $\limsup a_m/m \leq a + \epsilon$. This proves the lemma. □

The subadditivity property for entropy, Theorem I.6.3(c), gives

$$H(X_1^{n+m}) \leq H(X_1^n) + H(X_{n+1}^{n+m}).$$

If the process is stationary then $H(X_{n+1}^{n+m}) = H(X_1^m)$ so the subadditivity lemma with $a_n = H(X_1^n)$ implies that the limit superior in (2) is a limit.

An alternative formula for the process entropy in the stationary case is obtained by using the general addition law, $H(X, Y) = H(Y) + H(X|Y)$, to produce the formula

$$H(X_{-n}^0) - H(X_{-n}^{-1}) = H(X_0|X_{-n}^{-1}),\ n > 0.$$

The right-hand side is decreasing in n, from Lemma I.6.5, and the simple fact from analysis that if $a_n \geq 0$ and $a_{n+1} - a_n$ decreases to b then $a_n/n \to b$, can then be applied to give

$$H(\{X_n\}) = \lim_{n\to\infty} H\left(X_0|X_{-n}^{-1}\right), \tag{3}$$

a formula often expressed in the suggestive form

$$H(\{X_n\}) = H\left(X_0|X_{-\infty}^{-1}\right). \tag{4}$$

The next goal is to show that for ergodic processes, the process entropy H is the same as the entropy-rate h of the entropy theorem. Towards this end, the following lemma will be useful in controlling entropy on small sets.

Lemma I.6.8 (The subset entropy-bound.)

$$-\sum_{a\in B} p(a)\log p(a) \leq p(B)\log|B| - p(B)\log p(B),$$

for any $B \subset A$.

Proof. Let $p(\cdot\ |B)$ denote the conditional measure on B and use the trivial bound to obtain

$$-\sum_{a\in B} p(a|B)\log p(a|B) \leq \log|B|.$$

The left-hand side is the same as

$$\left(-\frac{1}{p(B)}\sum_{a\in B} p(a)\log p(a) + \log p(B)\right),$$

from which the result follows. □

Let μ be an ergodic process with alphabet A, and process entropy H, and let h be the entropy-rate of μ as given by the entropy theorem, that is

$$\lim_n \frac{1}{n}\log\frac{1}{\mu(x_1^n)} = h, \text{ a.e.}$$

Theorem I.6.9

Process entropy H and entropy-rate h are the same for ergodic processes.

Proof. First assume that $h > 0$ and fix ϵ such that $0 < \epsilon < h$. Define

$$G_n = \{x_1^n \colon 2^{-n(h+\epsilon)} \leq \mu(x_1^n) \leq 2^{-n(h-\epsilon)}\}$$

and let $B_n = A^n - G_n$. Then

$$\begin{aligned} H\left(X_1^n\right) &= -\sum_{A^n} \mu(x_1^n) \log \mu(x_1^n) \\ &= -\sum_{B_n} \mu(x_1^n) \log \mu(x_1^n) - \sum_{G_n} \mu(x_1^n) \log \mu(x_1^n). \end{aligned}$$

The two sums will be estimated separately.

The subset entropy-bound, Lemma I.6.8, gives

$$-\sum_{B_n} \mu(x_1^n) \log \mu(x_1^n) \leq n\mu(B_n) \log |A| - \mu(B_n) \log \mu(B_n), \tag{5}$$

since $|B| \leq |A|^n$. After division by n, this part must go to 0 as $n \to \infty$, since the entropy theorem implies that $\mu(B_n)$ goes to 0.

On the set G_n the following holds

$$n(h - \epsilon) \leq -\log \mu(x_1^n) \leq n(h + \epsilon),$$

so multiplication by $\mu(x_1^n)/n\mu(G_n)$ and summing produces

$$(h - \epsilon) \leq -\frac{1}{n\mu(G_n)} \sum_{G_n} \mu(x_1^n) \log \mu(x_1^n) \leq (h + \epsilon). \tag{6}$$

As $n \to \infty$ the measure of G_n approaches 1, from the entropy theorem, so the sum converges to $H(\{X_n\})$. Thus

$$h - \epsilon \leq H(\{X_n\}) \leq h + \epsilon,$$

which proves the theorem in the case when $h > 0$.

If $h = 0$ then define $G_n = \{x_1^n \colon \mu(x_1^n) \geq 2^{-n\epsilon}\}$. The bound (5) still holds, while only the upper bound in (6) matters and the theorem again follows. □

Remark I.6.10

Since for ergodic processes, the process entropy $H(\{X_n\})$ is the same as the entropy-rate h of the entropy theorem, both are often simply called the entropy. A detailed mathematical exposition and philosophical interpretation of entropy as a measure of information can be found in Billingsley's book, [4]. The Csiszár-Körner book, [7], contains an excellent discussion of combinatorial and communication theory aspects of entropy, in the i.i.d. case. The recent books by Gray, [18], and Cover and Thomas, [6], discuss many information-theoretic aspects of entropy for the general ergodic process.

I.6.c The entropy of i.i.d. and Markov processes.

Entropy formulas for i.i.d. and Markov processes can be derived from the entropy theorem by using the ergodic theorem to directly estimate $\mu(x_1^n)$, see Exercises 1 and 2 in Section I.5. Here they will be derived from the definition of process entropy.

Let $\{X_n\}$ be an i.i.d. process and let $p(x) = \text{Prob}(X_1 = x)$. The additivity of entropy for independent random variables, Theorem I.6.3(c), gives

$$H(X_1^n) = H(X_1) + H(X_2) + \ldots + H(X_n) = nH(X_1),$$

which yields the entropy formula

$$H(\{X_n\}) = H(X_1) = -\sum_x p(x)\log p(x). \tag{7}$$

An alternate proof can be given by using the conditional limit formula, (3), and the fact that $H(X|Y) = H(X)$ if X and Y are independent, to obtain

$$H(\{X_n\}) = \lim_n H\left(X_0|X_{-n}^{-1}\right) = H(X_0).$$

Now suppose $\{X_n\}$ is a Markov chain with stationary vector p and transition matrix M. The Markov property implies that

$$\text{Prob}\left(X_0 = a|X_{-n}^{-1}\right) = \text{Prob}\left(X_0 = a|X_{-1}\right),\ a \in A,$$

from which it follows that

$$H(X_0|X_{-n}^{-1}) = H(X_0|X_{-1}).$$

Thus the conditional entropy formula for the entropy of a process, (3), along with a direct calculation of $H(X_0|X_{-1})$ yields the following formula for the entropy of a Markov chain.

$$H(\{X_n\}) = H(X_0|X_{-1}) = -\sum_{i,j} p_i M_{ij}\log M_{ij}. \tag{8}$$

Recall that a process $\{X_n\}$ is Markov of order k if

$$\text{Prob}\left(X_0 = a|X_{-n}^{-1}\right) = \text{Prob}\left(X_0 = a|X_{-k}^{-1}\right),\ a \in A.$$

The argument used in the preceding paragraph shows that if $\{X_n\}$ is Markov of order k then

$$H(\{X_n\}) = H(X_0|X_{-k}^{-1}) = H(X_0|X_{-n}^{-1}),\ n \geq k.$$

This condition actually implies that the process is Markov of order k.

Theorem I.6.11 (The Markov order theorem.)
A stationary process $\{X_n\}$ is Markov of order k if and only if $H(X_0|X_{-k}^{-1}) = H(X_0|X_{-n}^{-1})$, $n \geq k$.

Proof. The conditional addition law gives

$$H((X_0, X_{-n}^{-k+1})|X_{-k}^{-1}) = H(X_{-n}^{-k+1}|X_{-k}^{-1}) + H(X_0|X_{-k}^{-1}, X_{-n}^{-k+1}).$$

The second term on the right can be replaced by $H(X_0|X_{-k}^{-1})$, provided $H(X_0|X_{-k}^{-1}) = H(X_0|X_{-n}^{-1})$, for $n \geq k$. The equality condition of the subadditivity principle, Theorem I.6.4(c), can then be used to conclude that X_0 and X_{-n}^{-k+1} are conditionally independent given X_{-k}^{-1}. If this is true for every $n \geq k$ then the process must be Markov of order k. This completes the proof of the theorem. □

In general the conditional entropy function $H_k = H(X_0|X_{-k}^{-1})$ is nonincreasing in k. To say that the process is Markov of some order is to say that H_k is eventually constant. If the true order is k^*, then $H_{k^*-1} > H_{k^*} = H_k,\ k \geq k^*$. This fact can be used to estimate the order of a Markov chain from observation of a sample path.

I.6.d Entropy and types.

The entropy of an empirical distribution gives the exponent for a bound on the number of sequences that could have produced that empirical distribution. This fact is useful in large deviations theory, and will be useful in some of the deeper interpretations of entropy to be given in later chapters.

The *empirical distribution* or *type* of a sequence $x_1^n \in A^n$ is the probability distribution $p_1 = p_1(\cdot\,|x_1^n)$ on A defined by the relative frequency of occurrence of each symbol in the sequence x_1^n, that is,

$$p_1(a|x_1^n) = \frac{|\{i\colon\ x_i = a\}|}{n},\quad a \in A.$$

Two sequences x_1^n and y_1^n are said to be *type-equivalent* if they have the same type, that is, if each symbol a appears in x_1^n the same number of times it appears in y_1^n. Type-equivalence classes are called *type classes.* The type class of x_1^n will be denoted by $\mathcal{T}(x_1^n)$ or by $\mathcal{T}_{p_1}^n$, where the latter stresses the type $p_1 = p_1(\cdot\,|x_1^n)$, rather than a particular sequence that defines the type. Thus, $x_1^n \in \mathcal{T}_{p_1}^n$ if and only if each symbol a appears in x_1^n exactly $np_1(a)$ times.

The *empirical (first-order) entropy* of x_1^n is the entropy of $p_1(\cdot\,|x_1^n)$, that is,

$$\widehat{H}(p_1) = \widehat{H}(p_1(\cdot\,|x_1^n)) = -\sum_a p_1(a|x_1^n)\log p_1(a|x_1^n).$$

The following purely combinatorial result bounds the size of a type class in terms of the empirical entropy of the type.

Theorem I.6.12 (The type-class bound.)
$\left|\mathcal{T}_{p_1}^n\right| \leq 2^{n\widehat{H}(p_1)}$, *for any type* p_1.

Proof. First note that if Q^n is a product measure on A^n then

$$Q^n(x_1^n) = \prod_{i=1}^n Q(x_i) = \prod_{a\in A} Q(a)^{np_1(a)},\quad x_1^n \in \mathcal{T}_{p_1}^n, \tag{9}$$

since $np_1(a)$ is the number of times the symbol a appears in a given $x_1^n \in \mathcal{T}_{p_1}^n$. In particular, a product measure is always constant on each type class.

A type p_1 defines a product measure $\widehat{P}^n$ on A^n by the formula

$$\widehat{P}^n(z_1^n) = \prod_i p_1(z_i), \ z_1^n \in A^n.$$

Replacing Q^n by $\widehat{P}^n$ in the product formula, (9), produces

$$\widehat{P}^n(x_1^n) = \prod_{a \in A} p_1(a)^{np_1(a)}, \ x_1^n \in \mathcal{T}_{p_1}^n,$$

which, after taking the logarithm and rewriting, yields

$$\widehat{P}^n(x_1^n) = 2^{-n\widehat{H}(p_1)}, \ x_1^n \in \mathcal{T}_{p_1}^n.$$

In other words, $\widehat{P}^n(x_1^n)$ has the constant value $2^{-n\widehat{H}(p_1)}$ on the type class of x_1^n, and hence $\widehat{P}^n(\mathcal{T}_{p_1}^n) = |\mathcal{T}_{p_1}^n| 2^{-n\widehat{H}(p_1)}$. But $\widehat{P}^n$ is a probability distribution so that $\widehat{P}^n(\mathcal{T}_{p_1}^n) \leq 1$. This establishes Theorem I.6.12. □

The bound $2^{n\widehat{H}}$, while not tight, is the correct asymptotic exponential bound for the size of a type class, as shown in the Csiszár-Körner book, [7]. The upper bound is all that will be used in this book.

Later use will also be made of the fact that there are only polynomially many type classes, as stated in the following theorem.

Theorem I.6.13 (The number-of-types bound.)
The number of possible types is at most $(n+1)^{|A|}$.

Proof. This follows from the fact that for each $a \in A$, the only possible values for $np_1(a|x_1^n)$ are $0, 1, \ldots, n$. □

The concept of type extends to the empirical distribution of overlapping k-blocks. The *empirical overlapping k-block distribution* or *k-type* of a sequence $x_1^n \in A^n$ is the probability distribution $p_k = p_k(\cdot\ |x_1^n)$ on A^k defined by the relative frequency of occurrence of each k-block in the sequence x_1^n, that is,

$$p_k(a_1^k|x_1^n) = \frac{\left|\{i \in [1, n-k+1]\colon x_i^{i+k-1} = a_1^k\}\right|}{n-k+1}, \ a_1^k \in A^k.$$

Two sequences x_1^n and y_1^n are said to be *k-type-equivalent* if they have the same k-type, and k-type-equivalence classes are called *k-type classes*. The k-type class of x_1^n will be denoted by $\mathcal{T}_{p_k}^n$, where $p_k = p_k(\cdot\ |x_1^n)$, that is, $x_1^n \in \mathcal{T}_{p_k}^n$ if and only if each block a_1^k appears in x_1^n exactly $(n-k+1)p_k(a_1^k)$ times.

The bound on the number of types, Theorem I.6.13, extends immediately to k-types.

Theorem I.6.14 (The number of k-types bound.)
The number of possible k-types is at most $(n-k+2)^{|A|^k}$.

Note, in particular, that if k is fixed the number of possible k-types grows polynomially in n, while if k is also growing, but satisfies $k \leq \alpha \log_{|A|} n$, with $\alpha < 1$, then the number of k-types is of lower order than $|A|^n$.

Estimating the size of a k-type class is a bit trickier, since the k-type measures the frequency of *overlapping* k-blocks. A suitable bound can be obtained, however, by considering the $(k-1)$-st order Markov measure defined by the k-type $p_k(\cdot\,|x_1^n)$, that is, the stationary $(k-1)$-st order Markov chain $\widehat{\mu}^{(k-1)}$ with the empirical transition function

$$\widehat{p}(a_k|a_1^{k-1}) = \frac{p_k(a_1^k|x_1^n)}{\sum_{b_k} p_k(a_1^{k-1}b_k|x_1^n)},$$

and with the stationary distribution given by $\widehat{p}(a_1^{k-1}) = \sum_{b_k} p_k(a_1^{k-1}b_k|x_1^n)$.

The Markov chain $\widehat{\mu}^{(k-1)}$ has entropy

$$\begin{aligned}\widehat{H}^{(k-1)} &= -\sum_{a_1^k} \widehat{p}(a_k|a_1^{k-1})\widehat{p}(a_1^{k-1}) \log \widehat{p}(a_k|a_1^{k-1}) \\ &= -\sum_{a_1^k} p_k(a_1^k|x_1^n) \log \widehat{p}(a_k|a_1^{k-1}),\end{aligned}$$

by formula (8). The entropy $\widehat{H}^{(k-1)}$ is called the *empirical* $(k-1)$*-st order Markov entropy* of x_1^n. Note that it is constant on the k-type class $\mathcal{T}_k(x_1^n) = \mathcal{T}_{p_k}$ of all sequences y_1^n that have the same k-type p_k as x_1^n.

Theorem I.6.15 (The k-type-class bound.)
$|\mathcal{T}_k(x_1^n)| \le (n-k)2^{(n-k)\widehat{H}^{(k-1)}}$.

Proof. First consider the case $k = 2$. If Q is Markov a direct calculation yields the formula

$$Q(x_1^n) = Q(x_1)\prod_{a,b} Q(b|a)^{(n-1)p_2(ab)}, \quad x_1^n \in \mathcal{T}_{p_2}.$$

This formula, with Q replaced by $\widehat{\mu}^{(2-1)} = \widehat{\mu}^{(1)}$ and after suitable rewrite using the definition of $\widehat{H}^{(1)}$ becomes

$$\widehat{\mu}^{(1)}(x_1^n) = \widehat{\mu}^{(1)}(x_1)2^{-(n-1)\widehat{H}^{(1)}}, \quad x_1^n \in \mathcal{T}_{p_2},$$

and hence

$$\widehat{\mu}^{(1)}(\mathcal{T}_{p_2}) \ge \frac{1}{n-1}2^{-(n-1)\widehat{H}^{(1)}}|\mathcal{T}_{p_2}|,$$

since $\widehat{\mu}^{(1)}(x_1) \ge 1/(n-1)$. This yields the desired bound for the first-order case, since $\widehat{\mu}^{(1)}(x_1^n)$ is a probability measure on A^n. The proof for general k is obtained by an obvious extension of the argument. □

I.6.e Exercises.

1. Prove the log-sum inequality: If $a_i \ge 0$, $b_i \ge 0$, $i = 1, 2, \ldots, n$, and $a = \sum a_i$, $b = \sum b_i$, then $\sum a_i \log(a_i/b_i) \ge a \log(a/b)$.

2. Let $\mathcal{P}$ be the time-0 partition for an ergodic Markov chain with period $d = 3$ and entropy $H > 0$.

   (a) Determine the entropies of the ergodic components of the $(T^3, \mathcal{P})$-process.

(b) Determine the entropies of the ergodic components of the $(T^3, \mathcal{P} \vee T\mathcal{P} \vee T^2\mathcal{P})$-process.

3. Let μ be an ergodic process and let α be a positive number. Show there is an $N = N(\alpha)$ such that if $n \geq N$ then there is a set F_n, measurable with respect to $X_{-\infty}^n$, such that $\mu(F_n) \geq 1 - \alpha$ and so that if $x_{-\infty}^n \in F_n$ then
$$2^{-\alpha n}\mu(x_1^n) \leq \mu(x_1^n | x_{-\infty}^0) \leq 2^{\alpha n}\mu(x_1^n).$$
(Hint: apply the ergodic theorem to $f(x) = -\log \mu(x_1 | x_{-\infty}^0)$, and use the equality of process entropy and entropy-rate.)

4. Let T be the shift on $X = \{-1, 1\}^Z$, and let μ be the product measure on X defined by $\mu_0(-1) = \mu(1) = 1/2$. Let $Y = X \times X$, with the product measure $\nu = \mu \times \mu$, and define $S(x, y) = (Tx, T^{x_0}y)$. Let $\mathcal{P}$ be the time-zero partition of X and let $\mathcal{Q} = \mathcal{P} \times \mathcal{P}$. Find the entropy of the $(S, \mathcal{Q})$-process, which is the "random walk with random scenery" process discussed in Exercise 8 of Section I.2. (Hint: recurrence of simple random walk implies that all the sites of y have been almost surely visited by the past of the walk.)

5. Prove that the entropy of an n-stationary process $\{Y_n\}$ is the same as the entropy the stationary process $\{X_n\}$ obtained by randomizing the start. (Hint: let S be uniformly distributed on $[1, n]$ and note that $H(X_1^N, S) = H(X_1^N) + H(S|X_1^N) = H(S) + H(X_1^N|S)$ for any $N > n$, so $H(X_1^N)/N \sim H(X_1^N|S)/N$. Then show that $H(X_1^N|S = s)$ is the same as the unshifted entropy of X_{s+1}^N.)

6. The divergence for k-set partitions is $D(\mathcal{P}\|\mathcal{Q}) = \sum_i \mu(P_i)\log(\mu(P_i)/\mu(Q_i))$.

   (a) Show that $D(\mathcal{P} \vee \mathcal{R} \;\|\mathcal{Q} \vee \mathcal{R}) \geq D(\mathcal{P}\|\mathcal{Q})$, for any partition $\mathcal{R}$.

   (b) Prove Pinsker's inequality, namely, that $D(\mathcal{P}\|\mathcal{Q}) \geq (1/2\ln 2)|\mathcal{P}-\mathcal{Q}|^2$, where $|\mathcal{P} - \mathcal{Q}| = \sum_i |\mu(P_i) - \mu_i(Q_i)|^2$. (Hint: use part (a) to reduce to the two set case, then use calculus.)

7. Show that the process entropy of a stationary process and its reversed process are the same. (Hint: use stationarity.)

Section I.7 Interpretations of entropy.

Two simple, useful interpretations of entropy will be discussed in this section. The first is an expression of the entropy theorem in exponential form, which leads to the concept of entropy-typical sequence, and to the related building-block concept. The second interpretation is the connection between entropy and expected code length for the special class of prefix codes.

I.7.a Entropy-typical sequences.

Let μ be an ergodic process of entropy h. For $\epsilon > 0$ and $n \geq 1$ define
$$\mathcal{T}_n(\epsilon) = \left\{x_1^n \colon 2^{-n(h+\epsilon)} \leq \mu(x_1^n) \leq 2^{-n(h-\epsilon)}\right\}.$$
The set $\mathcal{T}_n(\epsilon)$ is called the set of (n, ϵ)-*entropy-typical sequences,* or simply the set of entropy-typical sequences, if n and ϵ are understood. The entropy theorem may be expressed by saying that x_1^n is eventually almost surely entropy-typical, that is,

Theorem I.7.1 (The typical-sequence form of entropy.)
For each $\epsilon > 0$, $x_1^n \in \mathcal{T}_n(\epsilon)$, eventually almost surely .

The convergence in probability form, $\lim_n \mu(\mathcal{T}_n(\epsilon)) = 1$, is known as the *asymptotic equipartition property* or AEP, in information theory.

The phrase "typical sequence," has different meaning in different contexts. Sometimes it means entropy-typical sequences, as defined here, sometimes it means frequency-typical sequences, as defined in Section I.4, sometimes it means sequences that are both frequency-typical and entropy-typical, and sometimes it is just shorthand for those sequences that are likely to occur. The context usually makes clear the notion of typicality being used. Here the focus is on the entropy-typical idea.

The members of the entropy-typical set $\mathcal{T}_n(\epsilon)$ all have the lower bound $2^{-n(h+\epsilon)}$ on their probabilities. Since the total probability is at most 1, this fact yields an upper bound on the cardinality of $\mathcal{T}_n$, namely,

Theorem I.7.2 (The entropy-typical cardinality bound.)
The set of entropy-typical sequences satisfies $|\mathcal{T}_n(\epsilon)| \leq 2^{n(h+\epsilon)}$.

Thus, even though there are $|A|^n$ possible sequences of length n, the measure is eventually mostly concentrated on a set of sequences of the (generally) much smaller cardinality $2^{n(h+\epsilon)}$. This fact is of key importance in information theory and plays a major role in many applications and interpretations of the entropy concept.

The preceding theorem provides an upper bound on the cardinality of the set of typical sequences and depends on the fact that typical sequences have a lower bound on their probabilities. Typical sequences also have an upper bound on their probabilities, which leads to the fact that too-small sets cannot be visited too often.

Theorem I.7.3 (The too-small set principle.)
If $C_n \subset A^n$ and $|C_n| \leq 2^{n(h-\epsilon)}$, $n \geq 1$, then $x_1^n \notin C^n$, eventually almost surely.

Proof. Since $x_1^n \in \mathcal{T}_n(\epsilon/2)$, eventually almost surely, it is enough to show that $x_1^n \notin C_n \cap \mathcal{T}_n(\epsilon/2)$, eventually almost surely. The cardinality bound on C_n and the probability upper bound $2^{-n(h-\epsilon/2)}$ on members of $\mathcal{T}_n(\epsilon/2)$ combine to give the bound

$$\mu\left(C_n \cap \mathcal{T}_n(\epsilon/2)\right) \leq 2^{n(h+\epsilon)} 2^{-n(h-\epsilon/2)} \leq 2^{-n\epsilon/2}.$$

Thus $\mu(C_n \cap \mathcal{T}_n(\epsilon/2))$ is summable in n, and the theorem is established. □

Another useful formulation of entropy, suggested by the upper bound, Theorem I.7.2, on the number of entropy-typical sequences, expresses the connection between entropy and coverings. For $\alpha > 0$ define the *(n, α)-covering number* by

$$\mathcal{N}_n(\alpha) = \min\{|C|:\ C \subset A^n, \text{ and } \mu(C) \geq \alpha\},$$

that is, $\mathcal{N}_n(\alpha)$ is the minimum number of sequences of length n needed to fill an α-fraction of the total probability. A good way to think of $\mathcal{N}_n(\alpha)$ is given by the following algorithm for its calculation.

(A1) List the n-sequences in decreasing order of probability.

(A2) Count down the list until the first time a total probability of at least α is reached.

The covering number $\mathcal{N}_n(\alpha)$ is the count obtained in (A2).

The connection with entropy is given by the following theorem.

Theorem I.7.4 (The covering-exponent theorem.)
For each $\alpha > 0$, the covering exponent $(1/n) \log \mathcal{N}_n(\alpha)$ converges to h, as $n \to \infty$.

Proof. Fix $\alpha \in (0, 1)$ and $\epsilon > 0$. Since the measure of the set of typical sequences goes to 1 as $n \to \infty$, the measure $\mu(\mathcal{T}_n(\epsilon))$ eventually exceeds α. When this happens

$$\mathcal{N}_n(\alpha) \leq |\mathcal{T}_n(\epsilon)| \leq 2^{n(h+\epsilon)},$$

by Theorem I.7.2, and hence, $\limsup_n (1/n) \log \mathcal{N}_n(\alpha) \leq h$.

On the other hand, suppose, for each n, $\mu(C_n) \geq \alpha$ and $|C_n| = \mathcal{N}_n(\alpha)$. If n is large enough then $\mu(\mathcal{T}_n(\epsilon) \cap C_n) \geq \alpha(1 - \epsilon)$. The fact that $\mu(x_1^n) \leq 2^{-n(h-\epsilon)}$, for $x_1^n \in \mathcal{T}_n(\epsilon)$, then implies that

$$\mu(\mathcal{T}_n(\epsilon) \cap C_n) = \sum_{x_1^n \in \mathcal{T}_n(\epsilon) \cap C_n} \mu(x_1^n) \quad \leq \quad 2^{-n(h-\epsilon)} |\mathcal{T}_n(\epsilon) \cap C_n|,$$

and hence

$$\mathcal{N}_n(\alpha) = |C_n| \geq |\mathcal{T}_n(\epsilon) \cap C_n| \geq 2^{n(h-\epsilon)} \mu(\mathcal{T}_n(\epsilon) \cap C_n) \geq 2^{n(h-\epsilon)} \alpha(1 - \epsilon),$$

which proves the theorem. □

The covering exponent idea is quite useful as a tool for entropy estimation, see, for example, the proof in Section I.7.e, below, that a rotation process has entropy 0.

The connection between coverings and entropy also has as a useful approximate form, which will be discussed in the following paragraphs.

Let $d(a, b)$ be the (discrete) metric on A, defined by

$$d(a, b) = \begin{cases} 0 & \text{if } a = b \\ 1 & \text{otherwise,} \end{cases}$$

and extend to the metric d_n on A^n, defined by

$$d_n(x_1^n, y_1^n) = \frac{1}{n} \sum_{i=1}^{n} d(x_i, y_i).$$

The metric d_n is also known as the *per-letter Hamming distance.*

The distance $d_n(x_1^n, S)$ from x_1^n to a set $S \subset A^n$ is defined by

$$d_n(x_1^n, S) = \min\{d_n(x_1^n, y_1^n) \colon\ y_1^n \in S\},$$

and the *δ-neighborhood* or *δ-blowup* of S is defined by

$$[S]_\delta = \{x_1^n \colon\ d(x_1^n, S) \leq \delta\}.$$

It is important that a small blowup does not increase size by more than a small exponential factor. To state this precisely let $H(\delta) = -\delta \log \delta + (1 - \delta) \log(1 - \delta)$ denote the binary entropy function.

Lemma I.7.5 (The blowup bound.)
The δ-blowup of S satisfies

$$|[S]_\delta| \leq |S| 2^{nH(\delta)}(|A|-1)^{n\delta}.$$

In particular, given $\epsilon > 0$ there is a $\delta > 0$ such that $|[\mathcal{T}_n(\epsilon)]_\delta| \leq 2^{n(h+2\epsilon)}$, for all n.

Proof. Given x_1^n, there are at most $n\delta$ positions i in which x_i can be changed to create a member of $[\{x_1^n\}]_\delta$. Each such position can be changed in $|A|-1$ ways. Thus, the combinations bound, Lemma I.5.4, implies that $|[\{x_1^n\}]_\delta| \leq 2^{nH(\delta)}(|A|-1)^{n\delta}$, which yields the stated bound.

Since $|\mathcal{T}_n(\epsilon)| \leq 2^{n(h+\epsilon)}$, since $(|A|-1)^{n\delta} \leq 2^{n\delta \log|A|}$, and since $\delta \log|A|$ and $H(\delta)$ both go to 0 as $\delta \to 0$, it follows that if δ is small enough, then $|[\mathcal{T}_n(\epsilon)]_\delta| \leq 2^{n(h+2\epsilon)}$ will hold, for all n. This proves the blowup-bound lemma. □

The blowup form of the covering number is defined by

$$\mathcal{N}_n(\alpha, \delta) = \min\{|C|: C \subset A^n, \text{ and } \mu([C]_\delta) \geq \alpha\}, \tag{1}$$

that is, it is the minimum size of an n-set for which the δ-blowup covers an α-fraction of the probability. It is left to the reader to prove that the limit of $(1/n)\log \mathcal{N}_n(\alpha, \delta)$ exists as $n \to \infty$. An application of Lemma I.7.5, together with Theorem I.7.4 then establishes that

$$\lim_{\delta \to 0} \lim_{n \to \infty} \frac{1}{n} \log \mathcal{N}_n(\alpha, \delta) = h.$$

I.7.b The building-block concept.

An application of the ergodic theorem shows that frequency-typical sequences must consist mostly of n-blocks which are entropy-typical, provided only that n is not too small. In particular, the entropy-typical n-sequences can be thought of as the "building blocks," from which longer sequences are made by concatenating typical sequences, with occasional spacers inserted between the blocks. This idea and several useful consequences of it will now be developed.

Fix a collection $\mathcal{B}_n \subset A^n$, thought of as the set of building blocks. Also fix an integer $M > n$ and $\delta \geq 0$. A sequence x_1^M is said to be *$(1-\delta)$-built-up from the building blocks $\mathcal{B}_n$* if there is an integer $I = I(x_1^M)$ and a collection $\{[n_i, m_i]: i \leq I\}$, of disjoint n-length subintervals of $[1, M]$, such that

(a) $\sum_{i=1}^{I}(m_i - n_i + 1) \geq (1-\delta)M$

(b) $x_{n_i}^{m_i} \in \mathcal{B}_n, \quad i \in I.$

In the special case when $\delta = 0$ and M is a multiple of n, to say that x_1^M is 1-built-up from $\mathcal{B}_n$ is the same as saying that x_1^M is a concatenation of blocks from $\mathcal{B}_n$. If $\delta > 0$ then the notion of $(1-\delta)$-built-up requires that x_1^M be a concatenation of blocks from $\mathcal{B}_n$, with spacers allowed between the blocks, subject only to the requirement that the total length of the spacers be at most δM. Both the number I and the intervals $[n_i, m_i]$ for which $x_{n_i}^{m_i}$ is required to be a member of $\mathcal{B}_n$, are allowed to depend on the sequence x_1^M. The reader should also note the concepts of blowup and built-up are quite different. The blowup concept focuses on creating sequences by making a small density of otherwise arbitrary changes, while the built-up concept only allows changes in the spaces between the building blocks, but allows arbitrary selection of the blocks from a fixed collection.

An important fact about the building-block concept is that if δ is small and n is large, then the set of M-sequences that can be $(1-\delta)$-built-up from a given set $\mathcal{B}_n \subset A^n$ of building blocks cannot be exponentially much larger in cardinality than the set of all sequences that can be formed by selecting M/n sequences from $\mathcal{B}_n$ and concatenating them without spacers. The proof of this fact, which is stated as the following lemma, is similar in spirit to the proof of the key bound used to establish the entropy theorem, Lemma I.5.6, though simpler since now the blocks all have a fixed length. As usual, $H(\delta) = -\delta\log\delta - (1-\delta)\log(1-\delta)$ denotes the binary entropy function.

Lemma I.7.6 (The built-up set bound.)
Let D_M be the set of all sequences x_1^M that can be $(1-\delta)$-built-up from a given collection $\mathcal{B}_n \subset A^n$. Then

$$|D_M| \le |\mathcal{B}_n|^{\frac{M}{n}} 2^{MH(\delta)} |A|^{M\delta}.$$

In particular, if $B_n = \mathcal{T}_n(\epsilon)$, the set of entropy-typical n-sequences relative to ϵ, and if δ is small enough then $|D_M| \le 2^{M(h+2\epsilon)}$.

Proof. The number of ways to select a family $\{[n_i, m_i]\}$ of disjoint n-length subintervals that cover all but a δ-fraction of $[1, M]$ is upper bounded by the number of ways to select at most δM points from a set with M members, namely,

$$\sum_{j\le\delta M} \binom{M}{j},$$

which is, in turn, upper bounded by $2^{MH(\delta)}$, by the combinations bound, Lemma I.5.4. For a fixed configuration of locations, say $\{[n_i, m_i], i \in I\}$, for members of $\mathcal{B}_n$, the number ways to fill these with members of $\mathcal{B}_n$ is upper bounded by

$$|\mathcal{B}_n|^I \le |\mathcal{B}_n|^{M/n},$$

and the number of ways to fill the places that are not in $\cup_i [n_i, m_i]$ is upper bounded by $|A|^{\delta M}$. Thus

$$|D_M| \le 2^{MH(\delta)} |\mathcal{B}_n|^{M/n} |A|^{\delta M},$$

which is the desired bound. The bound for entropy-typical building blocks follows immediately from the fact that $|\mathcal{T}_n(\epsilon)| \le 2^{n(h+\epsilon)}$. This establishes the lemma. □

The building-block idea is closely related to the packing/covering ideas discussed in Section I.4.b. A sequence x_1^M is said to be $(1-\delta)$-*strongly-covered by* $\mathcal{B}_n \subseteq A^n$ if

$$|\{i \in [1, M-n+1]\colon x_i^{i+n-1} \notin \mathcal{B}_n\}| \le \delta(M-n+1).$$

The argument used to prove the packing lemma, Lemma I.3.3, can be used to show that almost strongly-covered sequences are also almost built-up, provided M is large enough. This is stated in precise form as the following lemma.

Lemma I.7.7 (The building-block lemma.)
If x_1^M is $(1-\delta/2)$-strongly-covered by $\mathcal{B}_n$, and if $M \geq 2n/\delta$, then x_1^M is $(1-\delta)$-built-up from $\mathcal{B}_n$.

Proof. Put $m_0 = 0$ and for $i > 0$, define n_i to be least integer $k > m_{i-1}$ such that $x_k^{k+n-1} \in \mathcal{B}_n$, stopping when within n of the end of x_1^M. The assumption that x_1^M is $(1-\delta/2)$-strongly-covered by $\mathcal{B}_n$ implies there are at most $\delta M/2$ indices $j \leq M-n+1$ which are not contained in one of the $[n_i, m_i]$, while the condition that $M \geq 2n/\delta$ implies there at most $\delta M/2$ indices $j \in [M-n+1, M]$ which are not contained in one of the $[n_i, m_i]$. The lemma is therefore established. □

Remark I.7.8
The preceding two lemmas are strictly combinatorial. In combination with the ergodic and entropy theorems they provide powerful tools for analyzing the combinatorial properties of partitions of sample paths, a subject to be discussed in the next chapter. In particular, if $B_n = \mathcal{T}_n(\epsilon)$, and n is large enough, then $\mu(\mathcal{T}_n(\epsilon)) \approx 1$. The almost-covering principle, Theorem I.4.7, implies that eventually almost surely x_1^M is almost strongly-covered by sequences from $\mathcal{T}_n(\epsilon)$, hence eventually almost surely mostly built-up from members of $\mathcal{T}_n(\epsilon)$. A suitable application of the built-up set lemma, Lemma I.7.6, implies that the set D_M of sequences that are mostly built-up from the building blocks $\mathcal{T}_n(\epsilon)$ will eventually have cardinality not exponentially much larger than $2^{M(h+\epsilon)}$. The sequences in D_M are not necessarily entropy-typical, for no upper and lower bounds on their probabilities are given; all that is known is that the members of D_M are almost built-up from entropy-typical n-blocks, where n is a *fixed* large number, and that x_1^M is eventually in D_M. This result, which can be viewed as the entropy version of the finite sequence form of the frequency-typical sequence characterization of ergodic processes, Theorem I.4.5, will be surprisingly useful later.

I.7.c Entropy and prefix codes.

In a typical data compression problem a given finite sequence x_1^n, drawn from some finite alphabet A, is to be mapped into a binary sequence $b_1^{\mathcal{L}} = b_1, b_2, \ldots, b_{\mathcal{L}}$, whose length $\mathcal{L}$ may depend on x_1^n, in such a way that the source sequence x_1^n is recoverable from knowledge of the encoded sequence $b_1^{\mathcal{L}}$. The goal is to use as little storage space as possible, that is, to make the code length as short as possible, at least in some average sense. Of course, there are many possible source sequences, often of varying lengths, so that a typical code must be designed to encode a large number of different sequences and accurately decode them. A standard model is to think of a source sequence as a (finite) sample path drawn from some ergodic process. For this reason, in information theory a stationary, ergodic process with finite alphabet A is usually called a *source.* In this section the focus will be on a special class of codes, known as prefix codes, for which there is a close connection between code length and source entropy.

In the following discussion B^* denotes the set of all finite-length binary sequences and $\ell(w)$ denotes the length of a member $w \in B^*$.

A *(binary) code on* A is a mapping $C: A \mapsto B^*$. A code is said to be *faithful* or *noiseless* if it is one-to-one. An image $C(a)$ is called a *codeword* and the range of C

is called the *codebook*. The function that assigns to each $a \in A$ the length of the code word $C(a)$ is called the *length function* of the code and denoted by $\mathcal{L}(\cdot|C)$ or by $\mathcal{L}$ if C is understood. Formally, it is the function defined by $\mathcal{L}(a|C) = \ell(C(a))$, $a \in C$.

The expected length of a code C relative to a probability distribution μ on A is $E_\mu(\mathcal{L}) = \sum_a \mathcal{L}(a|C)\mu(a)$. The entropy of μ is defined by the formula

$$H(\mu) = \sum_{a \in A} \mu(a) \log \frac{1}{\mu(a)},$$

that is, $H(\mu)$ is just the expected value of the random variable $-\log \mu(X)$, where X has the distribution μ. (As usual, base 2 logarithms are used.) Without further assumptions about the code there is little connection between entropy and expected code length. For codes that satisfy a simple prefix condition, however, entropy is a lower bound to code length, a lower bound which is almost tight.

To develop the prefix code idea some notation and terminology and terminology will be introduced. For $u = u_1^n \in B^*$ and $v = v_1^m \in B^*$, the *concatenation* uv is the sequence of length $n+m$ defined by

$$(uv)_i = \begin{cases} u_i & 1 \le i \le n \\ v_{i-n} & n < i \le n+m. \end{cases}$$

A nonempty word u is called a *prefix* of a word w if $w = uv$. The prefix u is called a *proper prefix* of w if $w = uv$ where v is nonempty. A nonempty set $W \subset B^*$ is *prefix free* if no member of W is a proper prefix of any member of W.

A prefix-free set W has the property that it can be represented as the leaves of the (labeled, directed) binary tree $T(W)$, defined by the following two conditions.

(i) The vertex set $V(W)$ of $T(W)$ is the set of prefixes of members of W, together with the empty word λ.

(ii) If $v \in V(W)$ has the form $v = ub$, where $b \in B$ then there is a directed edge from u to v, labeled by b. (See Figure I.7.9.)

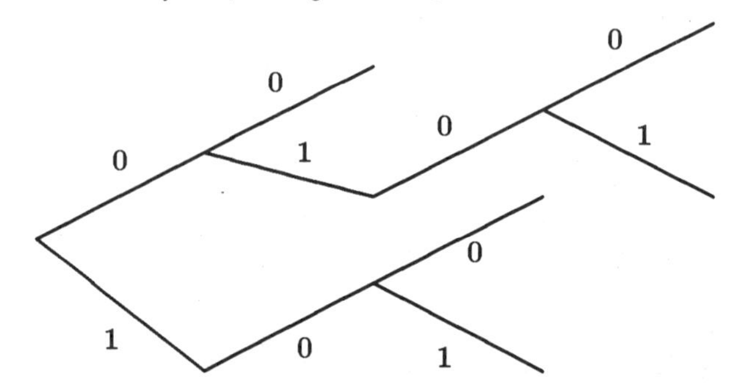

Figure I.7.9 The binary tree for $W = \{00, 0101, 101, 0100, 100\}$.

Note that the root of the tree is the empty word and the depth $d(v)$ of the vertex v is just the length of the word v. In particular, since W is prefix free, the words $w \in W$ correspond to leaves $L(T)$ of the tree $T(W)$. An important, and easily proved, property of binary trees is that the sum of $2^{-d(v)}$ over all leaves v of the tree can never exceed 1. For prefix-free sets W of binary sequences this fact takes the form

$$\sum_{w \in W} 2^{-\ell(w)} \le 1, \tag{2}$$

where $\ell(w)$ denotes the length of $w \in W$, a result known as the *Kraft inequality*. A geometric proof of this inequality is sketched in Exercise 10.

The Kraft inequality has the following converse.

Lemma I.7.10

If $1 \leq \ell_1 \leq \ell_2 \leq \ldots \leq \ell_t$ is a nondecreasing sequence of positive integers such that $\sum_i 2^{-\ell_i} \leq 1$ then there is a prefix-free set $W = \{w(i)\colon 1 \leq i \leq t\}$ such that $\ell(w(i)) = \ell_i, i \in [1, t]$.

Proof. Define i and j to be equivalent if $\ell_i = \ell_j$ and let $\{G_1, \ldots, G_s\}$ be the equivalence classes, written in order of increasing length $L(r) = \ell_i$, where $\ell_i \in G_r$. The Kraft inequality then takes the form

$$\sum_{r=1}^{s} |G_r| 2^{-L(r)} \leq 1.$$

Assign the indices in G_1 to binary words of length $L(1)$ in some one-to-one way, which is possible since $|G_1| \leq 2^{L(1)}$. Assign G_2 in some one-to-one way to binary words of length $L(2)$ that do not have already assigned words as prefixes. This is possible since $|G_1|2^{-L(1)} + |G_2|2^{-L(2)} \leq 1$, so that

$$|G_2| \leq 2^{L(2)} - |G_1| 2^{L(2)-L(1)},$$

where the second term on the right gives the number of words of length $L(2)$ that have prefixes already assigned. An inductive continuation of this assignment argument clearly establishes the lemma. □

A code C is a *prefix code* if $a = \tilde{a}$ whenever $C(a)$ is a prefix of $C(\tilde{a})$. Thus C is a prefix code if and only if C is one-to-one and the range of C is prefix-free. The codewords of a prefix code are therefore just the leaves of a binary tree, and, since the code is one-to-one, the leaf corresponding to $C(a)$ can be labeled with a or with $C(a)$. For prefix codes, the Kraft inequality

$$\sum_{a \in A} 2^{-\mathcal{L}(a|C)} \leq 1, \tag{3}$$

holds.

For prefix codes, entropy is always a lower bound on average code length, a lower bound that is "almost" tight. This fundamental connection between entropy and prefix codes was discovered by Shannon.

Theorem I.7.11 (Entropy and prefix codes.)

Let μ be a probability distribution on A.

(i) $H(\mu) \leq E_\mu(\mathcal{L}(\cdot|C))$, *for any prefix code C.*

(ii) *There is a prefix code C such that $E_\mu(\mathcal{L}(\cdot|C)) \leq H(\mu) + 1$.*

Proof. Let C be a prefix code. A little algebra on the expected code length yields

$$\begin{aligned} E_\mu(\mathcal{L}(\cdot|C)) &= \sum_a \mathcal{L}(a|C)\mu(a) = \sum_a \log 2^{\mathcal{L}(a|C)} \mu(a) \\ &= \sum_a \mu(a) \log \frac{\mu(a)}{2^{-\mathcal{L}(a|C)}} - \sum_a \mu(a) \log \mu(a). \end{aligned}$$

The second term is just $H(\mu)$, while the divergence inequality, Lemma I.6.2, implies that the first term is nonnegative, since the measure ν defined by $\nu(a) = 2^{-\mathcal{L}(a|C)}$ satisfies $\nu(A) \leq 1$, by the Kraft inequality. This proves (i).

To prove part (ii) define $\mathcal{L}(a) = \lceil -\log \mu(a) \rceil$, $a \in A$, where $\lceil x \rceil$ denotes the least integer $\geq x$, and note that

$$\sum_a 2^{-\mathcal{L}(a)} \leq \sum_a 2^{\log \mu(a)} = \sum_a \mu(a) = 1.$$

Lemma I.7.10 produces a prefix code $C(a) = w(a)$, $a \in A$, whose length function is $\mathcal{L}$. Since, however, $\mathcal{L}(a) \leq 1 - \log \mu(a)$, it follows that

$$\begin{aligned} \sum_a \mathcal{L}(a)\mu(a) &= \sum_a (1 - \log \mu(a))\mu(a) \\ &\leq 1 - \sum_a \mu(a) \log \mu(a) = 1 + H(\mu). \end{aligned}$$

Thus part (ii) is proved, which completes the proof of the theorem. □

Next consider n-codes, that is, mappings $C_n \colon A^n \mapsto B^*$ from source sequences of length n to binary words. In this case, a faithful n-code $C_n \colon A^n \mapsto B^*$ is a prefix n-code if and only if its range is prefix free. Of interest for n-codes is per-symbol code length $\mathcal{L}(a_1^n|C_n)/n$ and expected per-symbol code length $E(\mathcal{L}(\cdot|C_n))/n$. Let μ_n be a probability measure on A^n with per-symbol entropy $H_n(\mu_n) = H(\mu_n)/n$. Theorem I.7.11(i) takes the form

$$H_n(\mu_n) \leq \frac{1}{n} E_{\mu_n}(\mathcal{L}(\cdot|C_n)), \tag{4}$$

for any prefix n-code C_n, while Theorem I.7.11(ii) asserts the existence of a prefix n-code C_n such that

$$\frac{1}{n} E_{\mu_n}(\mathcal{L}(\cdot|C_n)) \leq H_n(\mu_n) + \frac{1}{n}. \tag{5}$$

A sequence $\{C_n\}$, such that, for each n, C_n is a prefix n-code with length function $\mathcal{L}(x_1^n) = \mathcal{L}(x_1^n|C_n)$, will be called a *prefix-code sequence*. The *(asymptotic) rate* of such a code sequence, relative to a process $\{X_n\}$ with Kolmogorov measure μ is defined by

$$\mathcal{R}_\mu(\{C_n\}) = \limsup_{n\to\infty} \frac{E_{\mu_n}(\mathcal{L}(\cdot|C_n))}{n}.$$

The two results, (4) and (5) then yield the following asymptotic results.

Theorem I.7.12 (Process entropy and prefix-codes.)
If μ is a stationary process with process entropy H then

(a) *There is a prefix code sequence $\{C_n\}$ such that $\mathcal{R}_\mu(\{C_n\}) \leq H$.*

(b) *There is no prefix code sequence $\{C_n\}$ such that $\mathcal{R}_\mu(\{C_n\}) < H$.*

Thus "good" codes exist, that is, it is possible to compress as well as process entropy in the limit, and "too-good" codes do not exist, that is, no sequence of codes can asymptotically compress more than process entropy. In the next chapter, almost-sure versions of these two results will be obtained for ergodic processes.

Remark I.7.13

A prefix code with length function $\mathcal{L}(a) = \lceil -\log \mu(a) \rceil$ will be called a *Shannon code*. Shannon's theorem implies that a Shannon code is within 1 of being the best possible prefix code in the sense of minimizing expected code length. A somewhat more complicated coding procedure, called Huffman coding, produces prefix codes that minimize expected code length. Huffman codes have considerable practical significance, but for n-codes the per-letter difference in expected code length between a Shannon code and a Huffman code is at most $1/n$, which is asymptotically negligible. For this reason no use will be made of Huffman coding in this book, since, for Shannon codes, the code length function $\mathcal{L}(a) = \lceil -\log \mu(a) \rceil$ is closely related to the function $-\log \mu(a)$, whose expected value is entropy.

I.7.d Converting faithful codes to prefix codes.

Next it will be shown that, as far as asymptotic results are concerned, there is no loss of generality in assuming that a faithful code is a prefix code. The key to this is the fact that a faithful n-code can always be converted to a prefix code by adding a header (i.e, a prefix) to each codeword to specify its length length. This can be done in such a way that the length of the header is (asymptotically) negligible compared to total codeword length, so that asymptotic results about prefix codes automatically apply to the weaker concept of faithful code.

The following is due to Elias, [8].

Lemma I.7.14 (The Elias-code lemma.)

There is a prefix code $\mathcal{E}: \{1, 2, \ldots\} \mapsto B^*$, such that $\ell(\mathcal{E}(n)) = \log n + o(\log n)$. Any prefix code with this property is called an Elias code.

Proof. The code word assigned to n is a concatenation of three binary sequences,

$$\mathcal{E}(n) = u(n)v(n)w(n).$$

The third part $w(n)$ is the usual binary representation of n, so that, for example, $w(12) = 1100$. The second part $v(n)$ is the binary representation of the length of $w(n)$, so that, for example, $v(12) = 100$. The first part $u(n)$ is just a sequence of 0's of length equal to the length of $v(n)$, so that, for example, $u(12)$=000. Thus $\mathcal{E}(12) = 0001001100$.

The code is a prefix code, for if

$$u(n)v(n)w(n) = u(m)v(m)w(m)w',$$

then, $u(n) = u(m)$, since both consist only of 0's and the first bit of both $v(n)$ and $v(m)$ is a 1. But then $v(n) = v(m)$, since both have length equal to the length of $u(n)$. This means that $w(n) = w(m)$, since both have length specified by $v(n)$, so that w' is empty and $n = m$.

The length of $w(n)$ is $\lceil \log(n+1) \rceil$, the length of the binary representation of n, while both $u(n)$ and $v(n)$ have length equal to

$$\lceil \log(1 + \lceil \log(n+1) \rceil) \rceil,$$

the length of the binary representation of $\lceil \log(n+1) \rceil$. The desired bound $\ell(\mathcal{E}(n)) = \log n + o(\log n)$ follows easily, so the lemma is established. □

Given a faithful n-code C_n with length function $\mathcal{L}(\cdot|C_n)$, a prefix n-code C_n^* is obtained by the formula

(6) $$C_n^*(x_1^n) = \mathcal{E}(\mathcal{L}(x_1^n|C_n))C_n(x_1^n),\ x_1^n \in A^n,$$

where $\mathcal{E}$ is an Elias prefix code on the integers. For example, if $C_n(x_1^n) = 001001100111$, a word of length 12, then

$$C_n^*(x_1^n) = 0001001100,\ 001001100111,$$

where, for ease of reading, a comma was inserted between the header information, $\mathcal{E}(\mathcal{L}(x_1^n|C_n))$, and the code word $C_n(x_1^n)$. The decoder reads through the header information and learns where $C_n(x_1^n)$ starts and that it is 12 bits long. This enables the decoder to determine the preimage x_1^n, since C_n was assumed to be faithful. The code C_n^*, which is clearly a prefix code, will be called *an Elias extension of* C_n.

In the example, the header information is almost as long as $C_n(x_1^n)$, but header length becomes negligible relative to the length of $C_n(x_1^n)$ as codeword length grows, by Lemma I.7.14. Thus Theorem I.7.12(b) extends to the following somewhat sharper form. (The definition of faithful-code sequence is obtained by replacing the word "prefix" by the word "faithful," in the definition of prefix-code sequence.)

Theorem I.7.15

If μ is a stationary process with entropy-rate H there is no faithful-code sequence such that $\mathcal{R}_\mu(\{C_n\}) < H$.

Remark I.7.16

Another application of the Elias prefix idea converts a prefix-code sequence $\{C_n\}$ into a single prefix code $C: A^* \mapsto B^*$, where C is defined by

$$C(x_1^n) = \mathcal{E}(n)C_n(x_1^n),\ x_1^n \in A^n, n = 1, 2, \ldots$$

The header tells the decoder which codebook to apply to decode the received message.

I.7.e Rotation processes have entropy 0.

As an application of the covering-exponent interpretation of entropy it will be shown that ergodic rotation processes have entropy 0. Let T be the transformation defined by $T: x \mapsto x \oplus \alpha,\ x \in [0, 1)$, where α is irrational and $\oplus$ is addition modulo 1, and let $\mathcal{P}$ be a finite partition of the unit interval $[0, 1)$.

Proposition I.7.17

The $(T, \mathcal{P})$-process has entropy 0.

The proof for the special two set partition $\mathcal{P} = \{P_0, P_1\}$, where $P_0 = [0, 0.5)$, and $P_1 = [0.5, 1)$, will be given in detail. The argument generalizes easily to arbitrary partitions into subintervals, and, with a bit more effort, to general partitions (by approximating by partitions into intervals).

The proof will be based on direct counting arguments rather than the entropy formalism. Geometric arguments will be used to estimate the number of possible $(T, \mathcal{P})$-names of length n. These will produce an upper bound of the form $2^{n\epsilon_n}$, where $\epsilon_n \to 0$ as

$n \to \infty$. The result then follows from the covering-exponent interpretation of entropy, Theorem I.7.4.

The key to the proof is the following uniform distribution property of irrational rotations.

Proposition I.7.18 (The uniform distribution property.)

$$\lim_{n\to\infty} \frac{|\{j \in [1,n]:\ T^j x \in I\}|}{n} = \lambda(I),$$

for any interval $I \subseteq [0,1)$ *and any* $x \in [0,1)$, *where* λ *denotes Lebesgue measure.*

Ergodicity asserts the truth of the proposition for almost every x. The fact that rotations are rigid motions allows the almost-sure result to be extended to a result that holds for every x. The proof is left as an exercise.

The key to the proof to the entropy 0 result is that, given z and x, some power $y = T^n x$ of x must be close to z. But translation is a rigid motion so that $T^m y$ and $T^m z$ will be close for all m, and hence the $(T\text{-}\mathcal{P})$-names of y and z will agree most of the time. Thus every name is obtained by changing the name of some fixed z in a small fraction of places and shifting a bounded amount. In particular, there cannot be exponentially very many names.

To make the preceding argument precise define the metric

$$|x - y|_1 = \min\{|x-y|, |1+x-y|\},\quad x, y \in [0,1),$$

and the pseudometric

$$d_n(x,y) = d_n(x_1^n, y_1^n) = \frac{1}{n}\sum_1^n d(x_i, y_i),$$

where $d(a,b) = 0$, if $a = b$, and $d(a,b) = 1$, if $a \neq b$, and $\{x_n\}$ and $\{y_n\}$ denote the $(T,\mathcal{P})$-names of $x, y \in [0,1)$, respectively. The first lemma shows that $d_n(x,y)$ is continuous relative to $|x-y|_1$.

Lemma I.7.19

Given $\epsilon > 0$ *there is an* N *such that* $d_n(x,y) < \epsilon$, *if* $|x-y|_1 < \epsilon/4$ *and* $n \geq N$.

Proof. Suppose $|x-y|_1 < \epsilon/4$. Consider the case when $|x-y|_1 = |x-y|$. Without loss of generality it can be assumed that $x < y$. Let $I = [x,y]$. The names of x and y disagree at time j if and only if $T^j x$ and $T^j y$ belong to different atoms of the partition $\mathcal{P}$, which occurs when and only when either $T^{-j}0 \in I$ or $T^{-j}(0.5) \in I$. The uniform distribution property, Proposition I.7.18, can be applied to T^{-1}, which is rotation by $-\alpha$, to provide an $N = N(x,y)$ such that if $n \geq N$ then

$$\frac{|\{j \in [i,n] : T^{-j}0 \in I \text{ or } T^{-j}(0.5) \in I\}|}{n} < \epsilon$$

This implies that $d_n(x,y) < \epsilon$, $n \geq N$. The case when $|x-y|_1 = |1+x-y|$ can be treated in a similar manner. Compactness then shows that N can be chosen to be independent of x and y. □

The next lemma is just the formal statement that the name of any point x can be shifted by a bounded amount to obtain a sequence close to the name of $z = 0$ in the sense of the pseudometric d_n.

Lemma I.7.20

Given $\epsilon > 0$, there is an integer M and an integer $K > M$ such that if $x \in [0, 1)$ there is a $j = j(x) \in [1, M]$ such that if $z = 0$ and $y = T^j x$ then

$$d_k(z_1^k, y_1^k) < \epsilon, \ k \geq K.$$

Proof. Given $\epsilon > 0$, choose N from the preceding lemma so that if $n \geq N$ and $|x-y|_1 < \epsilon/4$ then $d_n(x, y) < \epsilon$. Given x, apply Kronecker's theorem, Proposition I.2.15, to obtain a least positive integer $j = j(x)$ such that $|T^j x - 0|_1 < \epsilon/4$. Since small changes in x do not change $j(x)$ there is a number M such that $j(x) \leq M$ for all x. With $K = M + N$ the lemma follows. □

Now for the final details of the proof that the entropy of the $(T, \mathcal{P})$-process is 0. Given $\epsilon > 0$, determine M and $K > M$ from the preceding lemma, let $n \geq K+M$ and let $z = 0$. The unit interval X can be partitioned into measurable sets $X_j = \{x : j(x) = j\}, \ j \leq M$, where $j(x)$ is given by the preceding lemma. Note that

$$d_{n-j}(z, T^j x) < \epsilon, \ x \in X_j.$$

The number of binary sequences of length $n - j$ that can be obtained by changing the $(n - j)$-name of z in at most $\epsilon(n - j)$ places is upper bounded by $2^{(n-j)h(\epsilon)}$, where

$$h(\epsilon) = -\epsilon \log \epsilon - (1 - \epsilon) \log(1 - \epsilon),$$

is the binary entropy function, according to the combinations bound, Lemma I.5.4. There are at most 2^j possible values for the first j places of any name. Thus an upper bound on the number of possible names of length n for members of X_j is

$$|\{x_1^n : x \in X_j\}| \leq 2^j 2^{(n-j)h(\epsilon)} \leq 2^M 2^{nh(\epsilon)}. \tag{7}$$

Since there only M possible values for j, the number of possible rotation names of length n is upper bounded by $M 2^M 2^{nh(\epsilon)}$, for $n \geq K + M$. Since $h(\epsilon) \to 0$ as $\epsilon \to 0$ this completes the proof that the $(T, \mathcal{P})$-process has entropy 0. □

I.7.f Exercises.

1. Give a direct proof of the covering-exponent theorem, Theorem I.7.4, based on the algorithm (A1,A2) given for the computation of the covering number. (Hint: what does the entropy theorem say about the first part of the list?)

2. Show that $\lim_n (1/n) \log \mathcal{N}_n(\alpha, \delta)$ exists, where $\mathcal{N}_n(\alpha, \delta)$ is defined in (1).

3. Let $\widehat{\mu}$ be the concatenated-block process defined by a measure μ on A^n. Let $H(\mu)$ denote the entropy of μ.

    (a) Show that $H_{mn}(\widehat{\mu}) \leq (m + 1)H(\mu) + \log n$.

    (b) Show that $H_{mn}(\widehat{\mu}) \geq mH(\mu)$. (Hint: there is at least one way and at most $n|A|^{2n}$ ways to express a given x_1^n in the form $uw(1)w(2) \ldots w(k - 1)v$, where each $w(i) \in A^n$ and u and v are the tail and head, respectively, of words $w(0), w(k) \in A^n$.)

4. Use the covering exponent concept to show that the entropy of an ergodic n-stationary process is the same as the entropy of the stationary process obtained by randomizing its start.

5. Let μ a shift-invariant ergodic measure on A^Z and let $B \subset A^Z$ be a measurable set of positive measure. Show, using the entropy theorem, that the entropy of the induced process is $H(\mu)/\mu(B)$.

6. Suppose the waiting times between occurrences of "1" for a binary process μ are independent and identically distributed. Find an expression for the entropy of μ, in terms of the entropy of the waiting-time distribution.

7. Let $\{Y_n\}$ be the process obtained from the stationary ergodic process $\{X_n\}$ by applying the block code C_N and randomizing the start. Suppose $\{Y_n\}$ is ergodic. Show that $H(\{Y_n\}) \leq H(\{X_n\})$.

8. Show that if $C_n: A^n \mapsto B^*$ is a prefix code with length function $\mathcal{L}$ then there is a prefix code $\widehat{C}_n: A^n \mapsto B^*$ with length function $\widehat{\mathcal{L}}$ such that

   (a) $\widehat{\mathcal{L}}(x_1^n) \leq \mathcal{L}(x_1^n) + 1$, for all $x_1^n \in A^n$.

   (b) $\widehat{\mathcal{L}}(x_1^n) \leq 2 + 2n \log |A|$, for all $x_1^n \in A^n$.

9. Prove Proposition I.7.18.

10. Let C be a prefix code on A with length function $\mathcal{L}$. Associate with each $a \in A$ the dyadic subinterval of the unit interval $[0, 1]$ of length $2^{-\mathcal{L}(a)}$ whose left endpoint has dyadic expansion $C(a)$. Show that the resulting set of intervals is disjoint, and that this implies the Kraft inequality (2). Use this idea to prove that if $\sum_a 2^{-\mathcal{L}(a)} \leq 1$, then there is prefix code with length function $\mathcal{L}(a)$.

11. Show, using Theorem I.7.4, that the entropy-rate of an ergodic process and its reversed process are the same.

Section I.8 Stationary coding.

Stationary coding was introduced in Example I.1.9. Two basic results will be established in this section. The first is that stationary codes can be approximated by finite codes. Often a property can be established by first showing that it holds for finite codes, then passing to a suitable limit. The second basic result is a technique for creating stationary codes from block codes. In some cases it is easy to establish a property for a block coding of a process, then extend using the block-to-stationary code construction to obtain a desired property for a stationary coding.

I.8.a Approximation by finite codes.

The definition and notation associated with stationary coding will be reviewed and the basic approximation by finite codes will be established in this subsection.

Recall that a stationary coder is a measurable mapping $F: A^Z \mapsto B^Z$ such that

$$F(T_A x) = T_B F(x), \ \forall x \in A^Z,$$

where T_A and T_B denote the shifts on the respective two-sided sequence spaces A^Z and B^Z. Given a (two-sided) process with alphabet A and Kolmogorov measure μ_A, the encoded process has alphabet B and Kolmogorov measure $\mu_B = \mu_A \circ F^{-1}$. The associated time-zero coder is the function $f: A^Z \mapsto B$ defined by the formula $f(x) = F(x)_0$.

Associated with the time-zero coder $f: A^Z \mapsto B$ is the partition $\mathcal{P} = \mathcal{P}(f) = \{P_b: b \in B\}$ of A^Z, defined by $P_b = \{x: f(x) = b\}$, that is,

$$x \in P_b \text{ if and only if } f(x) = b. \tag{1}$$

Note that if $y = F(x)$ then $y_n = f(T^n x)$ if and only if $T^n x \in P_{y_n}$. In other words, y is the $(T, \mathcal{P})$-name of x and the measure μ_B is the Kolmogorov measure of the $(T, \mathcal{P})$-process. The partition $\mathcal{P}(f)$ is called the *partition defined by the encoder* f or, simply, *the encoding partition.* Note also that a measurable partition $\mathcal{P} = \{P_b: b \in B\}$ defines an encoder f by the formula (1), such that $\mathcal{P} = \mathcal{P}(f)$, that is, $\mathcal{P}$ is the encoding partition for f. Thus there is a one-to-one correspondence between time-zero coders and measurable finite partitions of A^Z.

In summary, a stationary coder can be thought of as a measurable function $F: A^Z \mapsto B^Z$ such that $F \circ T_A = T_B \circ F$, or as a measurable function $f: A^Z \mapsto B$, or as a measurable partition $\mathcal{P} = \{P_b: b \in B\}$. The descriptions are connected by the relationships

$$f(x) = F(x)_0, \quad F(x)_n = f(T^n x), \quad P_b = \{x: f(x) = b\}.$$

A time-zero coder f is said to be finite (with window half-width w) if $f(x) = f(\tilde{x})$, whenever $x_{-w}^{w} = \tilde{x}_{-w}^{w}$. In this case, the notation $f(x_{-w}^{w})$ is used instead of $f(x)$. A key fact for finite coders is that

$$F^{-1}([y_m^n]) = \bigcup_{f(x_{m-w}^{n+w}) = y_m^n} [x_{m-w}^{n+w}],$$

and, as shown in Example I.1.9, for any stationary measure μ, cylinder set probabilities for the encoded measure $\nu = \mu \circ F^{-1}$ are given by the formula

$$\nu(y_1^n) = \mu(F^{-1}[y_1^n]) = \sum_{f(x_{1-w}^{n+w}) = y_1^n} \mu(x_{1-w}^{n+w}). \tag{2}$$

The principal result in this subsection is that time-zero coders are "almost finite", in the following sense.

Theorem I.8.1 (Finite-coder approximation.)

If $f: A^Z \mapsto B$ is a time-zero coder, if μ is a shift-invariant measure on A^Z, and if $\epsilon > 0$, then there is a finite time-zero encoder $\widetilde{f}: A^Z \mapsto B$ such that

$$\mu\left(\{x: f(x) \neq \widetilde{f}(x)\}\right) \leq \epsilon.$$

Proof. Let $\mathcal{P} = \{P_b: b \in B\}$ be the encoding partition for f. Since $\mathcal{P}$ is a finite partition of A^Z into measurable sets, and since the measurable sets are generated by the finite cylinder sets, there is a positive integer w and a partition $\widetilde{\mathcal{P}} = \{\widetilde{P}_b\}$ with the following two properties

(a) Each $\widetilde{P}_b$ is union of cylinder sets of the form $[a_{-w}^{w}]$.

(b) $\sum_b \mu(P_b \Delta \widetilde{P}_b) \leq \epsilon$.

Let $\widetilde{f}$ be the time-zero encoder defined by $\widetilde{f}(x) = b$, if $[x_{-w}^w] \subseteq \widetilde{P}_b$. Condition (a) assures that $\widetilde{f}$ is a finite encoder, and $\mu(\{x: f(x) \neq \widetilde{f}(x)\}) \leq \epsilon$ follows from condition (b). This proves the theorem. □

Remark I.8.2

The preceding argument only requires that the coded process be a finite-alphabet process. In particular, any stationary coding onto a finite-alphabet process of any i.i.d. process of finite or infinite alphabet can be approximated arbitrarily well by finite codes.

As noted earlier, the stationary coding of an ergodic process is ergodic. A consequence of the finite-coder approximation theorem is that entropy for ergodic processes is not increased by stationary coding. The proof is based on direct estimation of the number of sequences of length n needed to fill a set of fixed probability in the encoded process.

Theorem I.8.3 (Stationary coding and entropy.)

If $\nu = \mu \circ F^{-1}$ is a stationary encoding of an ergodic process μ then $h(\nu) \leq h(\mu)$.

Proof. First consider the case when the time-zero encoder f is finite with window half-width w. By the entropy theorem, there is an N such that if $n \geq N$, there is a set $C_n \subset A_{1-w}^{n+w}$ of measure at least 1/2 and cardinality at most $2^{(n+2w+1)(h(\mu)+\epsilon)}$. The image $f(C_n)$ is a subset of B_1^n of measure at least 1/2, by formula (2). Furthermore, because mappings cannot increase cardinality, the set $f(C_n)$ has cardinality at most $2^{(n+2w+1)(h(\mu)+\epsilon)}$. Thus

$$\frac{1}{n} \log |f(C_n)| \leq h(\mu) + \epsilon + \delta_n,$$

where $\delta_n \to 0$, as $n \to \infty$, since w is fixed. It follows that $h(\nu) \leq h(\mu)$, since entropy equals the asymptotic covering rate, Theorem I.7.4.

In the general case, given $\epsilon > 0$ there is a finite time-zero coder $\widetilde{f}$ such that $\mu\left(\{x: f(x) \neq \widetilde{f}(x)\}\right) \leq \epsilon^2$. Let F and $\widetilde{F}$ denote the sample path encoders and let $\nu = \mu \circ F^{-1}$ and $\widetilde{\nu} = \mu \circ \widetilde{F}^{-1}$ denote the Kolmogorov measures defined by f and $\widetilde{f}$, respectively. It has already been established that finite codes do not increase entropy, so that $h(\widetilde{\nu}) \leq h(\mu)$. Thus, there is an n and a collection $\widetilde{C} \subset B^n$ such that $\widetilde{\nu}(\widetilde{C}) \geq 1 - \epsilon$ and $|\widetilde{C}| \leq 2^{n(h(\mu)+\epsilon)}$.

Let $C = [\widetilde{C}]_\epsilon$ be the ϵ-blowup of $\widetilde{C}$, that is,

$$C = \left\{y_1^n: \; d_n(y_1^n, \widetilde{y}_1^n) \leq \epsilon, \text{ for some } \widetilde{y}_1^n \in \widetilde{C}\right\}.$$

The blowup bound, Lemma I.7.5, implies that

$$|C| \leq 2^{n(h(\mu)+\epsilon+\delta(\epsilon))},$$

where $\delta(\epsilon) = o(\epsilon)$. Let

$$\widetilde{D} = \{x: \; \widetilde{F}(x)_1^n \in \widetilde{C}\} \text{ and } D = \{x: \; F(x)_1^n \in C\}$$

be the respective pull-backs to A^Z. Since $\mu\left(\{x: f(x) \neq \widetilde{f}(x)\}\right) \leq \epsilon^2$, the Markov inequality implies that the set

$$G = \{x: \; d_n(F(x)_1^n, \widetilde{F}(x)_1^n) \leq \epsilon\}$$

has measure at least $1-\epsilon$, so that $\mu(G\cap\widetilde{D})\geq 1-2\epsilon$, since $\mu(\widetilde{D})=\widetilde{\nu}(\widetilde{C})\geq 1-\epsilon$. By definition of G and D, however, $G\cap\widetilde{D}\subset D$, so that

$$\nu(C)=\mu(D)\geq\mu(G\cap\widetilde{D})\geq 1-2\epsilon.$$

The bound $|C|\leq 2^{n(h(\mu)+\epsilon+\delta(\epsilon))}$, and the fact that entropy equals the asymptotic covering rate, Theorem I.7.4, then imply that $h(\nu)\leq h(\mu)$. This completes the proof of Theorem I.8.3. □

Example I.8.4 (Stationary coding preserves mixing.)

A simple argument, see Exercise 19, can be used to show that stationary coding preserves mixing. Here a proof based on approximation by finite coders will be given. While not as simple as the earlier proof, it gives more direct insight into why stationary coding preserves mixing and is a nice application of coder approximation ideas.

The σ-field generated by the cylinders $[a_m^n]$, for m and n fixed, that is, the σ-field generated by the random variables $X_m, X_{m+1},\ldots,X_n$, will be denoted by $\Sigma(X_m^n)$. As noted earlier, the i-fold shift of the cylinder set $[a_m^n]$ is the cylinder set $T^{-i}[a_m^n]=[c_{i+m}^{i+n}]$, where $c_{i+j}=a_j$, $m\leq j\leq n$.

Let $\nu=\mu\circ F^{-1}$ be a stationary encoding of a mixing process μ. First consider the case when the time-zero encoder f is finite, with window width $2w+1$, say. The coordinates y_1^n of $y=F(x)$ depend only on the coordinates x_{1-w}^{n+w}. Thus for $g>2w$, the intersection $[y_1^n]\cap[y_{m+g+1}^{m+g+n}]$ is the image under F of $C\cap T^{-m}D$ where C and D are both measurable with respect to $\Sigma(X_{1-w}^{n+w})$. If μ is mixing then, given $\epsilon>0$ and $n\geq 1$, there is an M such that

$$|\mu(C\cap T^{-m}D)-\mu(C)\mu(D)|\leq\epsilon,\quad C,D\in\Sigma(X_{1-w}^{n+w}),\ m\geq M,$$

which, in turn, implies that

$$|\nu([y_1^n]\cap[y_{m+g+1}^{m+g+n}])-\nu([y_1^n])\nu([y_{m+g+1}^{m+g+n}])|\leq\epsilon,\ m\geq M,\ g>2w.$$

Thus ν is mixing.

In the general case, suppose $\nu=\mu\circ F^{-1}$ is a stationary coding of the mixing process μ, with stationary encoder F and time-zero encoder f, and fix $n\geq 1$. Given $\epsilon>0$, and a_1^n and b_1^n, let δ be a positive number to be specified later and choose a finite encoder $\tilde{F}$, with time-zero encoder $\tilde{f}$, such that $\mu(\{x\colon f(x)\neq\tilde{f}(x)\})\leq\delta$. Thus,

$$\begin{aligned}\mu(\{x\colon\ F(x)_1^n\neq\tilde{F}(x)_1^n\}) &\leq \sum_{i=1}^n\mu(\{x\colon\ f(T^ix)\neq\tilde{f}(T^ix)\})\\ &\leq n\mu(\{x\colon\ f(x)\neq\tilde{f}(x)\}),\leq n\delta,\end{aligned}$$

by stationarity. But n is fixed, so that if δ is small enough then $\nu(a_1^n)$ will be so close to $\tilde{\nu}(a_1^n)$ and $\nu(b_1^n)$ so close to $\tilde{\nu}(b_1^n)$ that

$$|\nu(a_1^n)\nu(b_1^n)-\tilde{\nu}(a_1^n)\tilde{\nu}(b_1^n)|\leq\epsilon/3, \tag{3}$$

for all a_1^n and b_1^n. Likewise, for any $m\geq 1$,

$$\mu\left(\{x\colon\ F(x)_1^n\neq\tilde{F}(x)_1^n,\ \text{or}\ F(x)_{m+1}^{m+n}\neq\tilde{F}(x)_{m+1}^{m+n}\}\right)\leq 2n\delta,$$

so that

$$|\nu([a_1^n]\cap T^{-m}[b_1^n]) - \tilde{\nu}([a_1^n]\cap T^{-m}[b_1^n])| \le \epsilon/3, \tag{4}$$

provided only that δ is small enough, uniformly in m for all a_1^n and b_1^n.

Since $\tilde{\nu}$ is mixing, being a finite coding of μ, it is true that

$$|\tilde{\nu}([a_1^n]\cap T^{-m}[b_1^n]) - \tilde{\nu}([a_1^n])\tilde{\nu}([b_1^n])| \le \epsilon/3,$$

provided only that m is large enough, and hence, combining this with (3) and (4), and using the triangle inequality yields

$$|\nu([a_1^n]\cap T^{-m}[b_1^n]) - \nu([a_1^n])\nu([b_1^n])| \le \epsilon,$$

for all sufficiently large m, provided only that δ is small enough. Thus, indeed, stationary coding preserves the mixing property.

I.8.b From block to stationary codes.

As noted in Example I.1.10, an N-block code $C_N: A^N \mapsto B^N$ can be used to map an A-valued process $\{X_n\}$ into a B-valued process $\{Y_n\}$ by applying C_N to consecutive nonoverlapping blocks of length N,

$$Y_{jN+1}^{(j+1)N} = C_N\left(X_{jN+1}^{(j+1)N}\right),\ j = 0, 1, 2, \ldots$$

If $\{X_n\}$ is stationary, then a stationary process $\{\widetilde{Y}_n\}$ is obtained by randomizing the start, i.e., by selecting an integer $u \in [1, N]$ according to the uniform distribution and defining $\widetilde{Y}_i = Y_{u+i-1}$, $i = 1, 2, \ldots$. The final process $\{\widetilde{Y}_n\}$ is stationary, but it is not, except in rare cases, a stationary coding of $\{X_n\}$, and nice properties of $\{X_n\}$, such as mixing or even ergodicity, may get destroyed.

A method for producing stationary codes from block codes will now be described. The basic idea is to use an event of small probability as a signal to start using the block code. The block code is then applied to successive N-blocks until within N of the next occurrence of the event. If the event has small enough probability, sample paths will be mostly covered by nonoverlapping blocks of length exactly N to which the block code is applied.

Lemma I.8.5 (Block-to-stationary construction.)
Let μ be an ergodic measure on A^Z and let $C: A^N \mapsto B^N$ be an N-block code. Given $\epsilon > 0$ there is a stationary code $F = F_C: A^Z \mapsto A^Z$ such that for almost every $x \in A^Z$ there is an increasing sequence $\{n_i : i \in Z\}$, which depends on x, such that $Z = \cup[n_i, n_{i+1})$ and

(i) $n_{i+1} - n_i \le N$, $i \in Z$.

(ii) *If J_n is the set of indices i such that $[n_i, n_{i+1}) \subset [-n, n]$ and $n_{i+1} - n_i < N$, then $\limsup_n (1/2n) \sum_{i\in J_n} n_{i+1} - n_i \le \epsilon$, almost surely.*

(iii) *If $n_{i+1} - n_i = N$ then $y_{n_i}^{n_{i+1}-1} = C(x_{n_i}^{n_{i+1}-1})$, where $y = F(x)$.*

Proof. Let D be a cylinder set such that $0 < \mu(D) < \epsilon/N$, and let G be the set of all $x \in A^Z$ for which $T^m x \in D$ for infinitely many positive and negative values of m. The set G is measurable and has measure 1, by the ergodic theorem. For $x \in G$, define $m_0 = m_0(x)$ to be the least nonnegative integer m such that $T^x \in D$, then extend to obtain an increasing sequence $m_j = m_j(x)$, $j \in Z$ such that $T^m x \in D$ if and only if $m = m_j$, for some j.

The next step is to split each interval $[m_j, m_{j+1})$ into nonoverlapping blocks of length N, starting with m_j, plus a final remainder block of length shorter than N, in case $m_{j+1} - m_j$ is not exactly divisible by N. In other words, for each j let q_j and r_j be nonnegative integers such that

$$m_{j+1} - m_j = q_j N + r_j,\ 0 \leq r_j < N,$$

and form the disjoint collection $\mathcal{I}_x(m_j)$ of left-closed, right-open intervals

$$[m_j, m_j + N),\ [m_j + N, m_j + 2N),\ \ldots\ [m_j + q_j N, m_{j+1}).$$

All but the last of these have length exactly N, while the last one is either empty or has length $r_j < N$.

The definition of G guarantees that for $x \in G$, the union $\cup_j \mathcal{I}_x(m_j)$ is a partition of Z. The random partition $\cup_j \mathcal{I}_x(m_j)$ can then be relabeled as $\{[n_i, n_{i+1}),\ i \in Z\}$, where $n_i = n_i(x)$, $i \in Z$. If $x \notin G$, define $n_i = i$, $i \in Z$.

By construction, condition (i) certainly holds for every $x \in G$. Furthermore, the ergodic theorem guarantees that the average distance between m_j and m_{j+1} is at least N/ϵ, so that (ii) also holds, almost surely.

The encoder $F = F_C$ is defined as follows. Let b be a fixed element of B, called the *filler symbol*, and let b^j denote the sequence of length $1 \leq j < N$, each of whose terms is b. If x is sequence for which $Z = \cup[n_i, n_{i+1})$, then $y = F(x)$ is defined by the formula

$$y_{n_i}^{n_{i+1}-1} = \begin{cases} b^{n_{i+1}-n_i} & \text{if } n_{i+1} - n_i < N \\ C(x_{n_i}^{n_{i+1}-1}) & \text{if } n_{i+1} - n_i = N. \end{cases}$$

This definition guarantees that property (iii) holds for $x \in G$, a set of measure 1. For $x \notin G$, define $F(x)_i = b$, $i \in Z$. The function F is certainly measurable and satisfies $T_B^{-1} F(T_A x) = F(x)$, for all x. This completes the proof of Lemma I.8.5. □

The blocks of length N are called *coding blocks* and the blocks of length less than N are called *filler or spacer blocks.* Any stationary coding F of μ for which (i), (ii), and (iii) hold is called *a stationary coding with ϵ-fraction of spacers induced by the block code C_N*. There are, of course, many different processes satisfying the conditions of the lemma, since there are many ways to parse sequences so that properties (i), (ii), and (iii) hold, for example, any event of small enough probability can be used and how the spacer blocks are coded is left unspecified in the lemma statement. The terminology applies to any of these processes.

Remark I.8.6

Lemma I.8.5 was first proved in [40], but it is really only a translation into process language of a theorem about ergodic transformations first proved by Rohlin, [60]. Rohlin's theorem played a central role in Ornstein's fundamental work on the isomorphism problem for Bernoulli shifts, [46, 63].

I.8.c A string-matching example.

The block-to-stationary code construction of Lemma I.8.5 provides a powerful tool for making counterexamples. As an illustration of the method a string matching example will be constructed. The string-matching problem is of interest in DNA modeling and is defined as follows. For $x_1^n \in A^n$ let $L(x_1^n)$ be the length of the longest block appearing at least twice in x_1^n, that is,

$$L(x_1^n) = \max\{k\colon x_{s+1}^{s+k} = x_{t+1}^{t+k}, \text{ for some } 0 \le s < t \le n-k\}.$$

The problem is to determine the asymptotic behavior of $L(x_1^n)$ along sample paths drawn from a stationary, finite-alphabet ergodic process. For i.i.d. and Markov processes it is known that $L(x_1^n) = O(\log n)$, see [26, 2]. A proof for the i.i.d. case using coding ideas is suggested in Exercise 2, Section II.1, and extended to a larger class of processes in Theorem II.5.5. A question arose as to whether such a $\log n$ type bound could hold for the general ergodic case, at least if the process is assumed to be mixing. A negative solution to the question was provided in [71], where it was shown that for any ergodic process $\{X_n\}$ and any positive function $\lambda(n)$ for which $n^{-1}\lambda(n) \to 0$, there is a nontrivial stationary coding, $\{Y_n\}$, of $\{X_n\}$, and an increasing unbounded sequence $\{n_i\}$, such that

$$\limsup_{i\to\infty} \frac{L(Y_1^{n_i})}{\lambda(n_i)} \ge 1, \text{ almost surely.}$$

In particular, if the start process $\{X_n\}$ is i.i.d, the $\{Y_n\}$ process will be mixing. It can, furthermore, be forced to have entropy as close to the entropy of $\{X_n\}$ as desired. The proof for the binary case and growth rate $\lambda(n) = n^{1/2}$ will be given here, as a simple illustration of the utility of block-to-stationary constructions.

The first observation is that if stationarity is not required, the problem is easy to solve merely by periodically inserting blocks of 0's to get bad behavior for one value of n, then nesting to obtain bad behavior for infinitely many n. To make this precise, for each $n \ge 64$, define the function $C_n\colon A^n \mapsto \{0, 1\}^n$ by changing the first $4n^{1/2}$ terms into all 0's and the final $4n^{1/2}$ terms into all 0's, that is, $C_n(x_1^n) = y_1^n$, where

$$y_i = \begin{cases} 0 & i \le 4n^{1/2} \text{ or } i > n - 4n^{1/2} \\ x_i & \text{otherwise.} \end{cases} \tag{5}$$

The n-block coding $\{Y_m\}$ of the process $\{X_m\}$ defined by

$$Y_{(j-1)n+1}^{jn} = C_n(X_{(j-1)n+1}^{jn}),\ j \ge 1,$$

clearly has the property

$$L(Y_{s+1}^{s+n}) \ge n^{1/2},$$

for any starting place s, since two disjoint blocks of 0's of length at least $n^{1/2}$ must appear in any set of n consecutive places.

The construction is iterated to obtain an infinite sequence of such n's, as follows. Let

$$\{X_m\} \xrightarrow{[C_n]} \{Y_m\},$$

indicate that $\{Y_m\}$ is the process obtained from $\{X_m\}$ by the encoding $Y_{(j-1)n+1}^{jn} = C_n(X_{(j-1)n+1}^{jn})$, $j \ge 1$, where C_n is the zero-inserter defined by the formula (5). Let

$\{n_i: i \geq 1\}$ be an increasing sequence of integers, to be specified later, and define the sequence of processes by starting with any process $\{Y(0)_m\}$ and inductively defining $\{Y(i)_m\}$ for $i \geq 1$ by the formula

$$\{Y(i-1)_m\} \xrightarrow{[C_{n_i}]} \{Y(i)_m\}.$$

For any start process $\{Y(0)_m\}$, the i-th iterate $\{Y(i)_m\}$ has the property

$$L(Y(i)_{s+1}^{s+n_j}) \geq n_j^{1/2}, \ 1 \leq j \leq i,$$

for all s, since 0's inserted at any stage are not changed in subsequent stages. In particular, each coordinate is changed at most once, which, in turn, guarantees that a limit process $\{Y(\infty)_m\}$ exists for which $L(Y(\infty)_{s+1}^{s+n_j}) \geq n_j^{1/2}$, $j \geq 1$. Furthermore, the final process has a positive frequency of 1's, provided the n_i increase rapidly enough.

The preceding construction is conceptually quite simple but clearly destroys stationarity. Randomizing the start at each stage will restore stationarity, but mixing properties are lost, and it is not easy to see how to pass to a limit process. An appropriate use of block-to-stationary coding, Lemma I.8.5, at each stage converts the block codes into stationary codings, which will, in turn, guarantee that the final limit process is a stationary coding of the starting process with the desired string-matching properties, hence, in particular, mixing is not destroyed.

The rigorous construction of a stationary coding is carried out as follows. For $n \geq 64$, let C_n be the zero-inserter code defined by (5), let $\{n_i\}$ be an increasing sequence of natural numbers, and let $\{\epsilon_i\}$ be a decreasing sequence of positive numbers, both to be specified later. Given the ergodic process $\{X_n\}$ (now thought of as a two-sided process, since stationary coding will be used), the sequence of processes $\{Y(i)\}$ is defined by setting $\{Y(0)_m\} = \{X_m\}$ and inductively defining $\{Y(i)_m\}$ for $i \geq 1$ by the formula

$$\{Y(i-1)_m\} \xrightarrow{(C_{n_i}, \epsilon_i)} \{Y(i)_m\},$$

where the notation indicates that $\{Y(i)_m\}$ is constructed from $\{Y(i-1)_m\}$ by stationary coding using the codebook C_{n_i} with a limiting ϵ_i-fraction of spacers. At each stage the same filler symbol, $b = 0$, will be used, so that the code at each stage never changes a 0 into a 1, that is, for all $m \in Z$ and $i \geq 1$,

(6) $$Y(i-1)_m = 0 \implies Y(i)_m = 0.$$

Furthermore, the choice of filler means that $L(Y(i)_1^{n_i}) \geq n_i^{1/2}$, since this always occurs if $Y(i)_1^{n_i}$ is the concatenation uv, where u is the end of a codeword and v the beginning of a codeword, and long matches are even more likely if $Y(i)_1^{n_i}$ is the concatenation ufv, where u is the end of a codeword, v is the beginning of a codeword, and f is filler. Combining this observation with the fact that 0's are not changed, (6), yields

(7) $$L(Y(i)_1^{n_j}) \geq n_j^{1/2}, \ j \leq i.$$

Since stationary codings of stationary codings are themselves stationary codings, each process $\{Y(i)_m\}$ is a stationary coding of the original process $\{X_m\}$. For each $i \geq 1$, let F_i denote the stationary encoder $F_i: \{0, 1\}^Z \mapsto \{0, 1\}^Z$ that maps $\{X_m\}$ to $\{Y(i)_m\}$. Property (6) guarantees that no coordinate is changed more than once, and hence there is a limit code F defined by the formula

$$[F(x)]_m = \lim_{i \to \infty} [F_i(x)]_m, \ m \in Z, \ x \in \{0, 1\}^Z$$

The limit process $\{Y(\infty)_m\}$ defined by $Y(\infty)_m = \lim_i Y(i)_m$, $m \in Z$ is the stationary coding of the initial process $\{X_m\}$ defined by the stationary encoder F. Since (7) hold for each i, the limit process has the property $L(Y(\infty)_1^{n_j}) \geq n_j^{1/2}$, $j \geq 1$. Furthermore, the limiting density of changes can be forced to be as small as desired, merely by making the n_i increase rapidly enough and the ϵ_i go to zero rapidly enough. In particular, the entropy of the final process will be as close to that of the original process as desired.

The above is typical of stationary coding constructions. First a sequence of block codes is constructed to produce a (nonstationary) limit process with the properties desired, then Lemma I.8.5 is applied to produce a stationary process with the same limiting properties.

I.8.d Exercises.

1. Show that if μ is mixing and $\delta > 0$, then the sequence $\{n_i(x)\}$ of Lemma I.8.5 can be chosen so that $\text{Prob}(x_{n_i}^{n_{i+1}-1} = a_1^N) < \delta$, for each $a_1^N \in A^N$. (Hint: replace D by $T^{-n}D$ for n large.)

2. Suppose $C\colon A^N \mapsto B^N$ is an N-code such that $E_\mu(d_N(x_1^N, C(x_1^N)) < \epsilon$, where μ is mixing. Show that there is a stationary code F such that $\lim_n E_\mu(d_n(x_1^n, F(x)_1^n) < 2\epsilon$ and such that $h(\mu \circ F^{-1}) \leq \epsilon + H(\mu_n \circ C^{-1})/n$. (Hint: use the preceding exercise.)

Section I.9 Process topologies.

Two useful ways to measure the closeness of stationary processes are described in this section. One concept, called the weak topology, declares two processes to be close if their joint distributions are close for a long enough time. The other concept, called the $\bar{d}$-topology, declares two ergodic processes to be close if only a a small limiting density of changes are needed to convert a typical sequence for one process into a typical sequence for the other process.

The weak topology is separable, compact, and easy to describe, but it has two important defects, namely, weak limits of ergodic processes need not be ergodic and entropy is not weakly continuous. On the other hand, entropy is $\bar{d}$-continuous and the class of ergodic processes is $\bar{d}$-closed, as are other classes of interest.

The $\bar{d}$-metric is not so easy to define, the $\bar{d}$-distance between processes is often difficult to calculate, and the $\bar{d}$-topology is nonseparable. Many of the deeper recent results in ergodic theory depend on the $\bar{d}$-metric, however, and it plays a central role in the theory of stationary codings of i.i.d. processes, a theory to be presented in Chapter 4.

I.9.a The weak topology.

The collection of (Borel) probability measures on a compact space X is denoted by $\mathcal{P}(X)$. Of primary interest are the two cases, $X = A^n$and $X = A^\infty$, where A is a finite set. The collection $\mathcal{P}(A^\infty)$ is just the collection of processes with alphabet A. The subcollection of stationary processes is denoted by $\mathcal{P}_s(A^\infty)$ and the subcollection of stationary ergodic processes is denoted by $\mathcal{P}_e(A^\infty)$.

A sequence $\{\mu^{(n)}\}$ of measures in $\mathcal{P}(A^\infty)$ converges *weakly* to a measure μ if

$$\lim_{n\to\infty} \mu^{(n)}(a_1^k) = \mu(a_1^k),$$

for all k and a_1^k. The weak topology on $\mathcal{P}(A^\infty)$ is the topology defined by weak convergence. It is the weakest topology for which each of the mappings $\mu \mapsto \mu(C)$ is continuous for each cylinder set C.

The weak topology is Hausdorff since a measure is determined by its values on cylinder sets. The weak topology is a metric topology, relative to the metric $D(\mu, \nu) = \sum 2^{-k}|\mu - \nu|_k$, where

$$|\mu - \nu|_k = \sum_{a_1^k} |\mu(a_1^k) - \nu(a_1^k)|,$$

denotes the k-th order distributional (variational) distance. If k is understood then $|\cdot|$ may be used in place of $|\cdot|_k$.

The weak topology is a compact topology. Indeed, for any sequence $\{\mu^{(n)}\}$ of probability measures and any a_1^k, the sequence $\{\mu^{(n)}(a_1^k)\}$ is bounded, hence has a convergent subsequence. Since there are only countably many a_1^k, the usual diagonalization procedure produces a subsequence $\{n_i\}$ such that $\{\mu^{(n_i)}(a_1^k)\}$ converges to, say, $\mu(a_1^k)$, as $i \to \infty$, for every k and every a_1^k. It is easy to check that the limit function μ is a probability measure. Furthermore, the limit μ is stationary if each $\mu^{(n)}$ is stationary, so the class of stationary measures is also compact in the weak topology.

The weak limit of ergodic measures need not be ergodic. One way to see this is to note that on the class of stationary Markov processes, weak convergence coincides with convergence of the entries in the transition matrix and stationary vector. In particular, if $\mu^{(n)}$ is the Markov process with transition matrix

$$M_n = \begin{bmatrix} 1 - 1/n & 1/n \\ 1/n & 1 - 1/n \end{bmatrix},$$

then $\mu^{(n)}$ converges weakly to the (nonergodic) process μ that puts weight 1/2 on each of the two sequences $(0, 0, \ldots)$ and $(1, 1, \ldots)$.

A second defect in the weak topology is that entropy is not weakly continuous. For example, for each n, let $x(n)$ be the concatenation of the members of $\{0, 1\}^n$ in some order and let $\mu^{(n)}$ be the concatenated-block process defined by the single $n2^n$-block $x(n)$. The process $\mu^{(n)}$ has entropy 0, yet

$$\lim_{n\to\infty} \mu^{(n)}(a_1^k) = 2^{-k}, \quad \forall\, k,\ a_1^k,$$

so that $\{\mu^{(n)}\}$ converges weakly to the coin-tossing process, which has entropy $\log 2$.

Entropy is weakly upper semicontinuous, however.

Theorem I.9.1

If $\mu^{(n)}$ converges weakly to μ, and each $\mu^{(n)}$ is stationary, then $\limsup_n H(\mu^{(n)}) \leq H(\mu)$.

Proof. For any stationary measure μ, let $H_\mu(X_0|X_{-k}^{-1})$ denote the conditional entropy of X_0 relative to X_{-k}^{-1}, where X_{-k}^{0} is distributed according to μ and recall that $H_\mu(X_0|X_{-k}^{-1})$ decreases to $H(\mu)$ as $k \to \infty$. Thus, given, $\epsilon > 0$ there is a k such that $H_\mu(X_0|X_{-k}^{-1}) \leq H(\mu) + \epsilon$.

Now suppose $\mu^{(n)}$ converges weakly to μ. Since $H_\mu(X_0|X_{-k}^{-1})$ depends continuously on the probabilities $\mu(a_1^{k+1})$, it must be true that

$$H_{\mu^{(n)}}(X_0|X_{-k}^{-1}) \leq H(\mu) + 2\epsilon,$$

for all sufficiently large n. But if this is true for some n, then $H(\mu^{(n)}) \leq H(\mu) + 2\epsilon$, since $H(\mu^{(n)}) \leq H_{\mu^{(n)}}(X_0|X_{-k}^{-1})$ is always true. This proves Theorem I.9.1. □

I.9.b The $\bar{d}$-metric.

The $\bar{d}$-distance between two ergodic processes is the minimal limiting (upper) density of changes needed to convert a typical sequence for one process into a typical sequence for the other. A precise formulation of this idea and some equivalent versions are given in this subsection.

The per-letter Hamming distance is defined by

$$d_n(x_1^n, y_1^n) = \frac{1}{n}\sum_{i=1}^{n} d(x_i, y_i), \text{ where } d(a, b) = \begin{cases} 0 & a = b \\ 1 & a \neq b. \end{cases}$$

It measures the fraction of changes needed to convert x_1^n into y_1^n. This extends to the pseudometric $\bar{d}(x, y)$ on A^∞, defined by

$$\bar{d}(x, y) = \limsup_{n\to\infty} d_n(x_1^n, y_1^n).$$

Thus $\bar{d}(x, y)$ is the limiting (upper) density of changes needed to convert the (infinite) sequence x into the (infinite) sequence y, that is, $\bar{d}(x, y)$ is the limiting per-letter Hamming distance between x and y.

Let $\mathcal{T}(\mu)$ denote the set of frequency-typical sequences for the stationary process μ, that is, the set of all sequences for which the limiting empirical measure is equal to μ. By the typical-sequence theorem, Theorem I.4.1, and its converse, Theorem I.4.2, the measure μ is ergodic if and only if $\mu(\mathcal{T}(\mu)) = 1$. The (minimal) limiting (upper) density of changes needed to convert y into a μ-typical sequence is given by

$$\bar{d}(y, \mathcal{T}(\mu)) = \inf\{\bar{d}(x, y) : x \in \mathcal{T}(\mu)\}.$$

Theorem I.9.2

If μ and ν are ergodic processes, then $\bar{d}(y, \mathcal{T}(\mu))$ is ν-almost surely constant.

Proof. The function $f(y) = \bar{d}(y, \mathcal{T}(\mu))$ is invariant since $\bar{d}(y, x) = \bar{d}(Ty, Tx)$, hence $f(y)$ is almost surely constant. □

The constant of the theorem is denoted by $\bar{d}(\mu, \nu)$ and is called *the $\bar{d}$-distance between the ergodic processes μ and ν.* Note, in particular, that for ν-almost every ν-typical sequence y, the $\bar{d}$-distance between μ and ν is the limiting density of changes needed to convert y into a μ-typical sequence. (The fact that $\bar{d}(\mu, \nu)$, as defined, is actually a metric on the class of ergodic processes is a consequence of an alternative description to be given later.)

Example I.9.3 (The $\bar{d}$-distance for i.i.d. processes.)

To illustrate the $\bar{d}$-distance concept the distance between two binary i.i.d. processes will be calculated. For $0 < p < 1$, let $\mu^{(p)}$ denote the binary i.i.d. process such that p is the probability that $X_1 = 1$. Suppose $0 < p < q < 1/2$, let x be a $\mu^{(p)}$-typical sequence, and let y be a $\mu^{(q)}$-typical sequence. The sequence x contains a limiting p-fraction of 1's while the sequence q contains a limiting q-fraction of 1's, so that x and y must disagree in at least a limiting $(q-p)$-fraction of places, that is, $\bar{d}(x, y) \geq q-p$. Thus it is enough to find, for each $\mu^{(p)}$-typical x, one $\mu^{(q)}$-typical y such that $\bar{d}(x, y) = q-p$.

Fix a $\mu^{(p)}$-typical x, and let r be the solution to the equation $q = p(1-r)+(1-p)r$. Let Z be an i.i.d. random sequence with the distribution of the $\mu^{(r)}$-process and let $Y = x \oplus Z$, where $\oplus$ is addition modulo 2. For any set S of natural numbers, the values $\{Z_s\colon s \in S\}$ are independent of the values $\{Z_s\colon s \notin S\}$, and hence, Y is $\mu^{(q)}$-typical, for almost any Z. Since Z is also almost surely $\mu^{(r)}$-typical, there is therefore at least one $\mu^{(r)}$-typical z such that $y = x \oplus z$ is $\mu^{(q)}$-typical. Thus there is indeed at least one $\mu^{(q)}$-typical y such that $\bar{d}(x, y) = q-p$, and hence $\bar{d}(\mu^{(p)}, \mu^{(q)}) = q-p$.

This result is also a simple consequence of the more general definition of $\bar{d}$-distance to be given later in this section, see Exercise 5.

Example I.9.4 (Weakly close, $\bar{d}$-far-apart Markov chains.)

As another example of the $\bar{d}$-distance idea it will be shown that the $\bar{d}$-distance between two binary Markov chains whose transition matrices are close can nevertheless be close to 1/2. Let μ and ν be the binary Markov chains with the respective transition matrices

$$M = \begin{bmatrix} p & 1-p \\ 1-p & p \end{bmatrix}, \quad N = \begin{bmatrix} 0 & 1 \\ 1 & 0 \end{bmatrix},$$

where p is small, say $p = 10^{-50}$. The ν-measure is concentrated on the alternating sequence $y = 101010\ldots$ and its shift Ty. On the other hand, a typical μ-sequence x has blocks of alternating 0's and 1's, separated by an extra 0 or 1 about every 10^{50} places. Thus $\bar{d}(x, y) \sim 1/2$, since about half the time the alternation in x is in phase with the alternation in y, which produces no errors, and about half the time the alternations are out of phase, which produces only errors. Likewise, $\bar{d}(x, Ty) \sim 1/2$, so that $\bar{d}(x, \mathcal{T}(\nu)) \sim 1/2$, for μ-almost all x. (With a little more effort it can be shown that $\bar{d}(\mu, \nu)$ is exactly 1/2; see Exercise 4.) These results show that the $\bar{d}$-topology is not the same as the weak topology, for μ converges weakly to ν as $p \to 0$. Later it will be shown that the class of mixing processes is $\bar{d}$-closed, hence the nonmixing process ν cannot be the $\bar{d}$-limit of any sequence of mixing processes.

I.9.b.1 The joining definition of $\bar{d}$.

The definition of $\bar{d}$-distance given by Theorem I.9.2 serves as a useful guide to intuition, but it is not always easy to use in proofs and it doesn't apply to nonergodic processes. To remedy these defects an alternative formulation of the $\bar{d}$-distance as a limit of a distance $\bar{d}_n(\mu_n, \nu_n)$ between the corresponding n-th order distributions will be developed. This new process distance will also be denoted by $\bar{d}(\mu, \nu)$. In a later subsection it will be shown that this new process distance is the constant given by Theorem I.9.2.

A key concept used in the definition of the $\bar{d}_n$-metric is the "join" of two measures. A *joining* of two probability measures μ and ν on a finite set A is a measure λ on $A \times A$ that has μ and ν as marginals, that is,

$$\mu(a) = \sum_b \lambda(a, b); \quad \nu(b) = \sum_a \lambda(a, b). \tag{1}$$

A simple geometric model is useful in thinking about the joining concept. First each of the two measures, μ and ν, is represented by a partition of the unit square into horizontal strips, with the width of a strip equal to the measure of the symbol it represents. The joining condition $\mu(a) = \sum_b \lambda(a, b)$ means that, for each a, the μ-rectangle corresponding to a can be partitioned into horizontal rectangles $\{R(a, b): b \in A\}$ such that $R(a, b)$ has area $\lambda(a, b)$. The second joining condition, $\nu(b) = \sum_a \lambda(a, b)$, means that the total mass of the rectangles $\{R(a, b): a \in A\}$ is exactly the ν-measure of the rectangle corresponding to b. (See Figure I.9.5.) In other words, a joining is just a rule for cutting each μ-rectangle into subrectangles and reassembling them to obtain the ν-rectangles.

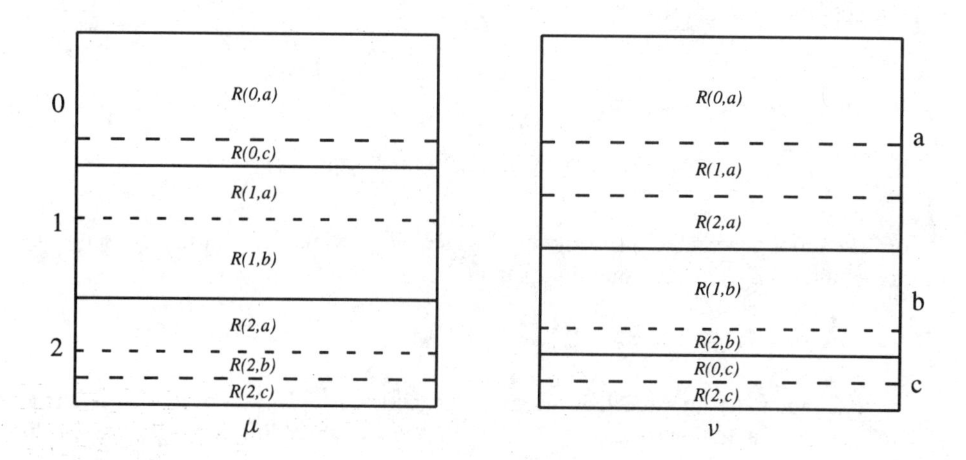

Figure I.9.5 Joining as reassembly of subrectangles.

Of course, one can also think of cutting up the ν-rectangles and reassembling to give the μ-rectangles. Also, the use of the unit square and rectangles is for simplicity only, for a finite distribution can always be represented as a partition of any nonatomic probability space into sets whose masses are given by the distribution. (A space is nonatomic if any subset of positive measure contains subsets of any smaller measure.) A representation of μ on a nonatomic probability space (Z, α) is just a measure-preserving mapping ϕ from (Z, α) to (A, μ), that is, a measurable mapping such that $\alpha(\phi^{-1}(a)) = \mu(a)$, $a \in A$. A joining λ can then be represented as a pair of measure-preserving mappings, ϕ from (Z, α) to (A, μ), and ψ from (Z, α) to (A, ν), such that λ is given by the formula

(2) $$\lambda(a_i, a_j) = \alpha(\phi^{-1}(a_i) \cap \psi^{-1}(a_j)).$$

Turning now to the definition of $\bar{d}_n(\mu, \nu)$, where μ and ν are probability measures on A^n, let $J_n(\mu, \nu)$ denote the set of joinings of μ and ν. The $\bar{d}_n$-metric is defined by

$$\bar{d}_n(\mu, \nu) = \min_{\lambda \in J_n(\mu,\nu)} \mathcal{E}_\lambda(d_n(x_1^n, y_1^n)),$$

where $\mathcal{E}_\lambda$ denotes expectation with respect to λ. The minimum is attained, since expectation is continuous with respect to the distributional distance $|\lambda - \tilde{\lambda}|_n$, relative to which $J_n(\mu, \nu)$ is a compact subset of the space $\mathcal{P}(A^n \times A^n)$. A measure λ on $A^n \times A^n$ is said to *realize* $\bar{d}_n(\mu, \nu)$ if it is a joining of μ and ν such that $E_\lambda(d_n(x_1^n, y_1^n)) = \bar{d}_n(\mu, \nu)$.

The function $\bar{d}_n(\mu, \nu)$ satisfies the triangle inequality and is strictly positive if $\mu \neq \nu$, hence is a metric on the class $\mathcal{P}(A^n)$ of measures on A^n, see Exercise 2.

The $\bar{d}$-distance between two processes is defined for the class $\mathcal{P}(A^\infty)$ of probability measures on A^∞ by passing to the limit, that is,

$$\bar{d}(\mu, \nu) = \limsup_{n\to\infty} \bar{d}_n(\mu_n, \nu_n).$$

Note that $\bar{d}(\mu, \nu)$ is a pseudometric on the class $\mathcal{P}(A^\infty)$, since $\bar{d}_n$ is a metric on $\mathcal{P}(A^n)$, and hence $\bar{d}$ defines a topology on $\mathcal{P}(A^\infty)$. In the stationary case, the limit superior is, in fact, both the limit and the supremum.

Theorem I.9.6

If μ and ν are stationary then

$$\bar{d}(\mu, \nu) = \sup_n \bar{d}_n(\mu_n, \nu_n) = \lim_n \bar{d}_n(\mu_n . \nu_n).$$

Proof. The proof depends on a superadditivity inequality, which is, in turn, a consequence of the definition of $\bar{d}_n$ as an minimum, namely,

$$n\ \bar{d}_n(\mu_1^n, \nu_1^n) + m\ \bar{d}_m(\mu_{n+1}^{n+m}, \nu_{n+1}^{n+m}) \le (n+m)\ \bar{d}_{n+m}(\mu_1^{n+m}, \nu_1^{n+m}), \tag{3}$$

where μ_i^j denotes the measure induced by μ on the set A_i^j. The proof of this is left to Exercise 3. In the stationary case, $\mu_{n+1}^{n+m} = \mu_1^m$, so the preceding inequality takes the form

$$n\bar{d}_n + m\bar{d}_m \le (n+m)\bar{d}_{n+m},$$

which implies that the limsup is in fact both a limit and the supremum. This establishes Theorem I.9.6. □

A consequence of the preceding result is that the $\bar{d}$-pseudometric is actually a metric on the class $\mathcal{P}_s(A^\infty)$ of stationary measures. Indeed, if $\bar{d}(\mu, \nu) = 0$ and μ and ν are stationary, then, since in the stationary case the limit superior is a supremum, $\bar{d}_n(\mu_n, \nu_n)$ must be 0, for all n, which implies that $\mu_n = \nu_n$ for all n, since $\bar{d}_n(\mu, \nu)$ is a metric, Exercise 2, which, in turn, implies that $\mu = \nu$.

The $\bar{d}$-distance between stationary processes is defined as a limit of n-th order $\bar{d}_n$-distance, which is, in turn, defined as a minimization of expected per-letter Hamming distance over joinings. It is useful to know that the $\bar{d}$-distance can be defined directly in terms of stationary joining measures on the product space $A^\infty \times A^\infty$. A joining λ of μ and ν is just a measure on $A^\infty \times A^\infty$ with μ and ν as marginals, that is,

$$\mu(B) = \lambda(B \times A^\infty),\ \ \nu(B) = \lambda(A^\infty \times B),\ \ B \in \Sigma.$$

The set of all stationary joinings of μ and ν will be denoted by $J_S(\mu, \nu)$.

Theorem I.9.7 (Stationary joinings and $\bar{d}$-distance.)

If μ and ν are stationary then

$$\bar{d}(\mu, \nu) = \min_{\lambda \in J_S(\mu,\nu)} \mathcal{E}_\lambda(d(x_1, y_1)). \tag{4}$$

Proof. The existence of the minimum follows from the fact that the set $J_S(\mu, \nu)$ of stationary joinings is weakly closed and hence compact, and from the fact that $\mathcal{E}_\lambda(d(x_1, y_1))$ is weakly continuous, for it depends only on the first order distribution. It is also easy to see that $\bar{d}$ cannot exceed the minimum of $\mathcal{E}_\lambda(d(x_1, y_1))$, for if $\lambda \in J_S(\mu, \nu)$ then stationarity implies that

$$\mathcal{E}_\lambda(d(x_1, y_1)) = \mathcal{E}_{\lambda_n}(d_n(x_1^n, y_1^n)),$$

and the latter dominates $\bar{d}_n(\mu, \nu)$, since λ_n belongs to $J_n(\mu_n, \nu_n)$.

It takes a bit more effort to show that $\bar{d}$ cannot be less than the right-hand side in (4). Towards this end, for each n choose $\lambda_n \in J_n(\mu_n, \nu_n)$, such that

$$\bar{d}_n(\mu_n, \nu_n) = \mathcal{E}_{\lambda_n}(d_n(x_1^n, y_1^n)).$$

The goal is to construct a stationary measure $\lambda \in J_S(\mu, \nu)$ from the collection $\{\lambda_n\}$, such that $\mathcal{E}_\lambda(d(x_1, y_1) = \lim_n \bar{d}_n(\mu.\nu)$. This is done by forming, for each n, the concatenated-block process defined by λ_n, then taking a weakly convergent subsequence to obtain the desired limit process λ. The details are given in the following paragraphs.

For each n, let $\widehat{\lambda}^{(n)}$ denote the concatenated-block process defined by λ_n. As noted in Exercise 8 of Section I.1, the measure $\widehat{\lambda}^{(n)}$ is given by the formula

$$\widehat{\lambda}^{(n)}(B) = \frac{1}{n}\sum_{i=1}^{n} \lambda^{(n)}(S^{-i}(B)), \tag{5}$$

where $\lambda^{(n)}$ is the measure on $(A \times A)^\infty$ defined by

$$\lambda^{(n)}((x_1^N, y_1^N)) = \\ \left(\prod_{k=0}^{K-1} \lambda_n((x_{kn+1}^{kn+n}, y_{kn+1}^{kn+n}))\right) \lambda_n((x_{Kn+1}^{Kn+r}, y_{Kn+1}^{Kn+r})),$$

if $N = Kn + r$, $0 \le r < n$. The definition of $\lambda^{(n)}$ implies that if $0 \le i \le n - j$ then

$$\lambda^{(n)}(S^{-i}([x_1^j] \times A^\infty)) = \lambda_n(T^{-i}[x_1^j] \times A^\infty) = \mu_n(T^{-i}[x_1^j]) = \mu([x_1^j]),$$

since λ_n has μ_n as marginal and μ is shift invariant. This, together with the averaging formula, (5), implies that $\lim_n \widehat{\lambda}^{(n)}(C \times A^\infty) = \mu(C)$, for any cylinder set C, which proves the first of the following equalities.

$$\begin{aligned} &\text{(a)} \quad \lim_{n\to\infty} \widehat{\lambda}^{(n)}(B \times A^\infty) = \mu(B),\ B \in \Sigma. \\ &\text{(b)} \quad \lim_{n\to\infty} \widehat{\lambda}^{(n)}(A^\infty \times B) = \mu(B),\ B \in \Sigma. \\ &\text{(c)} \quad \lim_{n\to\infty} \mathcal{E}_{\widehat{\lambda}^{(n)}}(d(x_1, y_1)) = \lim_n \bar{d}_n(\mu, \nu) = \bar{d}(\mu, \nu). \end{aligned} \tag{6}$$

The proof of (b) is similar to the proof of (a). To prove (c), let λ_i^* be projection of λ_n onto its i-th coordinate, that is, the measure on $A \times A$ defined by

$$\lambda_i^*(a, b) = \sum_{u,u' \in A^{i-1}, v,v' \in A^{n-i}} \lambda_n(uav, u'bv'),$$

and note that

$$\mathcal{E}_{\lambda_n}(d_n(x_1^n, y_1^n)) = \frac{1}{n}\sum_{x_i, y_i} d(x_i, y_i)\lambda_i^*(x_i, y_i) = \mathcal{E}_{\widehat{\lambda}^{(n)}}(d(x_1, y_1)),$$

since $\widehat{\lambda}^{(n)}(a, b) = (1/n)\sum_i \lambda_i^*(a, b)$. This proves (c), since $\bar{d}_n(\mu, \nu) = \mathcal{E}_{\lambda_n}(d_n(x_1^n, y_1^n))$.

To complete the proof select a convergent subsequence of $\widehat{\lambda}^{(n)}$. The limit λ must be stationary, and is a joining of μ and ν, by conditions (6a) and (6b). Thus $\lambda \in J_S(\mu, \nu)$. Condition (6c) guarantees that $\bar{d}(\mu, \nu) = \mathcal{E}_{\lambda}(d(x_1, y_1))$. □

I.9.b.2 Empirical distributions and joinings.

The pair (x_1^n, y_1^n) defines three empirical k-block distributions. Two of these, $p_k(a_1^k|x_1^n)$ and $p_k(b_1^k|y_1^n)$, are obtained by thinking of x_1^n and y_1^n as separate sequences, and the third, $p_k((a_1^k, b_1^k)|(x_1^n, y_1^n))$, is obtained by thinking of the pair (x_1^n, y_1^n) as a single sequence. A simple, but important fact, is that $p_k(\cdot\ |(x_1^n, y_1^n))$ is a joining of $p_k(\cdot\ |x_1^n)$ and $p_k(\cdot\ |y_1^n)$, since, for example, each side of

$$\sum_{b_1^k}(n-k+1)p((a_1^k, b_1^k)|(x_1^n, y_1^n)) = (n-k+1)p(a_1^k|\ x_1^n)$$

counts the number of times a_1^k appears in x_1^n. Note also, that

$$d_n(x_1^n, y_1^n) = E_{p_1(\cdot|(x_1^n, y_1^n))}(d(a, b)). \tag{7}$$

Limit versions of these finite ideas as $n \to \infty$ will be needed. One difficulty is that limits may not exist for a given pair (x, y). A diagonalization argument, however, yields an increasing sequence $\{n_j\}$ such that each of the limits

$$\lim_{j\to\infty} p(a_1^k|x_1^{n_j}),\ \lim_{j\to\infty} p(b_1^k|y_1^{n_j}),\ \lim_{j\to\infty} p((a_1^k, b_1^k)|(x_1^{n_j}, y_1^{n_j})), \tag{8}$$

exists for all k and all a_1^k and b_1^k. Fix such a $\{n_j\}$ and denote the Kolmogorov measures defined by the limits in (8) as $\widehat{\mu}(\cdot\ |x)$, $\widehat{\nu}(\cdot\ |y)$, and $\widehat{\lambda}(\cdot\ |(x, y))$, respectively. By dropping to a further subsequence, if necessary, it can be assumed that the limit

$$\widehat{d}(x, y) = \lim_{j\to\infty} d_{n_j}((x_1^{n_j}, y_1^{n_j}))$$

also exists. The limit results that will be needed are summarized in the following lemma.

Lemma I.9.8 (The empirical-joinings lemma.)
The measures $\widehat{\mu}$, $\widehat{\nu}$, and $\widehat{\lambda}$ are all stationary, and $\widehat{\lambda}$ is a joining of $\widehat{\mu}$ and $\widehat{\nu}$. Furthermore, $\widehat{d}(x, y) = E_{\widehat{\lambda}}(d(a_1, b_1))$.

Proof. Note that

$$0 \le (n-k+2)p(a_2^k|x_1^n) - (n+k-1)\sum_{a_1} p(a_1^k|x_1^n) \le 1,$$

since the only difference between the two counts is that a_2^k might be the first $k-1$ terms of x_1^n, hence is not counted in the second term. Passing to the limit it follows that

$\widehat{\mu}$ is stationary. Likewise, $\widehat{\nu}$ and $\widehat{\lambda}$ are stationary. The joining result follows from the corresponding finite result, while the $\bar{d}$-equality is a consequence of (7). □

The empirical-joinings lemma can be used in conjunction with the ergodic decomposition to show that the $\bar{d}$-distance between ergodic processes is actually realized by an ergodic joining, a strengthening of the stationary-joining theorem, Theorem I.9.7.

Theorem I.9.9 (Ergodic joinings and $\bar{d}$-distance.)
If μ and ν are ergodic then there is an ergodic joining λ of μ and ν such that $\bar{d}(\mu, \nu) = \mathcal{E}_\lambda(d(x_1, y_1))$.

Proof. Let λ be a stationary joining of μ and ν for which $\bar{d}(\mu, \nu) = \mathcal{E}_\lambda(d(x_1, y_1))$, and let

$$\lambda = \int \lambda_{\pi(x,y)}\, d\omega(\pi(x, y)) \tag{9}$$

be the ergodic decomposition of λ, as given by the ergodic decomposition theorem, Theorem I.4.10. In this expression π is the projection onto frequency-equivalence classes of pairs of sequences, ω is a probability measure on the set of such equivalence classes, $\lambda_{\pi(x,y)}$ is the empirical measure defined by $(x, y) \in A^\infty \times A^\infty$, and $\lambda_{\pi(x,y)}$ is ergodic for almost every (x, y). Furthermore, since λ is a stationary joining of μ and ν, it is concentrated on those pairs (x, y) for which x is μ-typical and y is ν-typical. Since, by definition of $\lambda_{\pi(x,y)}$, the pair (x, y) is $\lambda_{\pi(x,y)}$-typical, it follows that

$$x \in \mathcal{T}(\mu),\ \ y \in \mathcal{T}(\nu),\ \ (x, y) \in \mathcal{T}(\lambda_{\pi(x,y)})$$

for λ-almost every pair (x, y). But if this holds for a given pair, then $\lambda_{\pi(x,y)}$ must be a stationary joining of μ and ν, by the empirical-joinings lemma, Lemma I.9.8, and hence $\bar{d}(\mu, \nu) \le E_{\lambda_{\pi(x,y)}}(d(x_1, y_1))$. Thus,

$$\bar{d}(\mu, \nu) \le E_{\lambda_{\pi(x,y)}}(d(x_1, y_1)),\ \lambda\text{-almost surely.} \tag{10}$$

The integral expression (9) implies, however, that

$$\mathcal{E}_\lambda(d(x_1, y_1)) = \int \mathcal{E}_{\lambda_{\pi(x,y)}}(d(x_1, y_1))\, d\omega(\pi(x, y)),$$

since $d(x_1, y_1)$ depends only on the first order distribution and expectation is linear. This, combined with (10), means that

$$\bar{d}(\mu, \nu) = E_{\lambda_{\pi(x,y)}}(d(x_1, y_1)),\ \lambda\text{-almost surely,}$$

since it was assumed that $\bar{d}(\mu, \nu) = \mathcal{E}_\lambda(d(x_1, y_1))$. But, $\lambda_{\pi(x,y)}$ is ergodic and is a joining of μ and ν, for λ-almost every pair (x, y). This establishes Theorem I.9.9. □

Next it will be shown that the $\bar{d}$-distance defined as a minimization of expected per-letter Hamming distance over joinings, Theorem I.9.7, is, in the ergodic case, the same as the constant given by Theorem I.9.2.

Theorem I.9.10 (Typical sequences and $\bar{d}$-distance.)
If μ and ν are ergodic then $\bar{d}(\mu, \nu) = \bar{d}(y, \mathcal{T}(\mu))$, ν-almost surely.

Proof. First it will be shown that

$$(x, y) \in \mathcal{T}(\mu) \times \mathcal{T}(\nu) \Rightarrow \bar{d}(\mu, \nu) \le \bar{d}(x, y). \tag{11}$$

Let x be μ-typical and y be ν-typical, and choose an increasing sequence $\{n_j\}$ such that $p((a_1^k, b_1^k)|(x_1^{n_j}, y_1^{n_j}))$ converges to a limit $\widehat{\lambda}(a_1^k, b_1^k)$ for each k and each a_1^k and b_1^k, and so that $d_{n_j}(x_1^{n_j}, y_1^{n_j})$ converges to a limit $\widehat{d}(x, y)$, as $j \to \infty$. The empirical-joinings lemma implies that $\widehat{\lambda}$ is stationary and has μ and ν as marginals, and hence

$$\bar{d}(\mu, \nu) \le E_{\widehat{\lambda}}(d(X_1, Y_1)).$$

Since the latter is equal to $\widehat{d}(x, y)$, also by the empirical-joinings lemma, and $\widehat{d}(x, y) \le \bar{d}(x, y)$, the proof of (11) is finished.

Next it will be shown that there is a subset $X \subset A^\infty$ of ν-measure 1 such that

$$\bar{d}(y, \mathcal{T}(\mu)) = \bar{d}(\mu, \nu), \quad \forall y \in X. \tag{12}$$

Here is why this is so. Theorem I.9.9 implies that there is an ergodic joining λ for which $E_\lambda(d(x_1, y_1)) = \bar{d}(\mu, \nu)$. The ergodic theorem guarantees that for λ-almost all pairs (x, y), the following two properties hold.

(a) $d_n(x_1^n, y_1^n) = \frac{1}{n}\sum_{i=1}^{n} d(x_i, y_i) \to E_\lambda(d(x_1, y_1)) = \bar{d}(\mu, \nu)$.

(b) $(x, y) \in \mathcal{T}(\lambda) \subseteq \mathcal{T}(\mu) \times \mathcal{T}(\nu)$.

In particular, there is a set $X \subset A^\infty$ such that $\nu(X) = 1$ and such that for $y \in X$ there is at least one $x \in \mathcal{T}(\mu)$ for which both (a) and (b) hold. This establishes the desired result (12).

The lower bound result, (11), together with the almost sure limit result, (12), yields the desired result, Theorem I.9.10. □

I.9.b.3 Properties and interpretations of the $\bar{d}$-distance.

As earlier, the distributional (or variational) distance between n-th order distributions is given by

$$|\mu - \nu|_n = \sum_{a_1^n} |\mu(a_1^n) - \nu(a_1^n)|$$

and a measure is said to be T^N-invariant or N-stationary if it is invariant under the N-th power of the shift T. The vector obtained from x_1^n be deleting all but the terms with indices $j_1 < j_2 < \ldots < j_m$, will be denoted by $x_n(j_1^m)$, or by $x(j_1^m)$ when n is understood, that is,

$$x_n(j_1^m) = (x_{j_1}, x_{j_2}, \ldots, x_{j_m}).$$

For convenience, random variable notation will be used in stating $\bar{d}$-distance results, for example, $\bar{d}_n(X_1^n, Y_1^n)$ stands for $\bar{d}_n(\mu, \nu)$, if μ is the distribution of X_1^n and ν the distribution of Y_1^n. Another convenient notation is X_1^n/r to indicate the random vector X_1^n, conditioned on the value r of some auxiliary random variable R..

Lemma I.9.11 (Properties of the $\bar{d}$-distance.)

(a) *$\bar{d}_n(\mu, \nu) \le (1/2)|\mu - \nu|_n$, with equality when $n = 1$.*

(b) *$\bar{d}_n$ is a complete metric on the space of measures on A^n.*

(c) *$m\bar{d}_m(X(j_1^m), Y(j_1^m)) \le n\bar{d}_n(X_1^n, Y_1^n) \le m\bar{d}_m(X(j_1^m), Y(j_1^m)) + n - m$.*

(d) *If μ and ν are T^N-invariant, then $\bar{d}(\mu, \nu) = \lim \bar{d}_n(\mu, \nu)$.*

(e) *$\bar{d}(\mu, \nu) = \bar{d}(\mu \circ T^{-1}, \nu \circ T^{-1})$.*

(f) *If the blocks $\{X_{(j-1)n+1}^{jn}\}$ and $\{Y_{(j-1)n+1}^{jn}\}$ are independent of each other and of past blocks, then $m\bar{d}_{nm}(X_1^{nm}, Y_1^{nm}) = \sum_{j=1}^m \bar{d}_n(X_{(j-1)n+1}^{jn}, Y_{(j-1)n+1}^{jn})$.*

(g) *$\bar{d}_n(X_1^n, Y_1^n) \le \sum_r P(R = r)\bar{d}_n(X_1^n/r, Y_1^n)$, for any finite random variable R.*

Proof. A direct calculation shows that

$$|\mu - \nu|_n = 2 - 2\sum_{a_1^n} \min(\mu(a_1^n), \nu(a_1^n)),$$

and that

$$E_\lambda(d_n(a_1^n, b_1^n)) \le \lambda(\{(a_1^n, b_1^n)\colon\ a_1^n \ne b_1^n\}) = 1 - \sum_{a_1^n} \lambda(a_1^n, a_1^n),$$

for any joining λ of μ and ν. Moreover, there is always at least one joining λ^* of μ and ν for which $\lambda_n^*(a_1^n, a_1^n) = \min(\mu(a_1^n), \nu(a_1^n))$, see Exercise 8, and hence,

$$\bar{d}_n(\mu, \nu) \le E_{\lambda^*}(d_n(a_1^n, b_1^n)) \le \frac{1}{2}|\mu - \nu|_n.$$

In the case when $n = 1$, $E_\lambda(d(a, b)) \ge 1 - \sum_a \lambda(a, a)$, since $d(a, b) = 1$, unless $a = b$. This completes the proof of property (a).

The fact that $\bar{d}_n(\mu, \nu)$ satisfies the triangle inequality and is positive if $\mu \ne \nu$ is left as an exercise. Completeness follows from (a), together with the fact that the space of measures is compact and complete relative to the metric $|\mu - \nu|_n$.

The left-hand inequality in (c) follows from the fact that summing out in a joining of X_1^n, Y_1^n gives a joining of $X(j_1^m), Y(j_1^m)$, and $md_m(x(j_1^m), y(j_1^m)) \le nd_n(x_1^n, y_1^n)$.

The right-hand inequality in (c) is a consequence of the fact that any joining λ of $X(j_1^m)$ and $Y(j_1^m)$ can be extended to a joining of X_1^n and Y_1^n. To establish this extension property, let $\{i_1, i_2, \ldots, i_{n-m}\}$ denote the natural ordering of the complement $\{1, 2, \ldots, n\} - \{j_1, \ldots, j_m\}$ and for each $(x(j_1^m), y(j_1^m))$, let

$$\lambda_{(x(j_1^m), y(j_1^m))}(x(i_1^{n-m}), y(i_1^{n-m}))$$

be a joining of the conditional random variables, $X(i_1^{n-m})/x(j_1^m)$ and $Y(i_1^{n-m})/y(j_1^m)$. The function defined by

$$\lambda^*(x_1^n, y_1^n) = \lambda(x(j_1^m), y(j_1^m))\lambda_{(x(j_1^m), y(j_1^m))}(x(i_1^{n-m}), y(i_1^{n-m})),$$

is a joining of X_1^n and Y_1^n which extends λ. Furthermore,

$$nE_{\lambda^*}(d_n(x_1^n, y_1^n)) \le mE_\lambda(d_m(x(j_1^m), y(j_1^m)) + n - m,$$

since $d_{n-m}(x(i_1^{n-m}), y(i_1^{n-m})) \leq 1$. Minimizing over λ yields the right-hand inequality in property (c).

Property (d) for the case $N = 1$ was proved earlier; the extension to $N > 1$ is straightforward. Property (e) follows from (c).

To prove property (f), choose a joining λ_j of $\{X_{(j-1)n+1}^{jn}\}$ and $\{Y_{(j-1)n+1}^{jn}\}$, for each j, such that

$$E_{\lambda_j}(d_n(x_{(j-1)n+1}^{jn}, y_{(j-1)n+1}^{jn})) = \bar{d}_n(X_{(j-1)n+1}^{jn}, Y_{(j-1)n+1}^{jn}).$$

The product λ of the λ_j is a joining of X_1^{nm} and Y_1^{nm}, by the assumed independence of the blocks, and hence

$$m\bar{d}_{nm}(X_1^{nm}, Y_1^{nm}) \leq \sum_{j=1}^{m} \bar{d}_n(X_{(j-1)n+1}^{jn}, Y_{(j-1)n+1}^{jn}),$$

since

$$mnE_\lambda(d_{nm}(x_1^{nm}, y_1^{nm})) = \sum_{j=1}^{m} nE_{\lambda_j}(d_n(x_{(j-1)n+1}^{jn}, y_{(j-1)n+1}^{jn})).$$

The reverse inequality follows from the superadditivity property, (3), and hence property (f) is established.

The proof of Property (g) is left to the reader. This completes the discussion of Lemma I.9.11. □

Another metric, equivalent to the $\bar{d}_n$-metric, is obtained by replacing expected values by probabilities of large deviations. Let $\Delta_n(\epsilon) = \{(x_1^n, y_1^n): d_n(x_1^n, y_1^n) \leq \epsilon\}$, and define

$$d_n^*(\mu, \nu) = \min_{\lambda \in J_n(\mu,\nu)} \min\{\epsilon > 0\colon \lambda(\Delta_n(\epsilon)) \geq 1 - \epsilon\}. \tag{13}$$

The function $d_n^*(\mu, \nu)$ is a metric on the space $\mathcal{P}(A^n)$ of probability measures on A^n. It is one form of a metric commonly known as the Prohorov metric, and is uniformly equivalent to the $\bar{d}_n$-metric, with bounds given in the following lemma.

Lemma I.9.12

$\left[d_n^*(\mu, \nu)\right]^2 \leq \bar{d}_n(\mu, \nu) \leq 2d_n^*(\mu, \nu).$

Proof. Let λ be a joining of μ and ν such that $\bar{d}_n(\mu, \nu) = E_\lambda(d_n(x_1^n, y_1^n)) = \alpha$. The Markov inequality implies that $\lambda(\Delta_n(\sqrt{\alpha})) \geq 1 - \sqrt{\alpha}$, and hence $d_n^*(\mu, \nu) \leq \sqrt{\alpha}$. Likewise, if $\lambda \in J_n(\mu, \nu)$ is such that $\lambda(\Delta_n(\alpha)) \geq 1 - \alpha$ and $d_n^*(\mu, \nu) = \alpha$, then

$$E_\lambda(d_n(x_1^n, y_1^n)) \leq \sum_{(x_1^n, y_1^n) \in \Delta_n(\alpha)} d_n(x_1^n, y_1^n)\lambda(x_1^n, y_1^n) + \alpha \leq 2\alpha,$$

and hence $\bar{d}_n(\mu, \nu) \leq E_\lambda(d_n(x_1^n, y_1^n)) \leq 2\alpha$. This completes the proof of the lemma. □

The preceding lemma is useful in both directions; versions of each are stated as the following two lemmas. Part (b) of each version is based on the mapping interpretation of joinings, (2).

Lemma I.9.13

(a) *If there is a joining λ of μ and ν such that $d_n(x_1^n, y_1^n) \le \epsilon$, except for a set of λ-measure at most ϵ, then $\bar{d}_n(\mu, \nu) \le 2\epsilon$.*

(b) *If ϕ and ψ are measure-preserving mappings from a nonatomic probability space (Z, α) into (A^n, μ) and (A^n, ν), respectively, such that $d_n(\phi(z), \psi(z)) \le \epsilon$, except for a set of α-measure at most ϵ, then $\bar{d}_n(\mu, \nu) \le 2\epsilon$.*

Lemma I.9.14

(a) *If $\bar{d}_n(\mu, \nu) \le \epsilon^2$ and $\mu_n(B) > 1 - \epsilon$, then $\nu_n([B]_\epsilon) > 1 - 2\epsilon$, where $[B]_\epsilon = \{y_1^n : d_n(y_1^n, B) \le \epsilon\}$.*

(b) *If $\bar{d}_n(\mu, \nu) \le \epsilon^2$, and (Z, α) is a nonatomic probability space, there are measure-preserving mappings ϕ and ψ from (Z, α) into (A^n, μ) and (A^n, ν), respectively, such that $\alpha(\{z: d_n(\phi(z), \psi(z)) > \epsilon\}) \le \epsilon$.*

I.9.b.4 Ergodicity, entropy, mixing, and $\bar{d}$ limits.

In this section it will be shown that the $\bar{d}$-limit of ergodic (mixing) processes is ergodic (mixing) and that entropy is $\bar{d}$-continuous. The key to these and many other limit results is that $\bar{d}_n$-closeness implies that sets of large measure for one distribution must have large blowup with respect to the other distribution, which is just Lemma I.9.14. As before, for $\delta > 0$, let

$$[B]_\delta = \{y_1^n : d_n(y_1^n, B) \le \delta\},$$

denote the δ-neighborhood or δ-blow-up of $B \subset A^n$.

Theorem I.9.15 (Ergodicity and $\bar{d}$-limits.)
The $\bar{d}$-limit of ergodic processes is ergodic.

Proof. Suppose μ is the $\bar{d}$-limit of ergodic processes. By the ergodic theorem the relative frequency $p(a_1^k|x_1^n)$ converges almost surely to a limit $p(a_1^k|x)$. By Theorem I.4.2 it is enough to show that $p(a_1^k|x)$ is constant in x, almost surely, for every a_1^k. This follows from the fact that if ν is ergodic and close to μ, then a small blowup of the set of ν-typical sequences must have large μ-measure, and small blowups don't change empirical frequencies very much.

To show that small blowups don't change frequencies very much, fix k and note that if $y_i^{i+k-1} = a_1^k$ and $x_i^{i+k-1} \ne a_1^k$, then $y_j \ne x_j$ for at least one $j \in [i, i+k-1]$, and hence

$$(n-k+1)p(a_1^k|y_1^n) \le (n-k+1)p(a_1^k|x_1^n) + knd_n(x_1^n, y_1^n), \tag{14}$$

where the extra factor of k on the last term is to account for the fact that a j for which $y_j \ne x_j$ can produce as many as k indices i for which $y_i^{i+k-1} = a_1^k$ and $x_i^{i+k-1} \ne a_1^k$.

Let ϵ be a positive number and use (14) to choose a positive number $\delta \le \epsilon$, and N_1 so that if $n \ge N_1$ then

$$d_n(x_1^n, y_1^n) \le \delta \Rightarrow |p(a_1^k|x_1^n) - p(a_1^k|y_1^n)| < \epsilon. \tag{15}$$

Let ν be an ergodic measure such that $\bar{d}(\mu, \nu) \leq \delta^2$ and hence $\bar{d}_n(\mu_n, \nu_n) \leq \delta^2$, $n \geq N_1$, since for stationary measures $\bar{d}$ is the supremum of the $\bar{d}_n$. Because ν is ergodic there is an $N_2 \geq N_1$ such that if $n \geq N_2$, the set

$$B_n = \{x_1^n \colon |p(a_1^k|x_1^n) - \nu(a_1^k)| < \epsilon\}$$

has ν_n measure at least $1-\delta$. Fix $n \geq N_2$ and apply Lemma I.9.14 to obtain $\mu_n([B_n]_\delta) \geq 1 - 2\delta$. Note, however, that if $y_1^n \in [B_n]_\delta$ then there is a sequence $x_1^n \in B_n$ such that $d_n(x_1^n, y_1^n) \leq \delta$ and hence

$$|p(a_1^k|x_1^n) - p(a_1^k|y_1^n)| < \epsilon$$

by (15). Likewise if $\tilde{y}_1^n \in [B_n]_\delta$ then there is a sequence $\tilde{x}_1^n \in B_n$ such that

$$|p(a_1^k|\tilde{x}_1^n) - p(a_1^k|\tilde{y}_1^n)| < \epsilon.$$

The definition of B_n and the two preceding inequalities yield

$$|p(a_1^k|y_1^n) - p(a_1^k|\tilde{y}_1^n)| < 4\epsilon, \quad y_1^n, \tilde{y}_1^n \in [B_n]_\delta.$$

Since $[B_n]_\delta$ has large μ_n-measure, the finite sequence characterization of ergodicity, Theorem I.4.5, implies that μ is ergodic, completing the proof of Theorem I.9.15. □

Theorem I.9.16 (Entropy and $\bar{d}$-limits.)
Entropy is $\bar{d}$-continuous on the class of ergodic processes.

Proof. Now the idea is that if μ and ν are $\bar{d}$-close then a small blowup does not increase the set of μ-entropy-typical sequences by very much, and has large ν-measure. Let μ be an ergodic process with entropy $h = h(\mu)$. Given $\epsilon > 0$, choose N, and, for each $n \geq N$, choose $\mathcal{T}_n \subseteq A^n$ such that $|\mathcal{T}_n| \leq 2^{n(h+\epsilon)}$ and $\mu(\mathcal{T}_n) \to 1$ as $n \to \infty$. Let δ be a positive number such that

$$|[\mathcal{T}_n]_\delta| \leq 2^{n(h+2\epsilon)},$$

for all $n \geq N$. (Such a δ exists by the blowup-bound lemma, Lemma I.7.5.) Choose $N_1 \geq N$ such that $\mu(\mathcal{T}_n) \geq 1 - \delta$, for $n \geq N_1$.

Let ν be an ergodic process such that $\bar{d}(\mu, \nu) \leq \delta^2$. Lemma I.9.14 implies that $\nu([\mathcal{T}_n]_\delta) \geq 1 - 2\delta$, for all $n \geq N$, so the covering-exponent interpretation of entropy, Theorem I.7.4, yields $h(\nu) \leq h(\mu) + 2\epsilon$. Likewise, for sufficiently large n, ν gives measure at least $1 - \delta$ to a subset of A^n of cardinality at most $2^{n(h(\nu)+\epsilon)}$ and hence $h(\mu) \leq h(\nu) + 2\epsilon$. This completes the proof of Theorem I.9.16. □

Mixing is also preserved in the passage to $\bar{d}$-limits.

Theorem I.9.17 (Mixing and $\bar{d}$-limits.)
The $\bar{d}$-limit of mixing processes is mixing.

Proof. The proof is similar to the proof, via finite-coder approximation, that stationary coding preserves mixing, see Example 19. Suppose μ is the $\bar{d}$-limit of mixing processes, fix a_1^n and b_1^n, each of positive μ-measure, and fix $\epsilon > 0$. Given $\delta > 0$, let ν be a mixing process such that $\bar{d}(\mu, \nu) \leq \delta^2$, and let λ be a stationary joining of μ and ν such that

$$E_\lambda(d_1(x_1, y_1)) = E_\lambda(d_n(x_1^n, y_1^n)) \leq \delta^2.$$

This implies that $d_n(x_1^n, y_1^n) \leq \delta$ except for a set of λ measure at most δ, so that if $\delta < 1/n$ then the set $\{(x_1^n, y_1^n)\colon x_1^n \neq y_1^n\}$ has λ-measure at most δ. Thus by making δ even smaller it can be supposed that $\mu(a_1^n)$ is so close to $\nu(a_1^n)$ and $\mu(b_1^n)$ so close to $\nu(b_1^n)$ that

$$|\mu(a_1^n)\mu(b_1^n) - \nu(a_1^n)\nu(b_1^n)| \leq \epsilon/3.$$

Likewise, for any $m \geq 1$,

$$\begin{aligned}&\lambda(\{(x_1^{m+n}, y_1^{m+n})\colon x_1^n \neq y_1^n \text{ or } x_{m+1}^{m+n} \neq y_{m+1}^{m+n}\})\\ &\quad\leq \lambda(\{(x_1^n, y_1^n)\colon x_1^n \neq y_1^n\}) + \lambda(\{(x_{m+1}^{m+n}, y_{m+1}^{m+n})\colon x_{m+1}^{m+n} \neq y_{m+1}^{m+n}\}) \leq 2\delta,\end{aligned}$$

so that if $\delta < 1/2n$ is small enough, then

$$|\mu([a_1^n] \cap T^{-m}[b_1^n]) - \nu([a_1^n] \cap T^{-m}[b_1^n])| \leq \epsilon/3,$$

for all $m \geq 1$. Since ν is mixing, however,

$$|\nu(a_1^n)\nu(b_1^n) - \nu([a_1^n] \cap T^{-m}[b_1^n])| \leq \epsilon/3,$$

for all sufficiently large m, and hence, the triangle inequality yields

$$|\mu(a_1^n)\mu(b_1^n) - \mu([a_1^n] \cap T^{-m}[b_1^n])| \leq \epsilon,$$

for all sufficiently large m, provided only that δ is small enough. Thus μ must be mixing, which establishes the theorem. □

This section will be closed with an example, which shows, in particular, that the $\bar{d}$-topology is nonseparable.

Example I.9.18 (Rotation processes are generally $\bar{d}$-far-apart.)
Let S and T be the transformations of $X = [0, 1)$ defined, respectively, by

$$Sx = x \oplus \alpha, \quad Tx = x \oplus \beta,$$

where, as before, $\oplus$ denotes addition modulo 1.

Proposition I.9.19
Let λ be an $(S \times T)$-invariant measure on $X \times X$ which has Lebesgue measure μ as marginal on each factor. If α and β are rationally independent then λ must be the product measure.

Proof. "Rationally independent" means that $k\alpha + m\beta$ is irrational for any two rationals k and m with $(k, m) \neq (0, 0)$. Let C and D be measurable subsets of X. The goal is to show that

$$\lambda(C \times D) = \mu(C)\mu(D).$$

It is enough to prove this when C and D are intervals and $\mu(C) = 1/N$, where N is an integer. Given $\epsilon > 0$, let C_1 be a subinterval of C of length $(1 - \epsilon)/N$ and let

$$E = C_1 \times D, \quad F = X \times D.$$

Since α and β are rationally independent, the two-dimensional version of Kronecker's theorem, Proposition I.2.16, can be applied, yielding integers $m_1, m_2, \ldots, m_N$ such that if V denotes the transformation $S \times T$ then

$$V^{m_i}E \cap V^{m_j}E = \emptyset, \text{ if } i \neq j,$$

and

$$\lambda(F\Delta\bar{F}) < 2\epsilon, \quad \text{where } \bar{F} = \cup_{i=1}^{N} V^{m_i}E.$$

It follows that $\lambda(E) = \lambda(\bar{F})/N$ is within $2\epsilon/N$ of $\lambda(F)/N = \mu(C)\mu(D)$. Let $\epsilon \to 0$ to obtain

$$\lambda(C \times D) = \mu(C)\mu(D).$$

This completes the proof of Proposition I.9.19. □

Now let $\mathcal{P}$ be the partition of the unit interval that consists of the two intervals $P_0 = [0, 1/2)$, $P_1 = [1/2, 1)$. It is easy to see that the mapping that carries x into its $(T, \mathcal{P})$-name $\{x_n\}$ is an invertible mapping of the unit interval onto the space A^Z which carries Lebesgue measure onto the Kolmogorov measure. This fact, together with Proposition I.9.19, implies that the only joining of the $(T, \mathcal{P})$-process and the $(S, \mathcal{P})$-process is the product joining, and this, in turn, implies that the $\bar{d}$-distance between these two processes is 1/2. This shows in particular that the class of ergodic processes is not separable, for, in fact, even the translation (rotation) subclass is not separable. It can be shown that the class of all processes that are stationary codings of i.i.d. processes is $\bar{d}$-separable, see Exercise 3 in Section IV.2.

I.9.c Exercises.

1. A measure $\mu \in \mathcal{P}_s(X)$ is *extremal* if it cannot be expressed in the form $\mu = t\mu_1 + (1-t)\mu_2$, with $\mu_i \in P_s$, $0 < t < 1$, and $\mu_1 \neq \mu_2$.

   (a) Show that if μ is ergodic then μ is extremal. (Hint: if $\mu = t\mu_1 + (1-t)\mu_2$, apply the Radon-Nikodym theorem to obtain $\mu_i(B) = \int_B f_i \, d\mu$ and show that each f_i is T-invariant.)

   (b) Show that if μ is extremal, then it must be ergodic. (Hint: if $T^{-1}B = B$ then μ is a convex sum of the T-invariant conditional measures $\mu(\cdot|B)$ and $\mu(\cdot|X - B)$.)

2. Show that $\bar{d}_n(X_1^n, Y_1^n)$ is a complete metric by showing that

   (a) The triangle inequality holds. (Hint: if λ joins X_1^n and Y_1^n, and λ^* joins Y_1^n and Z_1^n, then $\sum_{y_1^n} \lambda(x_1^n, y_1^n)\lambda^*(z_1^n|y_1^n)$ joins X_1^n and Z_1^n.)

   (b) If $\bar{d}_n(X_1^n, Y_1^n) = 0$ then X_1^n and Y_1^n have the same distribution.

   (c) The metric $\bar{d}_n(X_1^n, Y_1^n)$ is complete.

3. Prove the superadditivity inequality (3).

4. Let μ and ν be the binary Markov chains with the respective transition matrices

$$M = \begin{bmatrix} p & 1-p \\ 1-p & p \end{bmatrix}, \quad N = \begin{bmatrix} 0 & 1 \\ 1 & 0 \end{bmatrix},$$

   Let $\tilde{\mu}$ be the Markov process defined by M^2.

(a) Show that if x is typical for μ, and $y_n = x_{2n}$, $n = 1, 2, \ldots$, then, almost surely, y is typical for $\tilde{\mu}$.

(b) Use the result of part (a) to show $\bar{d}(\mu, \nu) = 1/2$, if $0 < p < 1$.

5. Use Lemma I.9.11(f) to show that if μ and ν are i.i.d. then $\bar{d}(\mu, \nu) = (1/2)|\mu - \nu|_1$. (This is a different method for obtaining the $\bar{d}$-distance for i.i.d. processes than the one outlined in Example I.9.3.)

6. Suppose ν is ergodic and $\widehat{\mu}$ is the concatenated-block process defined by μ_n on A^n. Show that $\bar{d}(\nu, \widehat{\mu}) = \bar{d}_n(\nu_n, \mu_n)$. (Hint: $\widehat{\mu}$ is concentrated on shifts of sequences that are typical for the product measure on $(A^n)^\infty$ defined by μ_n.)

7. Prove property (d) of Lemma I.9.11.

8. Show there is a joining λ^* of μ and ν such that $\lambda_n^*(a_1^n, a_1^n) = \min(\mu(a_1^n), \nu(a_1^n))$.

9. Prove that $|\mu - \nu|_n = 2 - 2\sum_{a_1^n} \min(\mu(a_1^n), \nu(a_1^n))$.

10. Two sets $C, D \subseteq A^k$ are α-separated if $C \cap [D]_\alpha = \emptyset$. Show that if the supports of μ_k and ν_k are α-separated then $\bar{d}_k(\mu_k . \nu_k) \geq \alpha$.

11. Suppose μ and ν are ergodic and $\bar{d}(\mu, \nu) \geq \alpha$. Show that if $\epsilon > 0$ is given, then for sufficiently large n there are subsets C_n and D_n of A^n such that $\mu(C_n) > 1 - \epsilon$, $\nu(D_n) > 1 - \epsilon$, and $d_n(x_1^n, y_1^n) \geq \alpha - \epsilon$, for $x_1^n \in C_n$, $y_1^n \in D_n$. (Hint: if $p_k(\cdot|x_1^k) \sim \mu_k$ and $p_k(\cdot|y_1^k) \sim \nu_k$, and k is large enough, then $d_n(x_1^n, y_1^n)$ cannot be much smaller than α, by (7).)

12. Let (Y, ν) be a nonatomic probability space and suppose $\psi: Y \mapsto A^n$ is a measurable mapping such that $\bar{d}_n(\mu_n, \nu \circ \psi^{-1}) < \delta$. Show that there is a measurable mapping $\phi: Y \mapsto A^n$ such that $\mu_n = \nu \circ \phi^{-1}$ *and* such that $E_\nu(d_n(\phi(y), \psi(y))) < \delta$.

Section I.10 Cutting and stacking.

Concatenated-block processes and regenerative processes are examples of processes with block structure. Sample paths are concatenations of members of some fixed collection $\mathcal{S}$ of blocks, that is, finite-length sequences, with both the initial tail of a block and subsequent block probabilities governed by a product measure μ^* on $\mathcal{S}^\infty$. The assumption that μ^* is a product measure is not really necessary, for any stationary measure μ^* on $\mathcal{S}^\infty$ leads to a stationary measure μ on A^∞ whose sample paths are infinite concatenations of members of $\mathcal{S}$, provided only that expected block length is finite. Indeed, the measure μ is just the measure given by the tower construction, with base $\mathcal{S}^\infty$, measure μ^*, and transformation given by the shift on $\mathcal{S}^\infty$, see Section I.2.c.2.

It is often easier to construct counterexamples by thinking directly in terms of block structures, first constructing finite blocks that have some approximate form of the final desired property, then concatenating these blocks in some way to obtain longer blocks in which the property is approximated, continuing in this way to obtain the final process as a suitable limit of finite blocks. A powerful method for organizing such constructions will be presented in this section. The method is called "cutting and stacking," a name suggested by the geometric idea used to go from one stage to the next in the construction.

Before going into the details of cutting and stacking, it will be shown how a stationary measure μ^* on $\mathcal{S}^\infty$ gives rise to a sequence of pictures, called columnar representations, and how these, in turn, lead to a description of the measure μ on A^∞.

I.10.a The columnar representations.

Fix a set $\mathcal{S} \subset A^*$ of finite length sequences drawn from a finite alphabet A. The members of the initial set $\mathcal{S} \subseteq A^*$ are called *words* or *first-order blocks*. The length of a word $w \in \mathcal{S}$ is denoted by $\ell(w)$. The members w_1^n of the product space $\mathcal{S}^n$ are called *n-th order blocks*. The length $\ell(w_1^n)$ of an n-th order block w_1^n satisfies $\ell(w_1^n) = \sum_i \ell(w_i)$. The symbol w_1^n has two interpretations, one as the n-tuple $w_1, w_2, \ldots, w_n$ in $\mathcal{S}^n$, the other as the concatenation $w_1 w_2 \cdots w_n$ in A^L, where $L = \sum_i \ell(w_i)$. The context usually makes clear which interpretation is in use.

The space $\mathcal{S}^\infty$ consists of the infinite sequences $w_1^\infty = w_1, w_2, \ldots$, where each $w_i \in \mathcal{S}$. A (Borel) probability measure μ^* on $\mathcal{S}^\infty$ which is invariant under the shift on $\mathcal{S}^\infty$ will be called a *block-structure measure* if it satisfies the finite expected-length condition,

$$E(\ell(w)) = \sum_{w \in \mathcal{S}} \ell(w)\mu_1^*(w) < \infty.$$

Here μ_1^* denotes the projection of μ^* onto $\mathcal{S}$, while, in general, μ_n^* denotes the projection of μ^* onto $\mathcal{S}^n$. Note, by the way, that stationary gives

$$E(\ell(w_1^n)) = \sum_{w_1^n \in \mathcal{S}^n} \ell(w_1^n)\mu_n^*(w_1^n) = nE(\ell(w)).$$

Blocks are to be concatenated to form A-valued sequences, hence it is important to have a distribution that takes into account the length of each block. This is the probability measure λ on $\mathcal{S}$ defined by the formula

$$\lambda(w) = \frac{\ell(w)\mu^*(w)}{E(\ell_1)}, \quad w \in \mathcal{S},$$

where $E(\ell_1)$ denotes the expected length of first-order blocks. The formula indeed defines a probability distribution, since summing $\ell(w)\mu^*(w)$ over w yields the expected block length $E(\ell_1)$. The measure is called the *linear mass distribution of words* since, in the case when μ^* is ergodic, $\lambda(w)$ is the limiting fraction of the length of a typical concatenation $w_1 w_2 \ldots$ occupied by the word w. Indeed, using $f(w|w_1^n)$ to denote the number of times w appears in $w_1^n \in \mathcal{S}^n$, the fraction of the length of w_1^n occupied by w is given by

$$\frac{\ell(w)f(w|w_1^n)}{\ell(w_1^n)} = \ell(w)\frac{f(w|w_1^n)}{n}\frac{n}{\ell(w_1^n)} \longrightarrow \frac{\ell(w)\mu^*(w)}{E(\ell_1)} = \lambda(w), \text{ a. s.,}$$

since $f(w|w_1^n)/n \to \mu^*(w)$ and $\ell(w_1^n)/n \to E(\ell_1)$, almost surely, by the ergodic theorem applied to μ^*.

The ratio

$$\tau(w) = \mu^*(w)/E(\ell_1)$$

is called the *width* or *thickness* of w. Note that $\lambda(w) = \ell(w)\tau(w)$, that is, linear mass = length $\times$ width.

The unit interval can be partitioned into subintervals indexed by $w \in \mathcal{S}$ such that length of the subinterval assigned to w is $\lambda(w)$. Thus no harm comes from thinking of λ as Lebesgue measure on the unit interval. A more useful representation is obtained by subdividing the interval that corresponds to w into $\ell(w)$ subintervals of width $\tau(w)$, labeling the i-th subinterval with the i-th term of w, then stacking these subintervals into a column, called the column associated with w. This is called the *first-order columnar representation of* $(\mathcal{S}^\infty, \mu^*)$. Figure I.10.1 shows the first-order columnar representation of $\mathcal{S} = \{v, w\}$, where $v = 01$, $w = 011$, $\mu_1^*(v) = 1/3$, and $\mu_1^*(w) = 2/3$.

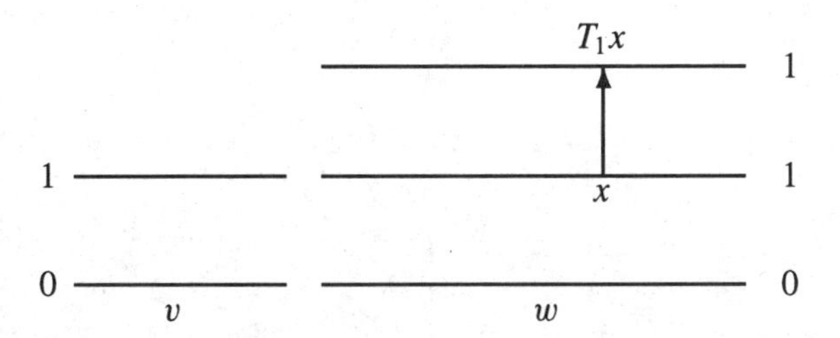

Figure I.10.1 The first-order columnar representation of $(\mathcal{S}, \mu_1^*)$.

In the columnar representation, shifting along a word corresponds to moving intervals one level upwards. This upward movement can be accomplished by a point mapping, namely, the mapping T_1 that moves each point one level upwards. This mapping is not defined on the top level of each column, but it is a Lebesgue measure-preserving map from its domain to its range, since a level is mapped linearly onto the next level, an interval of the same width. (This is also shown in Figure I.10.1.)

In summary, the columnar representation not only carries full information about the distribution $\lambda(w)$, or alternatively, $\mu_1^*(w) = \mu^*(w)$, but shifting along a block can be represented as the transformation that moves points one level upwards, a transformation which is Lebesgue measure preserving on its domain. The first-order columnar representation is determined by $\mathcal{S}$ and the first-order distribution μ_1^* (modulo, of course, the fact that there are many ways to partition the unit interval into subintervals and stack them into columns of the correct sizes.) Conversely, the columnar representation determines $\mathcal{S}$ and μ_1^*. The information about the final process is only partial since the picture gives no information about how to get from the top of a column to the base of another column, in other words, it does not tell how the blocks are to be concatenated. The first-order columnar representation is, of course, closely related to the tower representation discussed in Section I.2.c.2, the difference being that now the emphasis is on the width distribution and the partially defined transformation T that moves points upwards.

Information about how first-order blocks are concatenated to form second-order blocks is given by the columnar representation of the second-order distribution $\mu_2^*(w_1^2)$. Let $\tau(w_1^2) = \mu_2^*(w_1^2)/E(\ell_2)$ be the width of w_1^2, where $E(\ell_2)$ is the expected length of w_1^2, with respect to μ_2^*, and let $\lambda(w_1^2) = \ell(w_1^2)\tau(w_1^2)$ be the second-order linear mass. The second-order blocks w_1^2 are represented as columns of disjoint subintervals of the unit interval of width $\tau(w_1^2)$ and height $\ell(w_1^2)$, with the i-th subinterval labeled by the i-th term of the concatenation w_1^2.

A key observation, which gives the name "cutting and stacking" to the whole procedure that this discussion is leading towards, is that

Second-order columns can be be obtained by cutting each first-order column into subcolumns, then stacking these in pairs.

Indeed, the total mass in the second-order representation contributed by the first $\ell(w)$ levels of all the columns that start with w is

$$\sum_{w_2} \ell(w)\tau(ww_2) = \frac{\ell(w)}{E(\ell_2)} \sum_{w_2} \mu^*(ww_2) = \frac{\ell(w)\mu^*(w)}{2E(\ell_1)} = \frac{1}{2}\ell(w)\tau(w),$$

which is exactly half the total mass of w in the first-order columnar representation. Likewise, the total mass contributed by the top $\ell(w)$ levels of all the columns that end with w is

$$\sum_{w_1} \ell(w)\tau(w_1 w) = \frac{\ell(w)}{E(\ell_2)} \sum_{w_1} \mu^*(w_1 w) = \frac{\ell(w)\mu^*(w)}{2E(\ell_1)} = \frac{1}{2}\ell(w)\tau(w).$$

Thus half the mass of a first-order column goes to the top parts and half to the bottom parts of second-order columns, so, as claimed, it is possible to cut the first-order columns into subcolumns and stack them in pairs so as to obtain the second-order columnar representation. Figure I.10.2 shows how the second-order columnar representation of $\mathcal{S} = \{v, w\}$, where $\mu_2^*(vv) = 1/9$, $\mu_2^*(vw) = \mu_2^*(wv) = 2/9$, $\mu_2^*(ww) = 4/9$, can be built by cutting and stacking the first-order representation shown in Figure I.10.1.

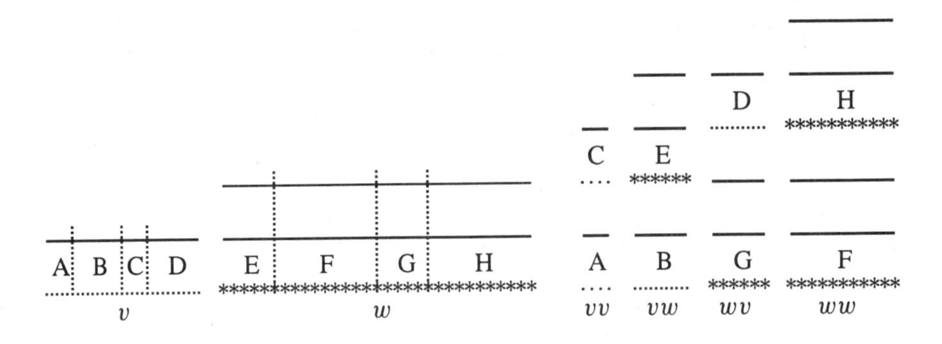

Figure I.10.2 The second-order representation via cutting and stacking.

The significance of the fact that second-order columns can be built from first-order columns by appropriate cutting and stacking is that this guarantees that the transformation T_2 defined for the second-order picture by mapping points directly upwards one level *extends the mapping* T_1 that was defined for the first-order picture by mapping points directly upwards one level. Indeed, if y is directly above x in some column of the first-order representation, then it will continue to be so in any second-order representation that is built from the first-order representation by cutting and stacking its columns. Note also the important property that the set of points where T_2 is undefined has only half the measure of the set where T_1 is defined, for the total width of the second-order columns is $1/E(\ell_2)$, which is just half the total width $1/E(\ell_1)$ of the first-order columns.

One can now proceed to define higher order columnar representations. In general, the $2m$-th order representation can be produced by cutting the columns of the m-th order columnar representation into subcolumns and stacking them in pairs. If this cutting and stacking method is used then the following two properties hold.

(a) The associated column transformation T_{2m} extends the transformation T_m.

(b) The set where T_{2m} is defined has one-half the measure of the set where T_m is undefined.

In particular, if cutting and stacking is used to go from stage to stage, then the set of transformations $\{T_{2^n}\}$ will have a common extension T defined for almost all x in the unit interval, and the transformation T will preserve Lebesgue measure. A general form of this fact, Theorem I.10.3, will be proved later. The labeling of levels by symbols from A provides a partition $\mathcal{P} = \{P_a : a \in A\}$ of the unit interval, where P_a is the union of all the levels labeled a. The $(T, \mathcal{P})$-process is the same as the stationary process μ defined by μ^*, see Corollary I.10.5.

The difference between the building of columnar representations and the general cutting and stacking method is really only a matter of approach and emphasis. The columnar representation idea starts with the final block-structure measure μ^* and uses it to construct a sequence of partial transformations of the unit interval each of which extends the previous ones. Their common extension is a Lebesgue measure-preserving transformation on the entire interval, which, in turn, gives the desired stationary A-valued process. Thus, in essence, it is a way of representing something already known.

The cutting and stacking idea focuses directly on the geometric concept of cutting and stacking labeled columns, using the desired goal (typically to make an example of a process with some sample path property) to guide how to cut up the columns of one stage and reassemble to create the next stage. The goal is still to produce a Lebesgue measure-preserving transformation on the unit interval. Of course, column structures define distributions on blocks, and cutting and stacking extends these to joint distributions on higher level blocks, but this all happens in the background while the user focuses on the combinatorial properties needed at each stage to produce the desired final process. Thus, in essence, it is a way of building something new.

The cutting and stacking language and theory will be rigorously developed in the following subsections. Some applications of its use to construct examples will be given in Chapter 3.

I.10.b The basic cutting and stacking vocabulary.

A *column* $\mathcal{C} = (L_1, L_2, \ldots, L_{\ell(\mathcal{C})})$ of *length*, or height, $\ell(\mathcal{C})$ is a nonempty, ordered, disjoint, collection of subintervals of the unit interval of equal positive *width* $\tau(\mathcal{C})$. The interval L_1 is called the *base* or *bottom*, the interval $L_{\ell(\mathcal{C})}$ is called the *top*, and the interval L_j is called the j-th *level* of the column. A *labeling* of $\mathcal{C}$ is a mapping $\mathcal{N}$ from $\{1, 2, \ldots, \ell(\mathcal{C})\}$ into the finite alphabet A. The vector $\mathcal{N}(\mathcal{C}) = (\mathcal{N}(1), \mathcal{N}(2), \ldots, \mathcal{N}(\ell(\mathcal{C})))$ is called the *name of* $\mathcal{C}$, and for each j, $\mathcal{N}(j) = \mathcal{N}(L_j)$ is called the *name* or *label* of L_j.

The *support* of a column $\mathcal{C}$ is the union of its levels L_j. Its *measure* $\lambda(\mathcal{C})$ is the Lebesgue measure λ of the support of $\mathcal{C}$, which is of course given by $\lambda(\mathcal{C}) = \ell(\mathcal{C})\tau(\mathcal{C})$. Two columns are said to be *disjoint* if their supports are disjoint.

A *column structure* $\mathcal{S}$ is a nonempty collection of mutually disjoint, labeled columns. The suggestive name, *column structure*, will be used in this book, rather than the non-informative name, *gadget*, which has been commonly used in ergodic theory. Note that the members of a column structure are labeled columns, hence, unless stated otherwise, terminology should be interpreted as statements about sets of labeled columns. Thus the union $\mathcal{S} \cup \mathcal{S}'$ of two column structures $\mathcal{S}$ and $\mathcal{S}'$ consists of the labeled columns of each, and $\mathcal{S} \subseteq \mathcal{S}'$ means that $\mathcal{S}$ is a substructure of $\mathcal{S}'$, that is, each labeled column in $\mathcal{S}$ is a labeled column in $\mathcal{S}'$ with the same labeling. An exception to this is the terminology *"disjoint column structures,"* which is taken as shorthand for *"column structures with disjoint support,"* where the support of a column structure is the union of the supports of its columns.

The *width* $\tau(\mathcal{S})$ of a column structure is the sum of the widths of its columns. The *base* or *bottom* of a column structure is the union of the bases of its columns and its *top* is the union of the tops of its columns. Note that the base and top have the same Lebesgue measure, namely, $\tau(\mathcal{S})$. The *width distribution* $\tilde{\tau}$ is the normalized column width

$$\tilde{\tau}(\mathcal{C}) = \frac{\tau(\mathcal{C})}{\sum_{\mathcal{D}\in\mathcal{S}} \tau(\mathcal{D})} = \frac{\tau(\mathcal{C})}{\tau(\mathcal{S})}.$$

The measure $\lambda(\mathcal{S})$ of a column structure is the Lebesgue measure of its support, in other words,

$$\lambda(\mathcal{S}) = \sum_{\mathcal{C}\in\mathcal{S}} \lambda(\mathcal{C}) = \sum_{\mathcal{C}\in\mathcal{S}} \ell(\mathcal{C})\tau(\mathcal{C}).$$

Note that $\lambda(\mathcal{S}) \leq 1$ since columns always consist of disjoint subintervals of the unit interval, and column structures have disjoint columns. In particular, this means that expected column height with respect to the width distribution is finite, for

$$\sum_{\mathcal{C}\in\mathcal{S}} \ell(\mathcal{C})\tilde{\tau}(\mathcal{C}) = \frac{\sum \ell(\mathcal{C})\tau(\mathcal{C})}{\tau(\mathcal{S})} = \frac{\sum \lambda(\mathcal{C})}{\tau(\mathcal{S})} \leq \frac{1}{\tau(\mathcal{S})} < \infty.$$

A column $\mathcal{C} = (L_1, L_2, \ldots, L_{\ell(\mathcal{C})})$ defines a transformation $T = T_{\mathcal{C}}$, called its *upward map.* If a column is pictured by drawing L_{j+1} directly above L_j then T maps points directly upwards one level, and is not defined on the top level. The precise definition is that T maps L_j in a linear and order-preserving manner onto L_{j+1}, for $1 \leq j < \ell(\mathcal{C})$. Note that T is one-to-one and its inverse maps points downwards.

The transformation $T = T_{\mathcal{S}}$ defined by a column structure $\mathcal{S}$ is the union of the upward maps defined by its columns. It is called the *mapping* or *upward mapping* defined by the column structure $\mathcal{S}$. Each point below the base of $\mathcal{S}$ is mapped one level directly upwards by T. T is not defined on the top of $\mathcal{S}$ and is a Lebesgue measure-preserving mapping from all but the top to all but the bottom of $\mathcal{S}$.

Columns are cut into subcolumns by slicing them vertically. This idea can be made precise in terms of subcolumns and column partitions as follows. A *subcolumn* of $\mathcal{C} = (L_1, L_2, \ldots, L_\ell)$ is a column $\mathcal{C}' = (L_1', L_2', \ldots, L_\ell')$, such that the following hold.

(a) For each j, L_j' is a subinterval of L_j with the same label as L_j.

(b) The distance from the left end point of L_j' to the left end point of L_j does not depend on j.

Note that implicit in the definition is that a subcolumn always has the same height as the column. (An alternative definition of subcolumn in terms of the upward map is given in Exercise 1.)

A *(column) partition* of a column $\mathcal{C}$ is a finite or countable collection

$$\{\mathcal{C}(i) = (L(i, 1), L(i, 2) \ldots, L(i, \ell(\mathcal{C})) \colon i \in I\}$$

of disjoint subcolumns of $\mathcal{C}$, the union of whose supports is the support of $\mathcal{C}$. Partitioning corresponds to cutting, thus *cutting $\mathcal{C}$ into subcolumns according to a distribution π* is the same as finding a column partition $\{\mathcal{C}(i)\}$ of $\mathcal{C}$ such that $\pi(i) = \tau(\mathcal{C}(i))/\tau(\mathcal{C})$, for each i.

The cutting idea extends to column structures. A *column partitioning* of a column structure $\mathcal{S}$ is a column structure $\mathcal{S}'$ with the same support as $\mathcal{S}$ such that each column of $\mathcal{S}'$ is a subcolumn of a column of $\mathcal{S}$. Thus, a column partitioning is formed by partitioning each column of $\mathcal{S}$ in some way, and taking the collection of the resulting subcolumns.

The stacking idea is defined precisely as follows. Let $\mathcal{C}_1 = (L(1, j)\colon 1 \leq j \leq \ell(\mathcal{C}_1))$ and $\mathcal{C}_2 = (L(2, j)\colon 1 \leq j \leq \ell(\mathcal{C}_2))$ be disjoint labeled columns of the same width. The *stacking of $\mathcal{C}_2$ on top of $\mathcal{C}_1$* is denoted by $\mathcal{C}_1 * \mathcal{C}_2$ and is the labeled column with levels L_j defined by

$$L_j = \begin{cases} L(1, j) & 1 \leq j \leq \ell(\mathcal{C}_1) \\ L(2, j - \ell(\mathcal{C}_1)) & \ell(\mathcal{C}_1) < j \leq \ell(\mathcal{C}_1) + \ell(\mathcal{C}_2), \end{cases}$$

with the label of $\mathcal{C}_1 * \mathcal{C}_2$ defined to be the concatenation vw of the label v of $\mathcal{C}_1$ and the label w of $\mathcal{C}_2$. Note that the width of $\mathcal{C}_1 * \mathcal{C}_2$ is the same as the width of $\mathcal{C}_1$, which is also the same as the width of $\mathcal{C}_2$. Longer stacks are defined inductively by

$$\mathcal{C}_1 * \mathcal{C}_2 * \cdots \mathcal{C}_k * \mathcal{C}_{k+1} = (\mathcal{C}_1 * \mathcal{C}_2 * \cdots \mathcal{C}_k) * \mathcal{C}_{k+1}.$$

A column structure $\mathcal{S}'$ is a *stacking* of a column structure $\mathcal{S}$, if they have the same support and each column of $\mathcal{S}'$ is a stacking of columns of $\mathcal{S}$. A column structure $\mathcal{S}'$ is *built by cutting and stacking* from a column structure $\mathcal{S}$ if it is a stacking of a column partitioning of $\mathcal{S}$. Thus, for example, the second-order columnar representation of $(\mathcal{S}, \mu^*)$, where $\mathcal{S} \subseteq A^*$, is built by cutting each of the first-order columns into subcolumns and stacking these in pairs. In general, the columns of $\mathcal{S}'$ are permitted to be stackings of variable numbers of subcolumns of $\mathcal{S}$.

The basic fact about stacking is that it extends upward maps. Indeed, the upward map $T_{\mathcal{C}_1 * \mathcal{C}_2}$ defined by stacking $\mathcal{C}_2$ on top of $\mathcal{C}_1$ agrees with the upward map $T_{\mathcal{C}_1}$, wherever it is defined, and with the upward map $T_{\mathcal{C}_2}$, wherever it is defined, and extends their union, since $T_{\mathcal{C}_1 * \mathcal{C}_2}$ is now defined on the top of $\mathcal{C}_1$. This is also true of the upward mapping $T = T_{\mathcal{S}}$ defined by a column structure $\mathcal{S}$, namely, it is extended by the upward map $T = T_{\mathcal{S}'}$ for any $\mathcal{S}'$ built by cutting and stacking from $\mathcal{S}$.

In summary, the key properties of the upward maps defined by column structures are summarized as follows.

(i) The domain of $T_{\mathcal{S}}$ is the union of all except the top of $\mathcal{S}$.

(ii) The range of $T_{\mathcal{S}}$ is the union of all except the bottom of $\mathcal{S}$.

(iii) The upward map $T_{\mathcal{S}}$ is a Lebesgue measure-preserving mapping from its domain to its range.

(iv) If $\mathcal{S}'$ is built by cutting and stacking from $\mathcal{S}$, then $T_{\mathcal{S}'}$ extends $T_{\mathcal{S}}$.

I.10.c The final transformation and process.

The goal of cutting and stacking operations is to construct a measure-preserving transformation on the unit interval. This is done by starting with a column structure $\mathcal{S}(1)$ with support 1 and applying cutting and stacking operations to produce a new structure $\mathcal{S}(2)$. Cutting and stacking operations are then applied to $\mathcal{S}(2)$ to produce a third structure $\mathcal{S}(3)$. Continuing in this manner a sequence $\{\mathcal{S}(m)\}$ of column structures is obtained for which each member is built by cutting and stacking from the preceding one. If the tops shrink to 0, then a common extension T of the successive upward maps is defined for almost all x in the unit interval and preserves Lebesgue measure. Together with the partition defined by the names of levels, the transformation produces a stationary process. A precise formulation of this result will be presented in this section, along with the basic formula for estimating the joint distributions of the final process from the sequence of column structures, and a condition for ergodicity.

A sequence $\{\mathcal{S}(m)\}$ of column structures is said to be *complete* if the following hold.

(a) For each $m \geq 1$, $\mathcal{S}(m+1)$ is built by cutting and stacking from $\mathcal{S}(m)$.

(b) $\lambda(\mathcal{S}(1)) = 1$ and the (Lebesgue) measure of the top of $\mathcal{S}(m)$ goes to 0 as $m \to \infty$.

Theorem I.10.3 (The complete-sequences theorem.)
If $\{\mathcal{S}(m)\}$ is complete then the collection $\{T_{\mathcal{S}(m)}\}$ has a common extension T defined for almost every $x \in [0, 1]$. Furthermore, T is invertible and preserves Lebesgue measure.

Proof. For almost every $x \in [0, 1]$ there is an $M = M(x)$ such that x lies in an interval below the top of $\mathcal{S}(M)$, since the top shrinks to 0 as $m \to \infty$. Further cutting and stacking preserves this relationship of being below the top and produces extensions of $T_{\mathcal{S}(M)}$, so that the transformation T defined by $Tx = T_{\mathcal{S}(M)}x$ is defined for almost all x and extends every $T_{\mathcal{S}(m)}$. Likewise, the common extension of the inverses of the $T_{\mathcal{S}(m)}$ is defined for almost all x; it is clearly the inverse of T.

If B is a subinterval of a level in a column of some $\mathcal{S}(m)$ which is not the base of that column, then $T^{-1}B = T^{-1}_{\mathcal{S}(m)}B$ is an interval of the same length as B. Since such intervals generate the Borel sets on the unit interval, it follows that T is measurable and preserves Lebesgue measure. This completes the proof of Theorem I.10.3. □

The transformation T is called the *transformation defined by the complete sequence* $\{\mathcal{S}(m)\}$. The label structure defines the partition $\mathcal{P} = \{P_a \colon a \in A\}$, where P_a is the union of all the levels of all the columns of $\mathcal{S}(1)$ that have the name a. The $(T, \mathcal{P})$-process $\{X_n\}$ and its Kolmogorov measure μ are, respectively, called the *process* and *Kolmogorov measure defined by* the complete sequence $\{\mathcal{S}(m)\}$. The process $\{X_n\}$ is described by selecting a point x at random according to the uniform distribution on the unit interval and defining $X_n(x)$ to be the index of the member of $\mathcal{P}$ to which $T^{n-1}x$ belongs. This is equivalent to picking a random x in the support of $\mathcal{S}(1)$, then choosing m so that x belongs to the j-th level, say L_j, of a column $\mathcal{C} \in \mathcal{S}(m)$ of height $\ell(\mathcal{C}) \geq j + n - 1$, and defining $X_n(x)$ to be the name of level L_{j+n-1} of $\mathcal{C}$. (Such an m exists eventually almost surely, since the tops are shrinking to 0.)

The k-th order joint distribution of the process μ defined by a complete sequence $\{\mathcal{S}(m)\}$ can be directly estimated by the relative frequency of occurrence of a_1^k in a column name, averaged over the Lebesgue measure of the column name. This estimate is exact for $k = 1$ for any m and for $k > 1$ is asymptotically correct as $m \to \infty$. The relative frequency of occurrence of a_1^k in a labeled column $\mathcal{C}$ is defined by

$$p_k(a_1^k|\mathcal{C}) = \frac{|\{i \in [1, \ell(\mathcal{C}) - k + 1]: \ x_i^{i+k-1} = a_1^k\}|}{\ell(\mathcal{C}) - k + 1},$$

where $x_1^{\ell(\mathcal{C})}$ is the name of $\mathcal{C}$, that is, $p_k(\cdot|\mathcal{C})$ is the empirical overlapping k-block distribution $p_k(\cdot|x_1^{\ell(\mathcal{C})})$ defined by the name of $\mathcal{C}$.

Theorem I.10.4 (Estimation of joint distributions.)
If μ is the measure defined by the complete sequence, $\{\mathcal{S}(m)\}$, then

$$\mu(a_1^k) = \lim_{m\to\infty} \sum_{\mathcal{C}\in\mathcal{S}(m)} p_k(a_1^k|\mathcal{C})\lambda(\mathcal{C}), \quad a_1^k \in A^k. \tag{1}$$

Proof. In this proof $\mathcal{C}$ will denote either a column or its support; the context will make clear which is intended. Let P_a be the union of all the levels of all the columns in $\mathcal{S}(1)$ that have the name a. Then, of course, $\mu(a) = \lambda(P_a)$, which is, in turn, given by

$$\lambda(P_a) = \sum_{\mathcal{C}\in\mathcal{S}(1)} p_1(a|\mathcal{C})\lambda(\mathcal{C}),$$

since $\ell(\mathcal{C})p_1(a|\mathcal{C})$ just counts the number of times a appears in the name of $\mathcal{C}$, and hence

$$\lambda(P_a \cap \mathcal{C}) = [\ell(\mathcal{C})p_1(a|\mathcal{C})]\,\tau(\mathcal{C}) = p_1(a|\mathcal{C})\lambda(\mathcal{C}).$$

The same argument applies to each $\mathcal{S}(m)$, establishing (1) for the case $k = 1$.

For $k > 1$, the only error in using (1) to estimate $\mu(a_1^k)$ comes from the fact that $p_k(a_1^k|\mathcal{C})$ ignores the final $k - 1$ levels of the column $\mathcal{C}$, a negligible effect when m is large for then most of the mass must be in long columns, since the top has small measure. To make this precise, for $x \in [0, 1]$, let $\{x_n\} \in A^Z$ be the sequence defined by the relation $T^n x \in P_{x_n}$, $n \in Z$, and let

$$P_{a_1^k} = \{x: \ x_1^k = a_1^k\}.$$

The quantity $(\ell(\mathcal{C}) - k + 1)p_k(a_1^k|\mathcal{C})$ counts the number of levels below the top $k - 1$ levels of $\mathcal{C}$ that are contained in $P_{a_1^k}$, and hence

$$|\lambda(P_{a_1^k} \cap \mathcal{C}) - p_k(a_1^k|\mathcal{C})\lambda(\mathcal{C})| \le 2(k-1)\tau(\mathcal{C}),$$

since the top $k - 1$ levels of $\mathcal{C}$ have measure $(k-1)\tau(\mathcal{C})$. The desired result (1) now follows since the sum of the widths $\tau(\mathcal{C})$ over the columns of $\mathcal{S}(m)$ was assumed to go to 0, as $n \to \infty$. This completes the proof of Theorem I.10.4. □

An application of the estimation formula (1) is connected to the earlier discussion of the sequence of columnar representations defined by a block-structure measure μ^* on $\mathcal{S}^\infty$, where $\mathcal{S} \subseteq A^*$. For each m let $\mathcal{S}(m)$ denote the 2^{m-1}-order columnar representation. Without loss of generality it can be assumed that $\mathcal{S}(m+1)$ is built by cutting and stacking from $\mathcal{S}(m)$. Furthermore, $\lambda(\mathcal{S}(1)) = 1$ and the measure of the top of $\mathcal{S}(m+1)$ is half the measure of the top of $\mathcal{S}(m)$, so that $\{\mathcal{S}(m)\}$ is a complete sequence. Let μ be the measure defined by this sequence. The sequence $\{\mathcal{S}(m)\}$ will be called the *standard cutting and stacking representation* of the measure μ.

Next let $\widehat{T}$ be the tower transformation defined by the base measure μ^*, with the shift S on $\mathcal{S}^\infty$ as base transformation and height function defined by $f(w_1, w_2, \ldots) = \ell(w_1)$. Also let $\widehat{\mu}$ be the Kolmogorov measure of the $(\widehat{T}, \widehat{\mathcal{P}})$-process, where $\widehat{\mathcal{P}} = \{\widehat{P}_a\}$ is the partition defined by letting $\widehat{P}_a$ be the set of all pairs (w_1^∞, j) for which $a_j = a$, where $w_1 = a_1^{\ell(w_1)}$.

Corollary I.10.5

The tower construction and the standard representation produce the same measures, that is, $\mu = \widehat{\mu}$.

Proof. Let $p_k(a_1^k|w_1^n)$ be the empirical overlapping k-block distribution defined by the sequence $w_1^n \in \mathcal{S}^n$, thought of as the concatenation $w_1 w_2 \cdots w_n \in A^L$, where $L = \sum_i \ell(w_i)$. Since this is clearly the same as $p_k(a_1^k|\mathcal{C})$ where $\mathcal{C}$ is the column with name w_1^n, it is enough to show that

$$\widehat{\mu}(a_1^k) = \lim_{n\to\infty} \sum_{w_1^n \in \mathcal{S}^n} p_k(a_1^k|w_1^n)\lambda(\mathcal{C}), \quad a_1^k \in A^k. \tag{2}$$

For $k = 1$ the sum is constant in n, since the start distribution for $\widehat{\mu}$ is obtained by selecting w_1 at random according to the distribution μ_1^*, then selecting the start position according to the uniform distribution on $[1, \ell(w_1)]$. The proof for $k > 1$ is left as an exercise. □

Next the question of the ergodicity of the final process will be addressed. One way to make certain that the process defined by a complete sequence is ergodic is to make sure that the transformation defined by the sequence is ergodic relative to Lebesgue measure. A condition for this, which is sufficient for most purposes, is that later stage structures become almost independent of earlier structures. These ideas are developed in the following paragraphs.

In the following discussion $\mathcal{C}$ denotes either a column or its support. For example, $\lambda(B|\mathcal{C})$ denotes the conditional measure $\lambda(B \cap \mathcal{C})/\lambda(\mathcal{C})$ of the intersection of the set B with the support of the column $\mathcal{C}$.

The column structures $\mathcal{S}$ and $\mathcal{S}'$ are ϵ-independent if

$$\sum_{\mathcal{C}\in\mathcal{S}} \sum_{\mathcal{D}\in\mathcal{S}'} |\lambda(\mathcal{C} \cap \mathcal{D}) - \lambda(\mathcal{C})\lambda(\mathcal{D})| \le \epsilon. \tag{3}$$

In other words, two column structures $\mathcal{S}$ and $\mathcal{S}'$ are ϵ-independent if and only if the partition into columns defined by $\mathcal{S}$ and the partition into columns defined by $\mathcal{S}'$ are ϵ-independent. The sequence $\{\mathcal{S}(m)\}$ of column structures is *asymptotically independent* if for each m and each $\epsilon > 0$, there is a $k \ge 1$ such that $\mathcal{S}(m)$ and $\mathcal{S}(m+k)$ are ϵ-independent. Note, by the way, that these independence concepts do not depend on how the columns are labeled, but only the column distributions. Related concepts are discussed in the exercises.

Theorem I.10.6 (Complete sequences and ergodicity.)
A complete asymptotically independent sequence defines an ergodic process.

Proof. Let T be the Lebesgue measure-preserving transformation of the unit interval defined by the complete, asymptotically independent sequence $\{\mathcal{S}(m)\}$. It will be shown that T is ergodic, which implies that the $(T, \mathcal{P})$-process is ergodic for any partition $\mathcal{P}$.

Let B be a measurable set of positive measure such that $T^{-1}B = B$. The goal is to show that $\lambda(B) = 1$. Towards this end note that since the top of $\mathcal{S}(m)$ shrinks to 0, the widths of its intervals must also shrink to 0, and hence the collection of all the levels of all the columns in all the $\mathcal{S}(m)$ generates the σ-algebra. Thus, given $\epsilon > 0$ there is an m and a level L, say level j, of some $\mathcal{C} \in \mathcal{S}(m)$ such that

$$\lambda(B \cap L) \geq (1 - \epsilon^2)\lambda(L).$$

This implies that the entire column $\mathcal{C}$ is filled to within $(1 - \epsilon^2)$ by B, that is,

$$\lambda(B \cap \mathcal{C}) \geq (1 - \epsilon^2)\lambda(\mathcal{C}), \tag{4}$$

since $T^k(B \cap L) = B \cap T^k L$ and $T^k L$ sweeps out the entire column as k ranges from $-j + 1$ to $\ell(W) - j$.

Fix $\mathcal{C}$ and choose M so large that $\mathcal{S}(m)$ and $\mathcal{S}(M)$ are $\epsilon\lambda(\mathcal{C})$-independent, so that, in particular, the following holds.

$$\sum_{\mathcal{D} \in \mathcal{S}(M)} |\lambda(\mathcal{D}|\, \mathcal{C}) - \lambda(\mathcal{D})| \leq \epsilon. \tag{5}$$

Let $\mathcal{F}$ be the set of all $\mathcal{D} \in \mathcal{S}(M)$ for which $\lambda(B|\, \mathcal{C} \cap \mathcal{D}) \geq 1 - \epsilon$. The set $\mathcal{C} \cap \mathcal{D}$ is a union of levels of $\mathcal{D}$, since $\mathcal{S}(M)$ is built by cutting and stacking from $\mathcal{S}(m)$, so if $\mathcal{D} \in \mathcal{F}$, then there must be at least one level of $\mathcal{D}$ which is at least $(1 - \epsilon)$ filled by B. The argument used to prove (4) then shows that the entire column $\mathcal{D}$ must be $(1 - \epsilon)$ filled by B. In summary,

$$\lambda(B \cap \mathcal{D}) \geq (1 - \epsilon)\lambda(\mathcal{D}),\ \mathcal{D} \in \mathcal{F}. \tag{6}$$

Since $\lambda(B|\, \mathcal{C}) = \sum_{\mathcal{D}} \lambda(B|\, \mathcal{C} \cap \mathcal{D})\lambda(\mathcal{D}|\, \mathcal{C})$, the Markov inequality and the fact that $\lambda(B|\, \mathcal{C}) \geq (1 - \epsilon^2)$, imply that

$$\sum_{\mathcal{D} \notin \mathcal{F}} \lambda(\mathcal{D}|\, \mathcal{C}) \leq \epsilon,$$

which together with the condition, (5), implies

$$\sum_{\mathcal{D} \notin \mathcal{F}} \lambda(\mathcal{D}) \leq 2\epsilon.$$

Thus summing over $\mathcal{D}$ and using (6) yields $\lambda(B) \geq 1 - 3\epsilon$. This shows that $\lambda(B) = 1$, and hence T must be ergodic. This completes the proof of Theorem I.10.6. □

If is often easier to make successive stages approximately independent than it is to force asymptotic independence, hence the following stronger form of the complete sequences and ergodicity theorem is quite useful.

Theorem I.10.7 (Complete sequences and ergodicity: strong form.)
If $\{\mathcal{S}(m)\}$ is a complete sequence such that for each m, $\mathcal{S}(m)$ and $\mathcal{S}(m+1)$ are ϵ_m-independent, where $\epsilon_m \to 0$, as $m \to \infty$, then $\{\mathcal{S}(m)\}$ defines an ergodic process.

Proof. Assume $T^{-1}B = B$ and $\lambda(B) > 0$. The only real modification that needs to be made in the preceding proof is to note that if $\{L_i\}$ is a disjoint collection of column levels such that

$$\lambda\left(B^c \cap (\cup L_i)\right) = \sum \lambda(B^c|L_i)\lambda(L_i) < \epsilon^2\delta, \tag{7}$$

where B^c is the complement of B, then by the Markov inequality there is a subcollection with total measure at least δ for which

$$\lambda(B \cap L_i) \geq (1-\epsilon^2)\lambda(L_i).$$

Let $\mathcal{S}'$ be the set of all columns $\mathcal{C}$ for which some level has this property. The support of $\mathcal{S}'$ has measure at least δ, and the sweeping-out argument used to prove (4) shows that

$$\lambda(B \cap \mathcal{C}) \geq (1-\epsilon^2)\lambda(\mathcal{C}), \quad \mathcal{C} \in \mathcal{S}'. \tag{8}$$

Thus, taking $\delta = \mu(B)/2$, and m so large that ϵ_m is smaller than both ϵ^2 and $(\mu(B)/2)^2$ and so that there is a set of levels of columns in $\mathcal{S}(m)$ such that (7) holds, it follows that there must be at least one $C \in \mathcal{S}(m)$ for which (8) holds and for which

$$\sum_{D \in \mathcal{S}(m+1)} |\lambda(D|C) - \lambda(D)| < \sqrt{\epsilon_m} \leq \epsilon.$$

The argument used in the preceding theorem then gives $\lambda(B) > 1 - 3\epsilon$. This proves Theorem I.10.7. □

The freedom in building ergodic processes via cutting and stacking lies in the arbitrary nature of the cutting and stacking rules. The user is free to vary which columns are to be cut and in what order they are to be stacked, as well as how substructures are to become well-distributed in later substructures. There are practical limitations, of course, in the complexity of the description needed to go from one stage to the next. A bewildering array of examples have been constructed, however, by using only a few simple techniques for going from one stage to the next. A few of these constructions will be described in later chapters of this book. The next subsection will focus on a simple form of cutting and stacking, known as independent cutting and stacking, which, in spite of its simplicity, can produce a variety of counterexamples when applied to substructures.

I.10.d Independent cutting and stacking.

Independent cutting and stacking is the geometric version of the product measure idea. In this discussion, as earlier, the same notation will be used for a column or its support, with the context making clear which meaning is intended.

A column structure $\mathcal{S}$ can be stacked independently on top of a labeled column $\mathcal{C}$ of the same width, provided that they have disjoint supports. This gives a new column structure denoted by $\mathcal{C} * \mathcal{S}$ and defined as follows.

(i) Partition $\mathcal{C}$ into subcolumns $\{\mathcal{C}_i\}$ so that $\tau(\mathcal{C}_i) = \tau(\mathcal{D}_i)$, where $\mathcal{D}_i$ is the ith column of $\mathcal{S}$.

(ii) Stack $\mathcal{D}_i$ on top of $\mathcal{C}_i$ to obtain the new column, $\mathcal{C}_i * \mathcal{D}_i$.

The new column structure $\mathcal{C} * \mathcal{S}$ consists of all the columns $\mathcal{C}_i * \mathcal{D}_i$. It is called the *independent stacking of $\mathcal{S}$ onto $\mathcal{C}$*. (See Figure I.10.8.)

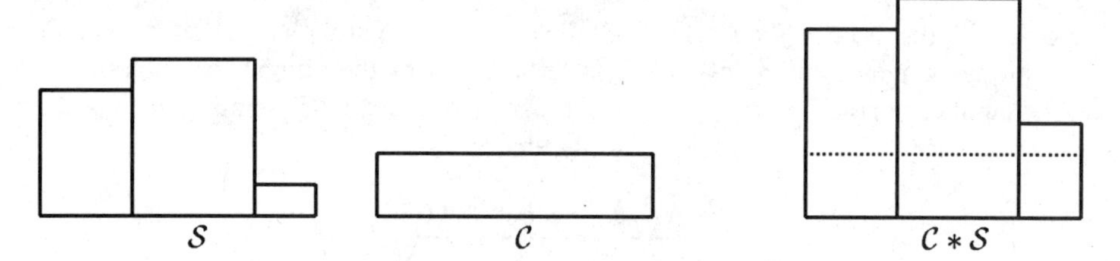

Figure I.10.8 Stacking a column structure independently onto a column.

To say precisely what it means to independently cut and stack one column structure on top of another column structure the concept of copy is useful.

A column structure $\mathcal{S}$ is said to be a *copy of size α* of a column structure $\mathcal{S}'$ if there is a one-to-one correspondence between columns such that corresponding columns have the same height and the same labeling, and the ratio of the width of a column of $\mathcal{S}$ to the width of its corresponding column in $\mathcal{S}'$ is α. In other words, a scaling of one structure is isomorphic to the other, where two column structures $\mathcal{S}$ and $\mathcal{S}'$ are said to be *isomorphic* if there is a one-to-one correspondence between columns such that corresponding columns have the same height, width, and name. Note, by the way, that a copy has the same width distribution as the original.

A column structure $\mathcal{S}$ can be *cut into copies* $\{\mathcal{S}_i\colon i \in I\}$ *according to a distribution* π on I, by partitioning each column of $\mathcal{S}$ according to the distribution π, and letting $\mathcal{S}_i$ be the column structure that consists of the i-th subcolumn of each column of $\mathcal{S}$.

Let $\mathcal{S}'$ and $\mathcal{S}$ be disjoint column structures with the same width. The *independent cutting and stacking of $\mathcal{S}'$ onto $\mathcal{S}$* is denoted by $\mathcal{S} * \mathcal{S}'$ and is defined for $\mathcal{S} = \{\mathcal{C}_i\}$ as follows. (See Figure I.10.9.)

(i) Cut $\mathcal{S}'$ into copies $\{\mathcal{S}'_i\}$ so that $\tau(\mathcal{S}'_i) = \tau(\mathcal{C}_i)$.

(ii) For each i, stack $\mathcal{S}'_i$ independently onto $\mathcal{C}_i$, obtaining $\mathcal{C}_i * \mathcal{S}'_i$.

The column structure $\mathcal{S} * \mathcal{S}'$ is the union of the column structures $\mathcal{C}_i * \mathcal{S}'_i$.

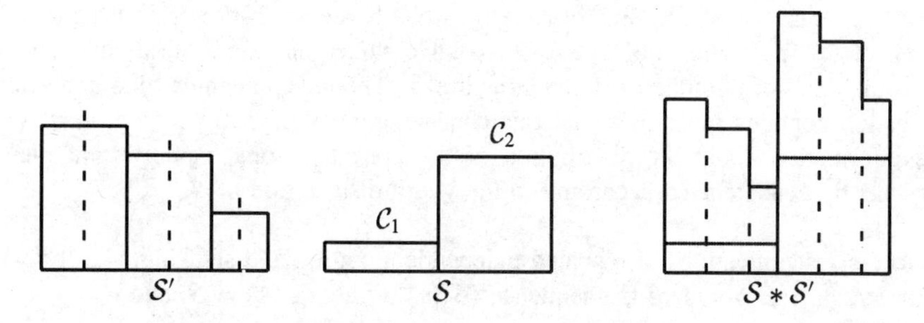

Figure I.10.9 Stacking a structure independently onto a structure.

An alternative description of the columns of $\mathcal{S} * \mathcal{S}'$ may given as follows.

(i) Cut each $C'_j \in \mathcal{S}'$ into subcolumns $\{C'_{ji}\}$ such that $\tau(C'_{ji}) = \tau(C_i)\tau(C'_j)/\tau(\mathcal{S}')$ for each $C_i \in \mathcal{S}$.

(ii) Cut each column $C_i \in \mathcal{S}$ into subcolumns $\{C_{ij}\}$, such that $\tau(C_{ij}) = \tau(C'_{ji})$.

The column structure $\mathcal{S} * \mathcal{S}'$ consists of all the $C_{ij} * C'_{ji}$. In particular, in the finite column case, the number of columns of $\mathcal{S} * \mathcal{S}'$ is the product of the number of columns of $\mathcal{S}$ and the number of columns of $\mathcal{S}'$. The key property of independent cutting and stacking, however, is that width distributions multiply, that is,

$$\frac{\tau(C_{ij} * C'_{ji})}{\tau(\mathcal{S} * \mathcal{S}')} = \frac{\tau(C_i)}{\tau(\mathcal{S})} \times \frac{\tau(C'_j)}{\tau(\mathcal{S}')},$$

since $\tau(\mathcal{S} * \mathcal{S}') = \tau(\mathcal{S})$. This formula expresses the probabilistic meaning of independent cutting and stacking: *cut and stack so that width distributions multiply.*

A column structure can be cut into copies of itself and these copies stacked to form a new column structure. The *M-fold independent cutting and stacking* of a column structure $\mathcal{S}$ is defined by cutting $\mathcal{S}$ into M copies $\{\mathcal{S}_m : m = 1, 2, \ldots, M\}$ of itself of equal width and successively independently cutting and stacking them to obtain

$$\mathcal{S}_1 * \mathcal{S}_2 * \cdots * \mathcal{S}_M,$$

where the latter is defined inductively by $\mathcal{S}_1 * \cdots * \mathcal{S}_M = (\mathcal{S}_1 * \cdots * \mathcal{S}_{M-1}) * \mathcal{S}_M$. Two-fold independent cutting and stacking is indicated in Figure I.10.10.

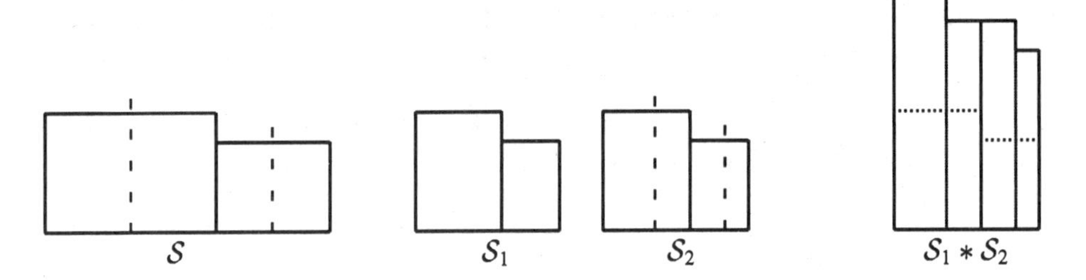

Figure I.10.10 Two-fold independent cutting and stacking.

Note that $\tau(\mathcal{S}_1 * \cdots * \mathcal{S}_M) = \tau(\mathcal{S})/M$, and, in the case when $\mathcal{S}$ has only finitely many columns, the number of columns of $\mathcal{S}_1 * \cdots * \mathcal{S}_M$ is the M-th power of the number of columns of $\mathcal{S}$. The columns of $\mathcal{S}_1 * \cdots * \mathcal{S}_M$ have names that are M-fold concatenations of column names of the initial column structure $\mathcal{S}$. The independent cutting and stacking construction contains more than just this concatenation information, for it carries with it the information about the distribution of the concatenations, namely, that they are independently concatenated according to the width distribution of $\mathcal{S}$.

Successive applications of repeated independent cutting and stacking, starting with a column structure $\mathcal{S}$ produces the sequence $\{\mathcal{S}(m)\}$, where $\mathcal{S}(1) = \mathcal{S}$, and

$$\mathcal{S}(m+1) = \mathcal{S}(m) * \mathcal{S}(m), \quad m \geq 1.$$

The sequence $\{\mathcal{S}(m)\}$ is called the sequence *built (or generated) from $\mathcal{S}$ by repeated independent cutting and stacking.* Note that $\mathcal{S}(m)$ is isomorphic to the 2^m-fold independent cutting and stacking of $\mathcal{S}$, so that its columns are just concatenations of 2^m subcolumns of the columns of $\mathcal{S}$, selected independently according to the width distribution of $\mathcal{S}$.

In the case when $\lambda(\mathcal{S}) = 1$, the $(T, \mathcal{P})$-process defined by the sequence generated from $\mathcal{S}$ by repeated independent cutting and stacking is called the *process built from $\mathcal{S}$ by repeated independent cutting and stacking.* This is, of course, just another way to define the regenerative processes, for they are the precisely the processes built by repeated independent cutting and stacking from the columnar representation of their first-order block-structure distributions, Exercise 8. Ergodicity is guaranteed by the following theorem.

Theorem I.10.11 (Repeated independent cutting and stacking.)
If $\{\mathcal{S}(m)\}$ is built from $\mathcal{S}$ by repeated independent cutting and stacking, then given $\epsilon > 0$, there is an m such that $\mathcal{S}$ and $\mathcal{S}(m)$ are ϵ-independent.

In particular, if $\lambda(\mathcal{S}) = 1$ then the process built from $\mathcal{S}$ by repeated independent cutting and stacking is ergodic.

Proof. The proof for the case when $\mathcal{S}$ has only a finite number of columns will be given here, the countable case is left to Exercise 2. The notation $\mathcal{D} = \mathcal{C}_1^k = \mathcal{C}_1\mathcal{C}_2\cdots\mathcal{C}_k$, will mean the column $\mathcal{D}$ was formed by taking, for each i, a subcolumn of $\mathcal{C}_i \in \mathcal{S}$, and stacking these in order of increasing i. For $k = 2^m$, $\mathcal{D} = \mathcal{C}_1\mathcal{C}_2\cdots\mathcal{C}_k \in \mathcal{S}(m)$, and $\mathcal{C} \in \mathcal{S}$, define

$$p(\mathcal{C}|\mathcal{C}_1^k) = \frac{|\{i \in [1, k]: \mathcal{C} = \mathcal{C}_i\}|}{k},$$

so that, in particular, $kp(\mathcal{C}|\mathcal{C}_1^k)$ is the total number of occurrences of $\mathcal{C}$ in the sequence $\mathcal{C}_1^k$. By the law of large numbers,

$$p(\mathcal{C}|\mathcal{C}_1^k) \to \tau(\mathcal{C})/\tau(\mathcal{S}), \tag{9}$$

and

$$\frac{1}{k}\ell(\mathcal{D}) = \frac{1}{k}\sum_{i=1}^{k}\ell(\mathcal{C}_i) \to E(\ell(\mathcal{S})), \tag{10}$$

both with probability 1, as $k \to \infty$, where $E(\ell(\mathcal{S}))$ is the expected height of the columns of $\mathcal{S}$, with respect to the width distribution.

Division of $\lambda(\mathcal{C} \cap \mathcal{D}) = kp(\mathcal{C}|\mathcal{C}_1^k)\ell(\mathcal{C})\tau(\mathcal{D})$ by the product of $\lambda(\mathcal{C}) = \ell(\mathcal{C})\tau(\mathcal{C})$ and $\lambda(\mathcal{D}) = \ell(\mathcal{D})\tau(\mathcal{D})$ yields

$$\frac{\lambda(\mathcal{C} \cap \mathcal{D})}{\lambda(\mathcal{C})\lambda(\mathcal{D})} = \frac{kp(\mathcal{C}|\mathcal{C}_1^k)}{\ell(\mathcal{D})\tau(\mathcal{C})} \to 1,$$

with probability 1, as $k \to \infty$, by (9) and (10), since $\tau(\mathcal{S})E(\ell(\mathcal{S})) = \lambda(\mathcal{S}) = 1$, by assumption. This establishes that indeed $\mathcal{S}$ and $\mathcal{S}(m)$ are eventually ϵ-independent.

Note that for each m_0, $\{\mathcal{S}(m): m \geq m_0\}$ is built from $\mathcal{S}(m_0)$ by repeated independent cutting and stacking, and hence the proof shows that $\{\mathcal{S}(m)\}$ is asymptotically independent, so that if $\lambda(\mathcal{S}) = 1$ then the process defined by $\{\mathcal{S}(m)\}$ is ergodic. This completes the proof of Theorem I.10.11. □

The following examples indicate how some standard processes can be built by cutting and stacking. More interesting examples of cutting and stacking constructions are given in Chapter 3. The simplest of these, an example of a process with an arbitrary rate of convergence for frequencies, see III.1.c, is recommended as a starting point for the reader unfamiliar with cutting and stacking constructions. It shows how repeated independent cutting and stacking on separate substructures can be used to make a process "look like another process on part of the space," so as to approximate the desired property, and how to "mix one part slowly into another," so as to guarantee ergodicity for the final process.

Example I.10.12 (I.i.d. processes.)

Let $\mathcal{S}$ be the column structure that consists of two columns each of height 1 and width 1/2, labeled as '0' and '1'. The process built from $\mathcal{S}$ by repeated independent cutting and stacking is just the coin-tossing process, that is, the binary i.i.d. process in which 0's and 1's are equally likely. By starting with a partition of the unit interval into disjoint intervals labeled by the finite set A, repeated independent cutting and stacking produces the A-valued i.i.d. process with the probability of a equal to the length of the interval assigned to a.

Example I.10.13 (Markov processes.)

The easiest way to construct an ergodic Markov chain $\{X_n\}$ via cutting and stacking is to think of it as the regenerative process defined by a recurrent state, say a. Let $\mathcal{S}$ be the set of all a_1^k for which $a_i \neq a$, $1 \leq i < k$, and $a_k = a$. Define

$$\mu_1^*(a_1^k) = \text{Prob}(X_1^k = a_1^k | X_0 = a), \quad a_1^k \in \mathcal{S},$$

and let μ^* be the product measure on $\mathcal{S}^\infty$ defined by μ_1^*. The process built from the first-order columnar representation of $\mathcal{S}$ by repeated independent cutting and stacking is the same as the Markov process $\{X_n\}$.

Example I.10.14 (Hidden Markov chains.)

A hidden Markov process $\{X_n\}$ is a function $X_n = f(Y_n)$ of a Markov chain. (These also known as finite-state processes.) If $\{Y_n\}$ is ergodic, then $\{X_n\}$ is regenerative, hence can be represented as the process built by repeated independent cutting and stacking. In fact, one can first represent the ergodic Markov chain $\{(X_n, Y_n)\}$ as a process built by repeated independent cutting and stacking, then drop the first coordinates of each label.

Example I.10.15 (Finite initial structures.)

If the initial structure $\mathcal{S}$ has only a finite number of columns, the process $\{X_n\}$ built from it by repeated independent cutting and stacking is a hidden Markov chain. To see why this is so, assume that the labeling set A does not contain the symbols 0 and 1 and relabel the column levels by changing the label $\mathcal{N}(L)$ of level L to $(\mathcal{N}(L), 0)$ if L is not the top of its column, and to $(\mathcal{N}(L), 1)$ if L is the top of its column. It is easy to see that the process $\{Y_n\}$ built from this new column structure by repeated independent cutting and stacking is Markov of order no more than the maximum length of its columns. Since dropping the second coordinates of each new label produces the old labels, it follows that X_n is a function of Y_n. Note, in particular, that if the original structure has two columns with heights differing by 1, then $\{Y_n\}$ is mixing Markov, and hence $\{X_n\}$ is a function of a mixing Markov chain.

Remark I.10.16

The cutting and stacking ideas originated with von Neumann and Kakutani. For a discussion of this and other early work see [12]. They have since been used to construct numerous counterexamples, some of which are mentioned [73]. The latter also includes several of the results presented here, most of which have long been part of the folklore of the subject. A sufficient condition for a cutting and stacking construction to produce a stationary coding of an i.i.d. process is given in [75].

I.10.e Exercises

1. Show that if T is the upward map defined by a column $\mathcal{C}$, then a subcolumn has the form $(D, TD, T^2D, \ldots, T^{\ell(\mathcal{C})-1}D)$, where D is a subset of the base B.

2. Prove Theorem I.10.11 for the case when $\mathcal{S}$ has countably many columns.

3. Show that formula (2) holds for $k > 1$. (Hint: the tower with base transformation S^n defines the same process $\widehat{\mu}$. Then apply the $k = 1$ argument.)

4. A column $\mathcal{C} \in \mathcal{S}$ is $(1-\epsilon)$-well-distributed in $\mathcal{S}'$ if $\sum_{\mathcal{D}\in\mathcal{S}'} |\lambda(\mathcal{D}|\,\mathcal{C}) - \lambda(\mathcal{D})| \leq \epsilon$. A sequence $\{\mathcal{S}(m)\}$ is asymptotically well-distributed if for each m, each $\mathcal{C} \in \mathcal{S}(m)$, and each $\epsilon > 0$, there is an $M = M(m, \mathcal{C}, \epsilon)$ such that $\mathcal{C}$ is $(1-\epsilon)$-well-distributed in $\mathcal{S}(n)$ for $n \geq M$.

   (a) Suppose $\mathcal{S}'$ is built by cutting and stacking from $\mathcal{S}$. Show that for $\mathcal{C} \in \mathcal{S}$ and $\mathcal{D} \in \mathcal{S}'$, the conditional probability $\lambda(\mathcal{D}|\,\mathcal{C})$ is the fraction of $\mathcal{C}$ that was cut into slices to put into $\mathcal{D}$.

   (b) Show that the asymptotic independence property is equivalent to the asymptotic well-distribution property.

   (c) Show that if $\mathcal{S}$ and $\mathcal{S}'$ are ϵ-independent, then $\sum_{\mathcal{C}\in\mathcal{S}} |\lambda(\mathcal{C}|\,\mathcal{D}) - \lambda(\mathcal{D})| \leq \sqrt{\epsilon}$, except for a set of $\mathcal{D} \in \mathcal{S}'$ of total measure at most $\sqrt{\epsilon}$.

   (d) Show that if $\sum_{\mathcal{C}\in\mathcal{S}} |\lambda(\mathcal{C}|\,\mathcal{D}) - \lambda(\mathcal{D})| \leq \epsilon$, except for a set of $\mathcal{D} \in \mathcal{S}'$ of total measure at most ϵ, then $\mathcal{S}'$ and $\mathcal{S}$ are 2ϵ-independent.

5. Let μ^* be a block-structure measure on $\mathcal{S}^\infty$, where $\mathcal{S} \subseteq A^*$, and let $\{\mathcal{S}(m)\}$ be the standard cutting and stacking representation of the stationary A-valued process defined by $(\mathcal{S}, \mu^*)$. Show that $\{\mathcal{S}(m)\}$ is asymptotically independent if and only if μ^* is totally ergodic. (Hint: for $\mathcal{S} = \mathcal{S}(m)$, Exercise 4c implies that μ^* is m-ergodic.)

6. Suppose $\mathcal{T}(m) \subset \mathcal{S}(m)$, for each m, and $\lambda(\mathcal{T}(m)) \to 1$ as $m \to \infty$.

   (a) Show that $\{\mathcal{S}(m)\}$ is asymptotically independent if and only if $\{\mathcal{T}(m)\}$ is asymptotically independent.

   (b) Suppose that for each m there exists $\mathcal{R}(m) \subset \mathcal{S}(m)$, disjoint from $\mathcal{T}(m)$, and an integer M_m such that $\mathcal{T}(m+1)$ is built by M_m-fold independent cutting and stacking of $\mathcal{T}(m) \cup \mathcal{R}(m)$. Suppose also that $\{\mathcal{S}(m\}$ is complete. Show that if $\{M_m\}$ increases fast enough then the process defined by $\{\mathcal{S}(m)\}$ is ergodic.

7. Suppose $\mathcal{S}$ has measure 1 and only finitely many columns.

   (a) Show that if the top of each column is labeled '1', with all other levels labeled '0', then the process built from $\mathcal{S}$ by repeated independent cutting and stacking is Markov of some order.

(b) Show that if two columns of $\mathcal{S}$ have heights differing by 1, then the process built from $\mathcal{S}$ by repeated independent cutting and stacking is a mixing finite-state process.

(c) Verify that the process constructed in Example I.10.13 is indeed the same as the Markov process $\{X_n\}$.

8. Show that process is regenerative if and only if it is the process built from a column structure $\mathcal{S}$ of measure 1 by repeated independent cutting and stacking.

Chapter II

Entropy-related properties.

Section II.1 Entropy and coding.

As noted in Section I.7.c, entropy provides an asymptotic lower bound on expected per-symbol code length for prefix-code sequences and faithful-code sequences, a lower bound which is "almost" tight, at least asymptotically in source word length n, see Theorem I.7.12 and Theorem I.7.15. Two issues left open by these results will be addressed in this section. The first issue is the universality problem. The sequence of Shannon codes compresses to entropy in the limit, but its construction depends on knowledge of the process, knowledge which may not be available in practice. It will be shown that there are universal codes, that is, code sequences which compress to the entropy in the limit, for almost every sample path from any ergodic process. The second issue is the almost-sure question. The entropy lower bound is an expected-value result, hence it does not preclude the possibility that there might be code sequences that beat entropy infinitely often on a set of positive measure. It will be shown that this cannot happen.

An *n-code* is a mapping $C_n\colon A^n \mapsto \{0, 1\}^*$, where $\{0, 1\}^*$ is the set of finite-length binary sequences. The *code length function* $\mathcal{L}(\cdot|C_n)$ of the code is the function that assigns to each x_1^n, the length $\ell(C_n(x_1^n))$ of the code word $C_n(x_1^n)$. When C_n is understood, $\mathcal{L}(x_1^n)$ will be used instead of $\mathcal{L}(x_1^n|C_n)$. A *code sequence* is a sequence $\{C_n\colon n = 1, 2, \ldots\}$, where each C_n is an n-code. If each C_n is one-to-one, the code sequence is called a faithful-code sequence, while if each C_n is a prefix code it is called a prefix-code sequence. The Shannon code construction provides a prefix-code sequence $\{C_n\}$ for which $\mathcal{L}(x_1^n) = \lceil -\log \mu(x_1^n) \rceil$, so that, in particular, $\mathcal{L}(x_1^n)/n \to h$, almost surely, by the entropy theorem. In general, however, if $\{C_n\}$ is a Shannon-code sequence for μ, and ν is some other ergodic process, then $\mathcal{L}(x_1^n)/n$ may fail to converge on a set of positive ν-measure, see Exercise 1.

A code sequence $\{C_n\}$ is said to be *universally asymptotically optimal* or, more simply, *universal* if

$$\limsup_{n\to\infty} \frac{\mathcal{L}(x_1^n)}{n} \leq h(\mu), \text{ almost surely,}$$

for every ergodic process μ, where $h(\mu)$ denotes the entropy of μ.

Theorem II.1.1 (Universal codes exist.)
There is a prefix-code sequence $\{C_n\}$ *such that* $\limsup_n \mathcal{L}(x_1^n)/n \le h(\mu)$, *almost surely, for any ergodic process* μ.

Theorem II.1.2 (Too-good codes do not exist.)
If $\{C_n\}$ *is a faithful-code sequence and* μ *is an ergodic measure with entropy* h, *then* $\liminf_n \mathcal{L}(x_1^n)/n \ge h$, *almost surely.*

In other words, it is possible to (universally) compress to entropy in the limit, almost surely for any ergodic process, but for no ergodic process is it possible to beat entropy infinitely often on a set of positive measure.

A counting argument, together with some results about entropy for Markov chains, will be used to establish the universal code existence theorem, Theorem II.1.1. The nonexistence theorem, Theorem II.1.2, will follow from the entropy theorem, together with a surprisingly simple lower bound on prefix-code word length. A second proof of the existence theorem, based on entropy estimation ideas of Ornstein and Weiss, will be discussed in Section II.3.d. A third proof of the existence theorem, based on the Lempel-Ziv coding algorithm, will be given in Section II.2.

Of interest, also, is the fact that both the existence and nonexistence theorems can be established using only the existence of process entropy, whose existence depends only on subadditivity of entropy as expected value, while the existence of decay-rate entropy is given by the much deeper entropy theorem. The proof of the existence theorem to be given actually shows the existence of codes that achieve at least the process entropy. A direct coding argument, which is based on a packing idea and is similar in spirit to some later proofs, will then be used to show that it is not possible to beat process entropy in the limit on a set of positive measure. These two results also provide an alternative proof of the entropy theorem. While no simpler than the earlier proof, they show clearly that the basic existence and nonexistence theorems for codes are, in essence, together equivalent to the entropy theorem, thus further sharpening the connection between entropy and coding. In addition, the direct coding construction extends to the case of semifaithful codes, for which a controlled amount of distortion is allowed, as shown in [49].

II.1.a Universal codes exist.

A prefix-code sequence $\{C_n\}$ will be constructed such that for any ergodic μ,

$$\limsup_n \mathcal{L}(x_1^n)/n \le H, \text{ a.s.,}$$

where $H = H_\mu$ denotes the process entropy of μ. Since process entropy H is equal to the decay-rate h given by the entropy theorem, this will show that good codes exist.

The code C_n utilizes the empirical distribution of k-blocks, that is, the k-type, for suitable choice of k as a function of n, with code performance established by using the type counting results discussed in Section I.6.d. The code is a two-part code, that is, the codeword $C_n(x_1^n)$ is a concatenation of two binary blocks. The first block gives the index of the k-type of the sequence x_1^n, relative to some enumeration of possible k-types; it has fixed length which depends only on the number of type classes. The second block gives the index of the particular sequence x_1^n in its k-type class, relative to some enumeration of the k-type class; its length depends on the size of the type class. Distinct words, x_1^n and y_1^n, either have different k-types, in which case the first blocks of $C_n(x_1^n)$ and $C_n(y_1^n)$

will be different, or they have the same k-types but different indices in their common k-type class, hence the second blocks of $C_n(x_1^n)$ and $C_n(y_1^n)$ will differ. Since the first block has fixed length, the code C_n is a prefix code.

If k does not grow too rapidly with n, say, $k \sim (1/2)\log_{|A|} n$, then the log of the number of possible k-types is negligible relative to n, so asymptotic code performance depends only on the asymptotic behavior of the cardinality of the k-type class of x_1^n. An entropy argument shows that the log of this class size cannot be asymptotically larger than nH.

A slight modification of the definition of k-type, called the circular k-type, will be used as it simplifies the final entropy argument, yet has negligible effect on code performance. Given a sequence x_1^n and an integer $k \le n$, let $\tilde{x}_1^{n+k-1} = x_1^n x_1^{k-1}$, the concatenation of x_1^n with x_1^{k-1}, that is, x_1^n is extended periodically for $k-1$ more terms. The *circular k-type* is the measure $\tilde{P}_k = \tilde{P}_k(\cdot|x_1^n)$ on A^k defined by the relative frequency of occurrence of each k-block in the sequence $\tilde{x}_1^{n+k-1}$, that is,

$$\tilde{P}_k(a_1^k|x_1^n) = \frac{\left|\{i \in [1,n]\colon\ \tilde{x}_i^{i+k-1} = a_1^k\}\right|}{n}, \quad a_1^k \in A^k.$$

The circular k-type is just the usual k-type of the sequence $\tilde{x}_1^{n+k-1}$, so that the bounds, Theorem I.6.14 and Theorem I.6.15, on the number of k-types and on the size of a k-type class yield the following bounds

$$\tilde{N}(k,n) \le (n+1)^{|A|^k}, \tag{1}$$

$$|\tilde{\mathcal{T}}_k(x_1^n)| \le (n-1)2^{(n-1)\tilde{H}_{k-1,x_1^n}} \tag{2}$$

on the number $\tilde{N}(k,n)$ of circular k-types that can be produced by sequences of length n, and on the cardinality $|\tilde{\mathcal{T}}_k(x_1^n)|$ of the set of all n-sequences that have the same circular k-type as x_1^n, where $\tilde{H}_{k-1,x_1^n}$ denotes the entropy of the $(k-1)$-order Markov chain defined by $\tilde{P}_k(\cdot|x_1^n)$. The gain in using the circular k-type is compatibility in k, that is,

$$\tilde{P}_{k-1}(a_1^{k-1}|x_1^n) = \sum_{a_k} \tilde{P}_k(a_1^k|x_1^n),$$

which, in turn, implies the entropy inequality

$$\tilde{H}_{k-1,x_1^n} \le \tilde{H}_{i,x_1^n}, \quad i \le k-1. \tag{3}$$

Now suppose $k = k(n)$ is the greatest integer in $(1/2)\log_{|A|} n$ and let C_n be the two-part code defined by first transmitting the index of the circular k-type of x_1^n, then transmitting the index of x_1^n in its circular k-type class, that is, $C_n(x_1^n) = b_1^m b_{m+1}^t$, where b_1^m is a fixed length binary sequence specifying the index of the circular k-type of x_1^n, relative to some fixed enumeration of the set of possible circular k-types, and b_{m+1}^t is a variable-length binary sequence specifying the index of the particular sequence x_1^n relative to some particular enumeration of $\tilde{\mathcal{T}}_k(x_1^n)$. (Variable length is needed for the second part, because the size of the type class depends on the type.) Total code length satisfies

$$\mathcal{L}(x_1^n) \le \log \tilde{N}(k,n) + \log|\tilde{\mathcal{T}}_k(x_1^n)| + 2,$$

where the two extra bits are needed in case the logarithms are not integers.

The bound (1) on the number of circular k-types, together with the assumption that $k \leq (1/2)\log_{|A|} n$, implies that it takes at most

$$1 + \sqrt{n}\log(n+1)$$

bits to specify a given circular $k(n)$-type, relative to some fixed enumeration of circular $k(n)$-types. This is, of course, negligible relative to n. The bound (2) on the size of a type class implies that at most

$$1 + (n-1)\widetilde{H}_{k-1,x_1^n} + \log(n-1)$$

bits are needed to specify a given member of $\widetilde{T}_k(x_1^n)$, relative to a given enumeration of the type class. Thus,

$$\limsup_n \frac{\mathcal{L}(x_1^n)}{n} \leq \limsup_n \widetilde{H}_{k(n)-1,x_1^n},$$

so that to show the existence of good codes it is enough to show that for any ergodic measure μ,

$$\limsup_n \widetilde{H}_{k(n)-1,x_1^n} \leq H, \text{ almost surely,} \tag{4}$$

where H is the process entropy of μ.

Fix an ergodic measure μ with process entropy H. Given $\epsilon > 0$, choose K such that $H_{K-1} \leq H + \epsilon$, where, in this discussion, $H_{K-1} = H(X_K|X_1^{K-1})$ denotes the process entropy of the Markov chain of order $K-1$ defined by the conditional probabilities $\mu(a_1^K)/\mu(a_1^{K-1})$. Since K is fixed, the ergodic theorem implies that $\widetilde{P}_K(a_1^K|x_1^n) \to \mu(a_1^K)$, almost surely, and hence $\widetilde{H}_{K-1,x_1^n} \to H_{K-1}$, almost surely, as $n \to \infty$. In particular, for almost every x there is an integer $N = N(x,\epsilon)$ such that

$$\widetilde{H}_{K-1,x_1^n} \leq H + 2\epsilon, \quad n \geq N.$$

For all sufficiently large n, $k(n)$ will exceed K. Once this happens the entropy inequality (3) combines with the preceding inequality to yield

$$\widetilde{H}_{k(n)-1,x_1^n} \leq H + 2\epsilon.$$

Since ϵ is arbitrary, this shows that $\limsup_n \widetilde{H}_{k(n),x_1^n} \leq H$, almost surely, establishing the desired bound (4).

The proof that good codes exist is now finished, for as noted earlier, process entropy H and decay-rate h are the same for ergodic processes. □

II.1.b Too-good codes do not exist.

A faithful-code sequence can be converted to a prefix-code sequence with no change in asymptotic performance by using the Elias header technique, described in Section I.7.d of Chapter 1. Thus it is enough to prove Theorem II.1.2 for the case when each C_n is a prefix code. Two quite different proofs of Theorem II.1.2 will be given. The first proof uses a combination of the entropy theorem and a simple lower bound on the pointwise behavior of prefix codes, a bound due to Barron, [3]. The second proof, which was developed in [49], and does not make use of the entropy theorem, is based on an explicit code construction which is closely connected to the packing ideas used in the proof of the entropy theorem.

The following lemma, due to Barron, is valid for any process, stationary or not.

Lemma II.1.3 (The almost-sure code-length bound.)

Let $\{C_n\}$ be a prefix-code sequence and let μ be a Borel probability measure on A^∞. If $\{\alpha_n\}$ is a sequence of positive numbers such that $\sum_n 2^{-\alpha_n} < \infty$, then

$$\mathcal{L}(x_1^n) + \log \mu(x_1^n) \geq -\alpha_n, \text{ eventually a.s.} \tag{5}$$

Proof. For each n define

$$B_n = \{x_1^n \colon \mathcal{L}(x_1^n) + \log \mu(x_1^n) \leq -\alpha_n\}.$$

Using the relation $\mathcal{L}(x_1^n) = \log 2^{\mathcal{L}(x_1^n)}$, this can be rewritten as

$$B_n = \{x_1^n \colon \mu(x_1^n) \leq 2^{-\mathcal{L}(x_1^n)} 2^{-\alpha_n}\},$$

which yields

$$\mu(B_n) = \sum_{x_1^n \in B_n} \mu(x_1^n) \leq 2^{-\alpha_n} \sum_{x_1^n \in B_n} 2^{-\mathcal{L}(x_1^n)} \leq 2^{-\alpha_n},$$

since $\sum_{x_1^n} 2^{-\mathcal{L}(x_1^n)} \leq 1$, by the Kraft inequality for prefix codes. The lemma follows from the Borel-Cantelli lemma. □

Barron's inequality (5) with $\alpha_n = 2 \log_{|A|} n$, and after division by n, yields

$$\liminf_{n\to\infty} \frac{\mathcal{L}(x_1^n)}{n} \geq \liminf_{n\to\infty} \frac{-\log \mu(x_1^n)}{n},$$

eventually almost surely, for any ergodic measure μ. But $-(1/n) \log \mu(x_1^n)$ converges almost surely to h, the entropy of μ. This completes the proof of Theorem II.1.2. □

II.1.c Nonexistence: second proof.

A second proof of the nonexistence of too-good codes, Theorem II.1.2, will now be given, a proof which uses the existence of good codes but makes no use of the entropy theorem. The basic idea is that if a faithful-code sequence indeed compresses more than entropy infinitely often on a set of positive measure, then, with high probability, long enough sample paths can be partitioned into disjoint words, such that those words that can be coded too well cover at least a fixed fraction of sample path length, while the complement can be compressed almost to entropy, thereby producing a code whose expected length per symbol is less than entropy by a fixed amount.

Let μ be a fixed ergodic process with process entropy H, and assume $\{C_n\}$ is a prefix-code sequence such that $\liminf_i \mathcal{L}(x_1^n|C_n)/n < H$, on a set of positive measure, with respect to μ. (As noted earlier, a faithful-code sequence with this property can be replaced by a prefix-code sequence with the same property, since adding headers to specify length has no effect on the asymptotics.) Thus there are positive numbers, ϵ and γ, such that if

$$B_j = \{x \colon \mathcal{L}(x_1^j|C_j) < i(H - \epsilon)\}$$

then $\mu(\cap_n \cup_{j\geq n} B_j) > \gamma$.

Let $\delta < 1/2$ be a positive number to be specified later. The existence theorem, Theorem II.1.1, provides a $k \geq 1/\delta$ and a prefix code $\widetilde{C}_k$, with length function $\widetilde{\mathcal{L}}(x_1^k) = \mathcal{L}(x_1^k|\widetilde{C}_k)$, such that the set

$$D_k = \left\{x_1^k\colon \widetilde{\mathcal{L}}(x_1^k) < k(H+\delta)\right\},$$

has measure at least $1-\delta$.

A version of the packing lemma will be used to show that long enough sample paths can, with high probability, be partitioned into disjoint words, such that the words that belong to $B_M \cup B_{M+1} \cup \ldots$, for some large enough M, cover at least a γ-fraction of total length, while the words that are neither in D_k nor in $B_M \cup B_{M+1} \cup \ldots$ cover at most a 3δ-fraction of total length. Sample paths with this property can be encoded by sequentially encoding the words, using the too-good code C_i on those words that belong to B_i for some $i \geq M$, using the good code $\widetilde{C}_k$ on those words that belong to D_k, and using a single-letter code on the remaining symbols, plus headers to tell which code is being used. If δ is small enough and such sample paths are long enough, this produces a single code with expected length smaller than expected entropy, contradicting Shannon's lower bound.

To make precise the desired sample path partition concept, let $M \geq 2k/\delta$ be a positive integer. For $n > M$, a concatentation

$$x_1^n = w(1)w(2)\cdots w(t)$$

is said to be a (γ, δ)-*good representation of* x_1^n if the following three conditions hold.

(a) Each $w(i)$ belongs to A, or belongs to D_k, or belongs to B_j for some $j \geq M$.

(b) The total length of the $w(i)$ that belong to A is at most $3\delta n$.

(c) The total length of the $w(i)$ that belong to $B_M \cup B_{M+1} \cup \ldots$ is at least γn.

Let $G(n)$ be the set of all sequences x_1^n that have a (γ, δ)-good representation. It will be shown later that

$$x_1^n \in G(n), \text{ eventually almost surely.} \tag{6}$$

Here it will be shown how a prefix n-code can be constructed by taking advantage of the structure of sequences in $G(n)$, a code that will eventually beat entropy for a suitable choice of δ and M, given that (6) holds. The coding result is stated as the following lemma.

Lemma II.1.4

For $n \geq M$, there is a prefix n-code C_n^* whose length function $\mathcal{L}^*(x_1^n)$ satisfies

$$\mathcal{L}^*(x_1^n) \leq n(H-\epsilon'),\quad x_1^n \in G(n), \tag{7}$$

for all sufficiently large n, for a suitable choice of $\epsilon' > 0$, $\delta > 0$, and $M \geq 2k/\delta$. Furthermore, $\mathcal{L}^*(x_1^n)/n$ will be uniformly bounded above by a constant on the complement of $G(n)$, independent of n.

The lemma is enough to contradict Shannon's lower bound on expected code length, for the uniform boundedness of $\mathcal{L}^*(x_1^n)/n$ on the complement of $G(n)$ combines with the properties (6) and (7) to yield the expected value bound

$$E(\mathcal{L}^*(X_1^n)) \leq n(H-\epsilon'/2),$$

for all sufficiently large n. But this, in turn, yields

$$E(\mathcal{L}^*(X_1^n)) < H(\mu_n),$$

for all sufficiently large n, since $(1/n)H(\mu_n) \to H$, thereby contradicting Shannon's lower bound on expected code length, Theorem I.7.11(i). □

Proof of Lemma II.1.4.

The code $C_n^*(x_1^n)$ is defined for sequences $x_1^n \in G(n)$ by first determining a (γ, δ)-good representation $x_1^n = w(1)w(2)\cdots w(t)$, then coding the $w(i)$ in order of increasing i, using the too-good code on those words that have length more than M, using the good code $\widetilde{C}_k$ on those words that have length k, and using some fixed code, say, $F : A \mapsto B^d$, where $d \geq \log|A|$, on the words of length 1. Appropriate headers are also inserted so the decoder can tell which code is being used, and, for $x_1^n \notin G(n)$, symbols are encoded separately using the fixed code F.

A precise definition of C_n^* is given as follows. For $x_1^n \notin G(n)$, the per-symbol code

$$C_n^*(x_1^n) = 0F(x_1)F(x_2)\cdots F(x_n),$$

is used. Note that $\mathcal{L} * (x_1^n)/n$ is uniformly bounded on the complement of $G(n)$. If $x_1^n \in G(n)$, then a (γ, δ)-good representation $x_1^n = w(1)w(2)\cdots w(t)$ is determined and the code is defined by

$$C_n^*(x_1^n) = 1v(1)v(2)\cdots v(t),$$

where

$$v(i) = \begin{cases} 0F((w(i)) & \text{if } w(i) \text{ has length } 1 \\ 10\widetilde{C}_k(w(i)) & \text{if } w(i) \in D_k \\ 11\mathcal{E}(j)C_j(w(i)) & \text{if } w(i) \in B_j \text{ for some } j \geq M. \end{cases} \tag{8}$$

Here $\mathcal{E}(\cdot)$ is an Elias code, that is, a prefix code on the natural numbers for which the length of $\mathcal{E}(j)$ is $\log j + o(\log j)$.

The code C_n^* is clearly a prefix code, since the initial header specifies whether $x_1^n \in G(n)$ or not, while the other headers specify which of one of the prefix-code collection $\{F, \widetilde{C}_k, C_M, C_{M+1}, \ldots\}$ is being applied to a given word. The goal is to show that on sequences in $G(n)$ the code C_n^* compresses more than entropy by a fixed amount, for suitable choice of δ and $M \geq 2k/\delta$. This is so because, as shown in the following paragraphs, the principal contribution to code length comes from the encodings $C_{\ell(w(i))}(w(i))$ of the $w(i)$ that belong to $\cup_{j \geq M} B_j$ and the encodings $\widetilde{C}_k(w(i))$ of the $w(i)$ that belong to D_k.

By the definition of the B_j, the code word $C_j(w(i))$ is no longer than $j(H - \epsilon)$, so that if L is total length of those $w(i)$ that belong to $B_M \cup B_{M+1} \ldots$, then it requires at most $L(H - \epsilon)$ bits to encode all such $w(i)$. Likewise, by the definition of D_k, a code word $\widetilde{C}_k(w(i))$ has length at most $k(h + \delta)$, so that, since the total length of such $w(i)$ is at most $N - L$, it takes at most

$$L(H - \epsilon) + (N - L)(H + \delta)$$

to encode all those $w(i)$ that belong to either D_k or to $\cup_{j \geq M} B_j$. But $L \geq \gamma n$, by property (c) in the definition of $G(n)$, so that, ignoring the headers needed to specify which code is being used as well as the length of the words that belong to $\cup_{j \geq M} B_j$, the contribution to total code length from encoding those $w(i)$ that belong to either D_k or to $\cup_{j\geq M} B_j$ is upper bounded by

$$n(H+\delta-\gamma(\epsilon+\delta)) \leq n(H+\delta-\gamma\epsilon). \tag{9}$$

Each word $w(i)$ of length 1 requires d bits to encode, and there are at most $3\delta n$ such words, by property (b) in the definition of $G(n)$, so that, still ignoring header lengths, the total length needed to encode all the $w(i)$ is upper bounded by

$$n(H+\delta-\gamma\epsilon+3d\delta). \tag{10}$$

If δ is small enough, then $\delta-\gamma\epsilon+3d\delta$ will be negative, so essentially all that remains is to show that the headers require few bits, relative to n, provided δ is small enough and M is large enough.

The 1-bit headers used to indicate that $w(i)$ has length 1 require a total of at most $3\delta n$ bits, since the total number of such words is at most $3\delta n$, by property (b) of the definition of $G(n)$. There are at most n/k words $w(i)$ the belong to D_k or to $\cup_{j\geq M} B_j$, since $k \leq M$ and hence such words must have length at least k. By assumption, $k \geq 1/\delta$, so these 2-bits headers contribute at most $2\delta n$ to total code length.

The words $w(i)$ need an additional header to specify their length and hence which C_j is to be used. Since an Elias code is used to do this, there is a constant K such that at most

$$K \sum_{w(i)\in\cup_{j\geq M} B_j} \log \ell(w(i))$$

bits are required to specify all these lengths. But $(1/x)\log x$ is a decreasing function for $x \geq e$, so that

$$K \sum_{w(i)\in\cup_{j\geq M} B_j} \log \ell(w(i)) \leq K\frac{\log M}{M}n,$$

since each such $w(i)$ has length $M \geq 2k/\delta$, which exceeds e since it was assumed that $\delta < 1/2$ and $k \geq 1/\delta$. Thus, if it assumed that $\log M \leq \delta M$, then at most $K\delta n$ bits are needed to specify the lengths of the $w(i)$ that belong to $\cup_{j\geq M} B_j$.

In summary, ignoring the one bit needed to tell that $x_1^n \in G(n)$, which is certainly negligible, the total contribution of all the headers as well as the length specifiers is upper bounded by

$$3\delta n + 2\delta n + K\delta n,$$

for any choice of $0 < \delta < 1/2$, provided $k \geq 1/\delta$, $M \geq 2k/\delta$ and $\log M \leq \delta M$. Adding this bound to the bound (10) obtained by ignoring the headers gives the bound

$$\mathcal{L}^*(x_1^n) \leq n[H+\delta(6+3d+K)-\gamma\epsilon)],$$

on total code length for members of $G(n)$. Since d and K are constants, it follows that if δ is chosen small enough, then

$$\mathcal{L}^*(x_1^n) \leq n(H-\epsilon'), \quad x_1^n \in G(n),$$

for suitable choice of δ, ϵ', and $M \geq 2k/\delta$, which establishes the lemma. □

Of course, the lemma depends on the truth of the assertion that $x_1^n \in G(n)$, eventually almost surely, provided only that $M \geq 2k/\delta$. To establish this, choose $L > M$ so large that if $B = \cup_{j=M}^{L} B_j$, then $\mu(B) > \gamma$. The partial packing lemma, Lemma I.3.8, provides, eventually almost surely, a γ-packing of x_1^n by blocks drawn from B. On the other hand, by the ergodic theorem, eventually almost surely there are at least δn indices $i \in [1, n-k+1]$ that are starting places of blocks from D_k, that is, x_1^n is $(1-\delta)$-strongly-covered by D_k. But the building-block lemma, Lemma I.7.7, then implies that x_1^n is eventually almost surely $(1-2\delta)$-packed by k-blocks drawn from D_k. In particular, eventually almost surely, x_1^n has both a γ-packing by blocks drawn from B and a $(1-2\delta)$-packing by blocks drawn from D_k.

To say that x_1^n has both a γ-packing by blocks drawn from B and a $(1-2\delta)$-packing by blocks drawn from D_k, is to say that there is a disjoint collection $\{[u_i, v_i]: i \in I\}$ of k-length subintervals of $[1, n]$ of total length at least $(1-2\delta)n$ for which each $x_{u_i}^{v_i} \in D_k$ and a disjoint collection $\{[s_j, t_j]: j \in J\}$ of M-length or longer subintervals of $[1, n]$ of total length at least γn, for which each $x_{s_j}^{t_j} \in B$. Since each $[s_j, t_j]$ has length at least M, the cardinality of J is at most n/M, so the total length of those $[u_i, v_i]$ that meet the boundary of at least one $[s_j, t_j]$ is at most $2nk/M$, which is upper bounded by δn, since it was assumed that $M \geq 2k/\delta$. Thus if I^* is the set of indices $i \in I$ such that $[u_i, v_i]$ meets none of the $[s_j, t_j]$, then the collection

$$\{[s_j, t_j]: j \in J\} \cup \{[u_i, v_i]: i \in I^*\}$$

is disjoint and covers at least a $(1-3\delta)$-fraction of $[1, n]$. (This argument is essentially just the proof of the two-packings lemma, Lemma I.3.9, suggested in Exercise 5, Section I.3.) The words

$$\{x_{s_j}^{t_j}: j \in J\} \cup \{x_{u_i}^{v_i}: i \in I^*\}$$

together with the length 1 words $\{x_s\}$ defined by those

$$s \notin \bigcup_j [s_j, t_j] \cup \bigcup_{i \in I^*} [u_i, v_i]\},$$

provide a representation $x_1^n = w(1)w(2)\cdots w(t)$ for which the properties (a), (b), and (c) hold.

The preceding paragraph can be summarized by saying that if x_1^n has both a γ-packing by blocks drawn from B and a $(1-2\delta)$-packing by blocks drawn from D_k then x_1^n *must belong to* $G(n)$. But this completes the proof that $x_1^n \in G(n)$, eventually almost surely, since, eventually almost surely, x_1^n has both a γ-packing by blocks drawn from B and a $(1-2\delta)$-packing by blocks drawn from D_k.

This completes the second proof that too-good codes do not exist. □

II.1.d Another proof of the entropy theorem.

The entropy theorem can be deduced from the existence and nonexistence theorems connecting process entropy and faithful-code sequences, as follows. Fix an ergodic measure μ with process entropy H, and let $\{C_n\}$ be a prefix-code sequence such that $\limsup_n \mathcal{L}(x_1^n)/n \leq H$, a.s. Fix $\epsilon > 0$ and define

$$D_n = \left\{x_1^n: \mathcal{L}(x_1^n) < n(H+\epsilon)\right\},$$

so that $x \in D_n$, eventually almost surely, and $|D_n| \leq 2^{n(H+\epsilon)}$, since C_n is invertible. Thus, if

$$B_n = \left\{x_1^n: \mu(x_1^n) \leq 2^{-n(H+2\epsilon)}\right\},$$

then

$$\mu(B_n \cap D_n) \leq |D_n| 2^{-n(H+2\epsilon)} \leq 2^{-n\epsilon}.$$

Therefore $x_1^n \notin B_n \cap D_n$, eventually almost surely, and so $x_1^n \notin B_n$, eventually almost surely, establishing the upper bound

$$\limsup_n \frac{1}{n} \log \frac{1}{\mu(x_1^n)} \leq H, \text{ a.s.}$$

For the lower bound define $U_n = \{x_1^n \colon \mu(x_1^n) \geq 2^{-n(H-\epsilon)}\}$, and note that $|U_n| \leq 2^{n(H-\epsilon)}$. Thus there is a one-to-one function ϕ from U_n into binary sequences of length no more than $2 + n(H - \epsilon)$, such that the first symbol in $\phi(x_1^n)$ is always a 0. Let $\widetilde{C}_n$ be the code obtained by adding the prefix 1 to the code C_n. Define $\bar{C}_n(x_1^n) = \phi(x_1^n)$, $x_1^n \in U_n$ and $\bar{C}_n(x_1^n) = \widetilde{C}_n(x_1^n)$, $x_1^n \notin U_n$. The resulting code $\bar{C}_n$ is invertible so that Theorem II.1.2 guarantees that $x_1^n \notin U_n$, eventually almost surely. This proves that

$$\liminf_n \frac{1}{n} \log \frac{1}{\mu(x_1^n)} \geq H, \text{ a.s.},$$

and completes this proof of the entropy theorem. □

II.1.e Exercises

1. This exercise explores what happens when the Shannon code for one process is used on the sample paths of another process.

   (a) For each n let μ_n be the projection of the Kolmogorov measure of unbiased coin tossing onto A^n and let C_n be a Shannon code for μ_n. Show that if $\nu \neq \mu$ is ergodic then $\mathcal{L}(x_1^n)/n$ cannot converge in probability to the entropy of ν.

   (b) Let $D(\mu_n \| \nu_n) = \sum_{x_1^n} \mu_n(x_1^n) \log(\mu_n(x_1^n)/\nu_n(x_1^n))$ be the divergence of μ_n from ν_n, and let $\mathcal{L}(x_1^n)$ be the length function of a Shannon code with respect to ν_n. Show that the expected value of $\mathcal{L}(x_1^n)$ with respect to μ_n is $H(\mu_n) + D(\mu_n \| \nu_n)$.

   (c) Show that if μ is the all 0 process, then there is a renewal process ν such that $\limsup_n D(\mu_n \| \nu_n)/n = \infty$ and $\liminf_n D(\mu_n \| \nu_n)/n = 0$.

2. Let $L(x_1^n)$ denote the length of the longest string that appears twice in x_1^n. Show that if μ is i.i.d. then there is a constant C such that $\limsup_n L(x_1^n)/\log n \leq C$, almost surely. (Hint: code the second occurrence of the longest string by telling how long it is, where it starts, and where it occurred earlier. Code the remainder by using the Shannon code. Add suitable headers to each part to make the entire code into a prefix code and apply Barron's lemma, Lemma 5.)

3. Suppose C_n^* is a one-to-one function defined on a subset S of A^n, whose range is prefix free. Show that there is prefix n-code C_n whose length function satisfies $\mathcal{L}(x_1^n) \leq Kn$ on the complement of S, where K is a constant independent of n, and such that $C_n(x_1^n) = 1C_n^*(x_1^n)$, for $x_1^n \in S$. Such a code C_n is called a *bounded extension of C_n^* to A^n*.

Section II.2 The Lempel-Ziv algorithm.

An important coding algorithm was invented by Lempel and Ziv in 1975, [91]. In finite versions it is the basis for many popular data compression packages, and it has been extensively analyzed in various finite and limiting forms. Ziv's proof, [90], that the Lempel-Ziv (LZ) algorithm compresses to entropy in the limit will be given in this section. A second proof, which is due to Ornstein and Weiss, [53], and uses several ideas of independent interest, is given in Section II.4.

The LZ algorithm is based on a parsing procedure, that is, a way to express an infinite sequence x as a concatenation

$$x = w(1)w(2)\cdots$$

of variable-length blocks, called *words*. In its simplest form, called here *(simple) LZ parsing*, the word-formation rule can be summarized by saying:

The next word is the shortest new word.

To be precise, x is parsed inductively according to the following rules.

(a) The first word $w(1)$ consists of the single letter x_1.

(b) Suppose $w(1)\ldots w(j) = x_1^{n_j}$.

(i) If $x_{n_j+1} \notin \{w(1), \ldots, w(j)\}$ then $w(j+1)$ consists of the single letter x_{n_j+1}.

(ii) Otherwise, $w(j+1) = x_{n_j+1}^{m+1}$, where m is the least integer larger than n_j such that $x_{n_j+1}^{m} \in \{w(1), \ldots, w(j)\}$ and $x_{n_j+1}^{m+1} \notin \{w(1), \ldots, w(j)\}$.

Thus, for example, 11001010001000100... parses into

$$1, 10, 0, 101, 00, 01, 000, 100, \ldots$$

where, for ease of reading, the words are separated by commas.

Note that the parsing is sequential, that is, later words have no effect on earlier words, and therefore an initial segment of length n can be expressed as

$$x_1^n = w(1)w(2)\ldots w(C)\, v, \tag{1}$$

where the final block v is either empty or is equal to some $w(j)$, for $j \leq C$. The parsing defines a prefix n-code C_n, called the *(simple) Lempel-Ziv (LZ) code* by noting that each new word is really only new because of its final symbol, hence it is specified by giving a pointer to where the part before its final symbol occurred earlier, together with a description of its final symbol.

To describe the LZ code C_n precisely, let $\lceil\cdot\rceil$ denote the least integer function and let $f\colon \{0, 1, 2, \ldots, n\} \mapsto B^{\lceil \log n\rceil}$ and $g\colon A \mapsto B^{\lceil \log |A|\rceil}$ be fixed one-to-one functions. If x_1^n is parsed as in (1), the LZ code maps it into the concatenation

$$C_n(x_1^n) = b(1)b(2)\ldots b(C)b(C+1)$$

of the binary words $b(1), b(2), \ldots, b(C), b(C+1)$, defined according to the following rules.

(a) If $j \leq C$ and $w(j)$ has length 1, then $b(j) = 0\, g(w(j))$.

(b) If $j \leq C$ and $i < j$ is the least integer for which $w(j) = w(i)a$, $a \in A$, then $b(j) = 1\, f(i)\, 0\, g(a)$.

(c) If v is empty, then $b(C+1)$ is empty, otherwise $b(C+1) = 1\, f(i)$, where i is the least integer such that $v = w(i)$.

Part (a) requires at most $|A|(\lceil \log |A| \rceil + 1)$ bits, part (b) requires at most $C(\lceil \log n \rceil + \lceil \log |A| \rceil + 2)$ bits, and part (c) requires at most $\lceil \log n \rceil + 1$ bits, so total code length $\mathcal{L}(x_1^n)$ is upper bounded by

$$(C+1)\log n + \alpha C + \beta,$$

where α and β are constants. The dominant term is $C \log n$, so to establish universality it is enough to prove the following theorem of Ziv, [90], in which $C_n(x_1^n)$ now denotes the number of new words in the simple parsing (1).

Theorem II.2.1 (The LZ convergence theorem.)
If μ is an ergodic process with entropy h, then $(1/n)C(x_1^n)\log n \to h$, almost surely.

Of course, it is enough to prove that entropy is an upper bound, since no code can beat entropy in the limit, by Theorem II.1.2. Ziv's proof that entropy is the correct almost-sure upper bound is based on an interesting extension of the entropy idea to individual sequences, together with a proof that this individual sequence entropy is an asymptotic upper bound on $(1/n)C(x_1^n)\log n$, for *every sequence*, together with a proof that, for ergodic processes, individual sequence entropy almost surely upper bounds the entropy given by the entropy theorem.

Ziv's concept of entropy for an individual sequence begins with a simpler idea, called topological entropy, which is the growth rate of the number of observed strings of length k, as $k \to \infty$. The *k-block universe* of x is the set $\mathcal{U}_k(x)$ of all a_1^k that appear as a block of consecutive symbols in x, that is,

$$\mathcal{U}_k(x) = \{a_1^k \colon x_i^{i+k-1} = a_1^k, \text{ for some } i \geq 1\}.$$

The *topological entropy* of x is defined by

$$h(x) = \lim_{k\to\infty} \frac{1}{k} \log |\mathcal{U}_k(x)|,$$

a limit which exists, by subadditivity, Lemma I.6.7, since $|\mathcal{U}_{mk}(x)| \leq |\mathcal{U}_m(x)| \cdot |\mathcal{U}_k(x)|$. (The topological entropy of x, as defined here, is the same as the usual topological entropy of the orbit closure of x, see [84, 34] for discussions of this more general concept.)

Topological entropy takes into account only the number of k-strings that occur in x, but gives no information about frequencies of occurrence. For example, if x is a typical sequence for the binary i.i.d. process with p equal to the probability of a 1 then every finite string occurs with positive probability, hence its topological entropy $h(x)$ is $\log 2 = 1$. The entropy given by the entropy theorem is $h(p) = -p\log p - (1-p)\log(1-p)$, which depends strongly on the value of p, for it takes into account the frequency with which strings occur.

The concept of Ziv-entropy is based on the observation that strings with too small frequency of occurrence can be eliminated by making a small limiting density of changes. The natural distance concept for this idea is $\bar{d}(x, y)$, defined as before by

$$\bar{d}(x, y) = \limsup_{n\to\infty} d_n(x_1^n, y_1^n),$$

where

$$d_n(x_1^n, y_1^n) = \frac{1}{n}\sum_{i=1}^{n} d(x_i, y_i)$$

is per-letter Hamming distance. The *Ziv-entropy* of a sequence x is denoted by $H(x)$ and is defined by

$$H(x) = \lim_{\epsilon\to 0} \inf_{\bar{d}(x,y)<\epsilon} h(y).$$

Ziv established the LZ convergence theorem by proving the following two theorems, which also have independent interest.

Theorem II.2.2 (The LZ upper bound.)

$$\limsup_{n\to\infty} \frac{C(x_1^n)\log n}{n} \leq H(x), \quad x \in A^\infty.$$

Theorem II.2.3 (The Ziv-entropy theorem.)
If μ is an ergodic process with entropy h then $H(x) \leq h$, almost surely.

These two theorems show that $\limsup_n (1/n)C(x_1^n)\log n \leq h$, almost surely, which, combined with the fact it is not possible to beat entropy in the limit, Theorem II.1.2, leads immediately to the LZ convergence theorem, Theorem II.2.1.

The proof of the upper bound, Theorem II.2.2, will be carried out via three lemmas. The first gives a simple bound (which is useful in its own right), the second gives topological entropy as an upper bound, and the third establishes a $\bar{d}$-perturbation bound. The first lemma obtains a crude upper bound by a simple worst-case analysis.

Lemma II.2.4 (The crude bound.)
There exists $\delta_n \to 0$ such that

$$\frac{C(x_1^n)\log n}{n} \leq (1+\delta_n)\log|A|, \quad x \in A^\infty.$$

Proof. The most words occur in the case when the LZ parsing contains all the words of length 1, all the words of length 2,..., up to some length, say m, that is, when

$$n = \sum_{j=1}^{m} j|A|^j \text{ and } C = \sum_{j=1}^{m} |A|^j.$$

In this case, $n \sim (m+1)|A|^{m+1}/(|A|-1)$ and $C \sim |A|^{m+1}/(|A|-1)$, and hence $C\log n \sim n\log|A|$.

To establish the general case in the form stated, first choose m such that

$$\sum_{j=1}^{m} j|A|^j \leq n < \sum_{j=1}^{m+1} j|A|^j.$$

The desired upper bound then follows from the bound $\sum_1^m t^j \leq t^{m+1}/(t-1)$, and the bounds

$$\left(\frac{t}{t-1}\right) mt^m \left(1 - \frac{1}{m(t-1)}\right) \leq \sum_1^m jt^j \leq \left(\frac{t}{t-1}\right) mt^m, \tag{2}$$

all of which are valid for $t > 1$ and are simple consequences of the well-known formula $\sum_1^m t^j = (t^{m+1} - t)/(t-1)$. □

The second bounding result extends the crude bound by noting that short words don't matter in the limit, and hence the rate of growth of the number of words of length k, as $k \to \infty$, that is, the topological entropy, must give an upper bound.

Lemma II.2.5 (The topological entropy bound.)
For each $x \in A^\infty$, there exists $\delta_n \to 0$, such that

$$\frac{C(x_1^n)\log n}{n} \leq (1+\delta_n)h(x).$$

Proof. Let $h_k(x) = (1/k)\log|\mathcal{U}_k(x)|$, for $k \geq 1$, and choose m such that

$$\sum_{j=1}^{m} j2^{jh_j(x)} \leq n < \sum_{j=1}^{m+1} j2^{jh_j(x)}.$$

The bounds (2), together with the fact that

$$h_j(x) - \epsilon_j \leq h(x) \leq h_j(x) + \epsilon_j,$$

where $\epsilon_j \to 0$ as $j \to \infty$, yield the desired result. □

The third lemma asserts that the topological entropy of any sequence in a small $\bar{d}$-neighborhood of x is (almost) an upper bound on code performance. At first glance the result is quite unexpected, for the LZ parsing may be radically altered by even a few changes in x.

Lemma II.2.6 (The perturbation bound.)
Given x and $\epsilon > 0$, there is a $\delta > 0$ such that if $\bar{d}(x, y) < \delta$, then

$$\limsup_{n\to\infty} \frac{C(x_1^n)\log n}{n} \leq h(y) + \epsilon.$$

Proof. Define the word-length function by $\ell(w) = j$, for $w \in A^j$. Fix $\delta > 0$ and suppose $\bar{d}(x, y) < \delta$. Let $x = w(1)w(2)\ldots$ be the LZ parsing of x and parse y as $y = v(1)v(2)\ldots$, where $\ell(v(i)) = \ell(w(i))$, $i = 1, 2, \ldots$. The word $w(i)$ will be called *well-matched* if

$$d_{\ell(w(i))}(w(i), v(i)) < \sqrt{\delta},$$

otherwise *poorly-matched.* Let $C_1(x_1^n)$ be the number of poorly-matched words and let $C_2(x_1^n)$ be the number of well-matched words in the LZ parsing of x_1^n.

By the Markov inequality, the poorly-matched words cover less than a limiting $\sqrt{\delta}$-fraction of x, and hence it can be supposed that for all sufficiently large n, the total length of the poorly-matched words in the LZ parsing of x_1^n is at most $n\sqrt{\delta}$. Thus, Lemma II.2.4 gives the bound

$$C_1(x_1^n) \le (1+\alpha_n)\frac{n\sqrt{\delta}}{\log n\sqrt{\delta}} \log|A| \tag{3}$$

where $\lim_n \alpha_n = 0$.

Now consider the well-matched words. For each k, let $G_k(x)$ be the set of well-matched $w(i)$ of length k, and let $G_k(y)$ be the set of all $v(i)$ for which $w(i) \in G_k(x)$. The cardinality of $G_k(y)$ is at most $2^{kh_k(y)}$, since this is the total number of words of length k in y. Since $d_k(w(i), v(i)) < \sqrt{\delta}$, for $w(i) \in G_k(x)$, the blowup-bound lemma, Lemma I.7.5, together with the fact that $h_k(y) \to h(y)$, yields

$$|G_k(x)| \le 2^{k(h(y)+f(\delta))}$$

where $f(\delta) \to 0$ as $\delta \to 0$. Lemma II.2.5 then gives

$$C_2(x_1^n) \le (1+\beta_n)\frac{n}{\log n}(h(y)+f(\delta)),$$

where $\lim_n \beta_n = 0$. This bound, combined with (3) and the fact that $f(\delta) \to 0$, implies the desired result, completing the proof of Lemma II.2.6. □

The LZ upper bound, Theorem II.2.2, follows immediately from Lemma II.2.6. □

The desired inequality, $H(x) \le h$, almost surely, is an immediate consequence of the following lemma.

Lemma II.2.7

Let μ be an ergodic process with entropy h. For any frequency-typical x and $\epsilon > 0$, there is a sequence y such that $\bar{d}(x, y) < \epsilon$ and $h(y) < h + \epsilon$.

Proof. Fix $\epsilon > 0$, and choose k so large that there is a set $\mathcal{T}_k \subseteq A^k$ of cardinality less than $2^{n(h+\epsilon)}$ and measure more than $1-\epsilon$. Fix a frequency-typical sequence x. The idea for the proof is that $x_i^{i+k-1} \in \mathcal{T}_k$ for all but a limiting $(1-\epsilon)$-fraction of indices i, hence the same must be true for nonoverlapping k-blocks for at least one shift $s \in [0, k-1]$. The resulting nonoverlapping blocks that are not in $\mathcal{T}_k$ can then be replaced by a single fixed block to obtain a sequence close to x with topological entropy close to h.

To make the preceding argument rigorous, first note that

$$\lim_{n\to\infty} \frac{|\{i \in [0, n-1]\colon\ x_{i+1}^{i+k} \in \mathcal{T}_k\}|}{n} > 1-\epsilon,$$

since x was assumed to be frequency-typical. Thus there must be an integer $s \in [0, k-1]$ such that

$$\liminf_{n\to\infty} \frac{|\{i \in [0, n-1]\colon\ x_{in+s+1}^{in+s+k} \in \mathcal{T}_k\}|}{n} > 1-\epsilon,$$

which can be expressed by saying that x is the concatenation

$$x = uw(1)w(2)\ldots,$$

where the initial block u has length s, each $w(i)$ has length k, and

$$\liminf_{n\to\infty} \frac{|\{i \leq n \colon w(i) \in \mathcal{T}_k\}|}{n} > 1 - \epsilon. \tag{4}$$

Fix $a \in A$ and let a^k denote the sequence of length k, all of whose members are a. The sequence y is defined as the concatenation

$$y = uv(1)v(2)\ldots,$$

where

$$v(i) = \begin{cases} w(i) & \text{if } w(i) \in \mathcal{T}_k \\ a^k & \text{otherwise.} \end{cases}$$

Condition (4) and the definition of y guarantee that $\bar{d}(x, y) < \epsilon$.

If M is large relative to k, a member of the m-block universe $\mathcal{U}_M(y)$ of y has the form

$$qv(j)v(j+1)\cdots v(j+m-1)r,$$

where q and r have length at most k, and $(M-2k)/k \leq m \leq M/k$. Since each $v(i) \in \mathcal{T}_k \cup \{a^k\}$, a set of cardinality at most $1 + 2^{k(h+\epsilon)}$, the built-up set bound, Lemma I.7.6, implies that

$$h_M(y) \leq h + \epsilon + \delta_M,$$

where $\delta_M \to 0$, and hence $h(y) \leq h + \epsilon$. This completes the proof of Lemma II.2.7, and, thereby, the proof of the Ziv-entropy theorem, Theorem II.2.3. □

Remark II.2.8

Variations in the parsing rules lead to different forms of the LZ algorithm. One version, which has also been extensively analyzed, defines a word to be new only if it has been seen nowhere in the past (recall that in simple parsing, "new" meant "not seen as a word in the past.") In this alternative version, new words tend to be longer, which improves code performance, but more bookkeeping is required, for now the location of both the start of the initial old word, as well as its length must be encoded. Nevertheless, the rate of growth in the number of new words still determines limiting code length and the upper bounding theorem, Theorem II.2.2, remains true, see Exercise 1.

Remark II.2.9

Another variation, also discussed in the exercises, includes the final symbol of the new word in the next word. This modification, known as the LZW algorithm, produces slightly more rapid convergence and allows some simplification in the design of practical coding algorithms. It is used in several data compression packages.

Remark II.2.10

For computer implementation of LZ-type algorithms some bound on memory growth is needed. One method is to use some form of the LZ algorithm until a fixed number, say C_0, of words have been seen, after which the next word is defined to be the longest block that appears in the list of previously seen words. Another version, called the sliding-window version, starts in much the same way, until it reaches the n_0-th term, then defines the next word as the longest block that appears somewhere in the past n_0 terms. Such algorithms cannot achieve entropy for every ergodic process, but do perform well for processes which have constraints on memory growth, and can be shown to approach optimality as window size or tree storage approach infinity, see [87, 88].

II.2.a Exercises

1. Show that the version of LZ discussed in Remark II.2.8 achieves entropy in the limit. (Hint: if the successive words lengths are $\ell_1, \ell_2, \ldots, \ell_C$, then $\sum \log \ell_i \leq C \log(n/C)$.)

2. Show that the LZW version discussed in Remark II.2.9 achieves entropy in the limit.

3. Show that simple LZ parsing defines a binary tree in which the number of words corresponds to the number of nodes in the tree.

Section II.3 Empirical entropy.

In many coding procedures a sample path x_1^n is partitioned into blocks, then each block is coded separately. The blocks may have fixed length, which can depend on n, or variable length. Furthermore, the coding of a block may depend only on the block, or it may depend on the past or even the entire sample path, e. g., through the empirical distribution of blocks in the sample path. In this section, a beautiful and deep result due to Ornstein and Weiss, [52], about the covering exponent of the empirical distribution of fixed-length blocks will be established. The theorem, here called the empirical-entropy theorem, suggests an entropy-estimation algorithm and a specific coding technique, both of which will also be discussed. Furthermore, the method of proof leads to useful techniques for analyzing variable-length codes, such as the Lempel-Ziv code, as well as other entropy-related ideas. These will be discussed in later sections.

The empirical-entropy theorem has a nonoverlapping-block form which will be discussed here, and an overlapping-block form, see Exercise 1. The nonoverlapping k-block distribution, is defined by

$$q_k(a_1^k|x_1^n) = \frac{|\{i \in [0, m-1]\colon x_{ik+1}^{ik+k} = a_1^k\}|}{m},$$

where $n = km + r$, $0 \leq r < k$. The theorem is concerned with the number of k-blocks it takes to cover a large fraction of a sample path, asserting that eventually almost surely only a small exponential factor more than 2^{kh} such k-blocks is enough, and that a small exponential factor less than 2^{kh} is not enough, provided only that $n \geq 2^{kh}$.

Theorem II.3.1 (The empirical-entropy theorem.)

Let μ be an ergodic measure with entropy $h > 0$. For each $\epsilon > 0$ and each k there is a set $\mathcal{T}_k(\epsilon) \subseteq A^k$ for which $|\mathcal{T}_k(\epsilon)| \leq 2^{k(h+\epsilon)}$, such that for almost every x there is a $K = K(\epsilon, x)$ such that if $k \geq K$ and $n \geq 2^{kh}$, then

(a) $q_k(\mathcal{T}_k(\epsilon)|x_1^n)) > 1 - \epsilon$.

(b) $q_k(B|x_1^n)) < \epsilon$, *for any* $B \subseteq A^k$ *for which* $|B| \leq 2^{k(h-\epsilon)}$.

The theorem is, in essence, an empirical distribution form of entropy as covering exponent, Theorem I.7.4, for the latter says that for given ϵ and all k large enough, there is a set $\mathcal{T}_k(\epsilon) \subseteq A^k$ such that $|\mathcal{T}_k(\epsilon)| \leq 2^{k(h+\epsilon)}$ and $\mu_k(\mathcal{T}_k(\epsilon)) > 1 - \epsilon$, and there is no set $B \subseteq A^k$ such that $|B| \leq 2^{k(h-\epsilon)}$ and $\mu(B) \geq \epsilon$.

The ergodic theorem implies that for fixed k and mixing μ, the empirical distribution $q_k(\cdot|x_1^n)$ is eventually almost surely close to the true distribution, μ_k, hence, in particular, for fixed k, the empirical and true distributions will eventually almost surely have approximately the same covering properties. The surprising aspect of the empirical-entropy theorem is its assertion that even if k is allowed to grow with n, subject only to the condition that $k \leq (1/h)\log n$, the empirical measure $q_k(\cdot|x_1^n)$ eventually almost surely has the same covering properties as μ_k, in spite of the fact that it may not otherwise be close to μ_k in variational or even $\bar{d}_k$-distance. For example, for unbiased coin-tossing the measure $q_k(\cdot|x_1^n)$ is not close to μ_k in variational distance for the case $n = k2^k$, see Exercise 3.

II.3.a Strong-packing.

Part (a) of the empirical-entropy theorem is a consequence of a very general result about parsings of "typical" sample paths. Eventually almost surely, if the short words cover a small enough fraction, then most of the path is covered by words that come from fixed collections of exponential size almost determined by entropy. A fixed-length version of the result is all that is needed for part (a) of the entropy theorem, but the more general result about parsings into variable-length blocks will be developed here, as it will be useful in later sections.

As before, a finite sequence w is called a word, word length is denoted by $\ell(w)$ and a parsing of x_1^n is an ordered collection $\mathcal{P} = \{w(1), w(2), \ldots, w(t)\}$ of words for which $x_1^n = w(1)w(2)\cdots w(t)$.

Fix a sequence of sets $\{\mathcal{T}_k\}$, such that $\mathcal{T}_k \subseteq A^k$, for $k \geq 1$. A parsing $x_1^n = w(1)w(2)\cdots w(t)$ is $(1-\epsilon)$-*built-up from* $\{\mathcal{T}_k\}$, or $(1-\epsilon)$-*packed by* $\{\mathcal{T}_k\}$, if the words that belong to $\cup\mathcal{T}_k$ cover at least a $(1-\epsilon)$-fraction of n, that is,

$$\sum_{w(i)\in\cup\mathcal{T}_k} \ell(w(i)) \geq (1-\epsilon)n.$$

A sequence x_1^n is (K, ϵ)-*strongly-packed by* $\{\mathcal{T}_k\}$ if *any* parsing $x_1^n = w(1)w(2)\cdots w(t)$ for which

$$\sum_{\ell(w(I))<K} \ell(w(i)) \leq \frac{\epsilon n}{2},$$

is $(1-\epsilon)$-built-up from $\{\mathcal{T}_k\}$.

Strong-packing captures the essence of the idea that if the long words of a parsing cover most of the sample path, those long words that come from the fixed collections

$\{\mathcal{T}_k\}$ must cover most of the sample path. The key result is the existence of collections for which the $\mathcal{T}_k$ are of exponential size determined by entropy and such that eventually almost surely x_1^n is almost strongly-packed.

Lemma II.3.2 (The strong-packing lemma.)

Let μ be an ergodic process of entropy h and let ϵ be a positive number. There is an integer $K = K(\epsilon)$ and, for each $k \geq K$, a set $\mathcal{T}_k = \mathcal{T}_k(\epsilon) \subseteq A^k$, such that both of the following hold.

(a) $|\mathcal{T}_k| \leq 2^{k(h+\epsilon)}$ *for* $k \geq K$.

(b) x_1^n *is eventually almost surely* (K, ϵ)*-strongly-packed by* $\{\mathcal{T}_k\}$.

Proof. The idea of the proof is quite simple. The entropy theorem provides an integer m and a set $C_m \subseteq A^m$ of measure close to 1 and cardinality at most $2^{m(h+\delta)}$. By the ergodic theorem, eventually almost surely most indices in x_1^n are starting places of m-blocks from C_m. But if such an x_1^n is partitioned into words then, by the Markov inequality, most of x_1^n must be covered by those words which themselves have the property that most of their indices are starting places of members of C_m. If a word is long enough, however, and most of its indices are starting places of members of C_m, then the word is mostly built-up from C_m, by the packing lemma. The collection $\mathcal{T}_k$ of words of length k that are mostly built-up from C_m has cardinality only a small exponential factor more than $2^{k(h+\delta)}$, by the built-up set lemma.

To make the outline into a precise proof, fix $\epsilon > 0$, and let δ be a positive number to be specified later. The entropy theorem provides an m for which the set

$$C_m = \{a_1^m\colon\ \mu(a_1^m) \geq 2^{-m(h+\delta)}\}$$

has measure at least $1 - \delta^2/4$. For $k \geq m$, let $\mathcal{T}_k$ be the set of sequences of length k that are $(1-\delta)$-built-up from C_m. By the built-up set lemma, Lemma I.7.6, it can be supposed that δ is small enough to guarantee that

$$|\mathcal{T}_k| \leq 2^{k(h+\epsilon)},\ k \geq m.$$

By making δ smaller, if necessary, it can be supposed that $\delta < \epsilon/2$.

It remains to show that eventually almost surely x_1^n is (K, ϵ)-strongly-packed by $\{\mathcal{T}_k\}$, for a suitable K. This is a consequence of the following three observations.

(i) The ergodic theorem implies that for almost every x there is an $N = N(x)$ such that for $n \geq N$, the sequence x_1^n has the property that $x_i^{i+m-1} \in C_m$ for at least $(1-\delta^2/2)n$ indices $i \in [1, n-m+1]$, that is, x_1^n is $(1-\delta^2/2)$-strongly-covered by C_m.

(ii) If x_1^n is $(1-\delta^2/2)$-strongly-covered by C_m and parsed as $x_1^n = w(1)w(2)\cdots w(t)$, then, by the Markov inequality, the words $w(i)$ that are not $(1-\delta/2)$-strongly-covered by C_m cannot have total length more than $\delta n \leq \epsilon n/2$.

(iii) If $w(i)$ is $(1-\delta/2)$-strongly-covered by C_m, and if $\ell(w(i)) \geq 2/\delta$, then, by the packing lemma, Lemma I.3.3, $w(i)$ is $(1-\delta)$-packed by C_m, that is, $w(i) \in \mathcal{T}_{\ell(w(i))}$.

From the preceding it is enough to take $K \geq 2/\delta$, for if this is so, if $n \geq N(x)$, and if the parsing $x_1^n = w(1)w(2)\cdots w(t)$ satisfies

$$\sum_{\ell(w(i))<K} \ell(w(i)) \leq \frac{\epsilon n}{2},$$

then the set of $w(i)$ for which $\ell(w(i)) \geq K$ and $w(i) \in \cup \mathcal{T}_k$ must have total length at least $(1-\epsilon)n$. This completes the proof of Lemma II.3.2. □

II.3.b Proof of the empirical-entropy theorem.

Proof of part (a).

Fix $\epsilon > 0$ and let $\delta < 1$ be a positive number to be specified later. The strong-packing lemma provides K and $\mathcal{T}_k \subset A^k$, such that $|\mathcal{T}_k| \leq 2^{k(h+\epsilon)}$ for $k \geq K$, and such that x_1^n is (K, δ)-strongly-packed by $\{\mathcal{T}_k\}$, eventually almost surely. Fix x such that x_1^n is (K, δ)-strongly-packed by $\{\mathcal{T}_k\}$ for $n \geq N(x)$, and choose $K(x) \geq K$ such that if $k \geq K(x)$ and $n \geq 2^{kh}$, then $n \geq N(x)$ and $k \leq \delta n$.

Fix $k \geq K(x)$ and $n \geq 2^{kh}$, and suppose

$$x_1^n = w(1)w(2)\cdots w(t)w(t+1),$$

where $\ell(w(i)) = k$, $i \leq t$, and $\ell(w(t+1)) < k$. All the blocks, except possibly the last one, are longer than K, while the last one has length less than δn, since $k \leq \delta n$.. The definition of strong-packing implies that

$$(n-k+1)q_k(\mathcal{T}_k|x_1^n) \geq (1-2\delta)n,$$

since the left side is just the fraction of x_1^n that is covered by members of $\mathcal{T}_k$. Thus, dividing by $n-k+1$ produces

$$q_k(\mathcal{T}_k|x_1^n) \geq (1-2\delta)(1-\delta) \geq (1-\epsilon),$$

for suitable choice of δ. This completes the proof of part (a) of the empirical-entropy theorem. □

Proof of part (b).

An informal description of the proof will be given first. Suppose x_1^n is too-well covered by a too-small collection B of k-blocks, for some k. A simple two-part code can be constructed that takes advantage of the existence of such sets B. The first part is an encoding of some listing of B; this contributes asymptotically negligible length if $|B|$ is exponentially smaller than 2^{kh} and $n \geq 2^{kh}$. The second part encodes successive k-blocks by giving their index in the listing of B, if the block belongs to B, or by applying a fixed good k-block code if the block does not belong to B. If B is exponentially smaller than 2^{kh} then fewer than kh bits are needed to code each block in B, so that if B covers too much the code will beat entropy.

The good k-code used on the blocks that are not in B comes from part (a) of the empirical-entropy theorem, which supplies for each k a set $\mathcal{T}_k \subset A^k$ of cardinality roughly 2^{kh}, such that eventually almost surely most of the k-blocks in an n-length sample path belong to $\mathcal{T}_k$, provided only that $n \geq 2^{kh}$. If a block belongs to $\mathcal{T}_k - B$ then its index in some ordering of $\mathcal{T}_k$ is transmitted; this requires roughly hk-bits. If the block does not belong to $\mathcal{T}_k \cup B$ then it is transmitted term by term using some fixed 1-block code; since such blocks cover at most a small fraction of the sample path this contributes little to overall code length.

To proceed with the rigorous proof, fix $\epsilon > 0$ and let δ be a positive number to be specified later. Part (a) of the theorem provides a set $\mathcal{T}_k \subset A^k$ of cardinality at most $2^{k(h+\delta)}$, for each k, and, for almost every x, an integer $K(x)$ such that if $k \geq K(x)$ and $n \geq 2^{kh}$ then $q_k(\mathcal{T}_k|x_1^n) \geq 1 - \delta$.

Let K be a positive integer to be specified later and for $n \geq 2^{Kh}$ let $\mathcal{B}(n)$ be the set of x_1^n for which there is some k in the interval $[K, (\log n)/h]$ for which there is a set $B \subset A^k$ such that the following two properties hold.

(i) $q_k(\mathcal{T}_k|x_1^n) \geq 1 - \delta$.

(ii) $|B| \leq 2^{k(h-\epsilon)}$ and $q_k(B|x_1^n) \geq \epsilon$.

By using the suggested coding argument it will be shown that if δ is small enough and K large enough then

$$x_1^n \notin \mathcal{B}(n), \text{ eventually almost surely.}$$

This fact implies part (b) of the empirical-entropy theorem. Indeed, the definition of $\{\mathcal{T}_k\}$ implies that for almost every x there is an integer $K(x) \geq K$ such that

$$q_k(\mathcal{T}_k|x_1^n) \geq 1 - \delta, \quad \text{for } K(x) \leq k \leq \frac{\log n}{h},$$

so that if $K(x) \leq k \leq (\log n)/h$ then either $x_1^n \in \mathcal{B}(n)$, or the following holds.

$$\text{(1)} \qquad \text{If } B \subseteq A^k \text{ and } |B| \leq 2^{k(h-\epsilon)}, \text{ then } q_k(B|x_1^n) < \epsilon.$$

If $x_1^n \notin \mathcal{B}(n)$, eventually almost surely, then for almost every x there is an $N(x)$ such that $x_1^n \notin \mathcal{B}(n)$, for $n \geq N(x)$, so that if k is enough larger than $K(x)$ to guarantee that $2^{kh} \geq N(x)$, then property (1) must hold for all $n \geq 2^{kh}$, which implies part (b) of the empirical-entropy theorem. □

It will be shown that the suggested code C_n beats entropy on $\mathcal{B}(n)$ for all sufficiently large n, if δ is small enough and K is large enough, which establishes that $x_1^n \notin \mathcal{B}(n)$, eventually almost surely, since no prefix-code sequence can beat entropy infinitely often on a set of positive measure.

Several auxiliary codes are used in the formal definition of C_n. Let $\lceil\cdot\rceil$ denote the upper integer function. A faithful single letter code, say, $F\colon A \mapsto \{0, 1\}^{\lceil \log |A| \rceil}$ is needed along with its extension to length $m \geq 1$, defined by

$$F_m(x_1^m) = F(x_1)F(x_2)\dots F(x_m).$$

Also needed for each k is a fixed-length faithful coding of $\mathcal{T}_k$, say

$$G_k\colon \mathcal{T}_k \mapsto \{0, 1\}^{\lceil k(h+\delta) \rceil},$$

and a fixed-length faithful code for each $B \subset A^k$ of cardinality at most $2^{k(h-\epsilon)}$, say,

$$M_{k,B}\colon B \mapsto \{0, 1\}^{\lceil k(h-\epsilon) \rceil}.$$

Finally, for each k and $B \subseteq A^k$, let $\phi_k(B)$ be the concatenation in some order of $\{F_k(x_1^k)\colon x_1^k \in B\}$, and let $\mathcal{E}$ be an Elias prefix code on the integers, that is, a prefix code such that the length of $\mathcal{E}(n)$ is $\log n + o(\log n)$.

The rigorous definition of C_n is as follows. If $x_1^n \notin \mathcal{B}(n)$ then $C_n(x_1^n) = 0F_n(x_1^n)$. If $x_1^n \in \mathcal{B}(n)$, then an integer $k \in [K, (\log n)/h]$ and a set $B \subset A^k$ of cardinality at most $2^{k(h-\epsilon)}$ are determined such that $q_k(B|x_1^n) \geq \epsilon$, and x_1^n is parsed as

$$x_1^n = w(1)w(2)\cdots w(t)w(t+1),$$

where $w(i)$ has length k for $i \leq t$, and $w(t+1)$ has length $r = [0, k)$. The code is defined as the concatenation

$$C(x_1^n) = 10^k 1\mathcal{E}(|B|)\phi_k(B)v(1)v(2)\cdots v(t)v(t+1),$$

where

$$v(i) = \begin{cases} 00M_{k,B}(w(i)) & w(i) \in B \\ 01G_k(w(i)) & w(i) \in \mathcal{T}_k - B \\ 11F_k(w(i)) & w(i) \notin \mathcal{T}_k \cup B, i \leq t \\ 11F_r(w(t+1) & i = t+1. \end{cases} \tag{2}$$

The code is a prefix code, since reading from the left, the first bit tells whether $x_1^n \in \mathcal{B}(n)$, and, given that $x_1^n \in \mathcal{B}(n)$, the block 0^k1 determines k. Once k is known, $\mathcal{E}(|B|)$ determines the size of B, and $\phi_k(B)$ determines B. The two-bit header on each $v(i)$ specifies whether $w(i)$ belongs to B, to $\mathcal{T}_k - B$, or to neither of these two sets, from which $w(i)$ can then be determined by using the appropriate inverse, since each of the codes $M_{k,B}$, G_k, F_k, and F_r are now known.

For $x_1^n \in \mathcal{B}(n)$, the principal contribution to total code length $\mathcal{L}(x_1^n) = \mathcal{L}(x_1^n|C_n)$ comes from the encoding of the $w(i)$ that belong to $\mathcal{T}_k \cup B$, a fact stated as the following lemma.

Lemma II.3.3

$$\mathcal{L}(x_1^n) \leq tkq_k(h-\epsilon) + tk(1-q_k)(h+\delta) + (\delta + \alpha_K)O(n), \quad x_1^n \in \mathcal{B}(n), \tag{3}$$

where $\alpha_K \to 0$ as $K \to \infty$, and $q_k = q_k(B|x_1^n)$, for the block length k and set $B \subset A^k$ used to define $C_n(x_1^n)$.

The lemma is sufficient to complete the proof of part (b) of the empirical-entropy theorem. This is because

$$tkq_k(h-\epsilon) + tk(1-q_k)(h+\delta) \leq n(h + \delta - \epsilon(\epsilon+\delta)),$$

since $q_k \geq \epsilon$, for $x_1^n \in \mathcal{B}(n)$, so that if K is large enough and δ small enough then

$$\mathcal{L}(x_1^n) \leq n(h - \epsilon^2/2), \quad x_1^n \in \mathcal{B}(n),$$

for all sufficiently large n. But this implies that $x_1^n \notin \mathcal{B}(n)$, eventually almost surely, since no sequence of codes can beat entropy infinitely often on a set of positive measure, and as noted earlier, this, in turn, implies part (b) of the empirical-entropy theorem. □

Proof of Lemma II.3.3.

Ignoring for the moment the contribution of the headers used to describe k and B and to tell which code is being applied to each k-block, as well as the extra bits that might be needed to round $k(h-\epsilon)$ and $k(h+\delta)$ up to integers, the number of bits used to encode the blocks in $\mathcal{T}_k \cup B$ is given by

$$tq_k(B|x_1^n)k(h-\epsilon) + \left(t - tq_k(B|x_1^n)\right) k(h+\delta),$$

since there are $tq_k(B|x_1^n)$ blocks that belong to B, each of which requires $k(h-\epsilon)$ bits, and $(t - tq_k(B|x_1^n))$ blocks in $\mathcal{T}_k - B$, each of which requires $k(h+\delta)$ bits. This gives the dominant terms in (3).

The header $10^k1\mathcal{E}(|B|)$ that describes k and the size of B has length

$$2 + k + \log|B| + o(\log|B|) \tag{4}$$

which is certainly $o(n)$, since $\log|B| \leq k(h-\epsilon)$ and $k \leq (\log n)/n$. It takes at most $k(1+\log|A|)$ bits to encode each $w(i) \notin \mathcal{T}_k \cup B$ as well as to encode the final r-block if $r > 0$, and there are at most $1 + \delta t$ such blocks, hence at most $\delta O(n)$ bits are needed to encode them. Adding this to the $o(n)$ bound of (4) then yields the $\delta O(n)$ term of the lemma.

The encoding $\phi_k(B)$ of B takes $k\lceil\log|A|\rceil|B| \leq k(1+\log|A|)2^{k(h-\epsilon)}$ bits, which is, in turn, upper bounded by

$$K(1+\log|A|)2^{-K\epsilon}n, \tag{5}$$

provided $K \geq (\epsilon \ln 2)^{-1}$, since $k \geq K$ and $n \geq 2^{kh}$. There are at most $t+1$ blocks, each requiring a two-bit header to tell which code is being used, as well as a possible extra bit to round up $k(h-\epsilon)$ and $k(h+\delta)$ to integers; together these contribute at most $3(t+1)$ bits to total length, a quantity which is at most $6n/K$, since $t \leq n/k$, and $n \geq k \geq K$. Thus with

$$\alpha_K = K(1+\log|A|)2^{-K\epsilon} + \frac{6}{K} = o(K),$$

the encoding of $\phi_k(B)$ and the headers contributes at most $n\alpha_K$ to total length. This completes completes the proof of Lemma II.3.3, thereby establishing part (b) of the empirical-entropy theorem. □

Remark II.3.4

The original proof, [52], of part (b) of the empirical-entropy theorem used a counting argument to show that the cardinality of the set $\mathcal{B}(n)$ must eventually be smaller than 2^{nh} by a fixed exponential factor. The argument parallels the one used to prove the entropy theorem and depends on a count of the number of ways to select the bad set B. The coding proof given here is simpler and fits in nicely with the general idea that coding cannot beat entropy in the limit.

II.3.c Entropy estimation.

A problem of interest is the entropy-estimation problem. Given a sample path $x_1, x_2, \ldots, x_n$ from an unknown ergodic process μ, the goal is to estimate the entropy h of μ. A simple procedure is to determine the empirical distribution of nonoverlapping k-blocks, and then take $H(q_k)/k$ as an estimate of h. If k is fixed and $n \to \infty$, then $H(q_k)/k$ will converge almost surely to $H(\mu_k)/k$, which, in turn, converges to h as

$k \to \infty$, at least if it is assumed that μ is totally ergodic. Thus, at least for a totally ergodic μ, there is some choice of $k = k(n)$ as a function of n for which the estimate will converge almost surely to h.

At first glance, the choice of $k(n)$ would appear to be very dependent on the measure μ, because, for example, there is no universal choice of $k = k(n)$ for which $k(n) \to \infty$ for which the empirical distribution $q_k(\cdot|x_1^n)$ is close to the true distribution μ_k for *every* mixing process μ. The empirical entropy theorem does imply a universal choice for the entropy-estimation problem, for example, the choice $k(n) \sim \log n$ works for any binary ergodic process. The general result may be stated as follows.

Theorem II.3.5 (The entropy-estimation theorem.)

If μ is an ergodic measure of entropy $h > 0$, if $k(n) \to \infty$, as $n \to \infty$, and if $k(n) \leq (1/h)\log n$, then

$$\lim_{n\to\infty} \frac{1}{k(n)} H(q_{k(n)}(\cdot|x_1^n)) = h, \ a.s. \tag{6}$$

In particular, if $k(n) \sim \log_{|A|} n$ then (6) holds for any ergodic measure μ with alphabet A, while if $k(n) \sim \log\log n$, then it holds for any finite-alphabet ergodic process.

Proof. Fix $\epsilon > 0$ and let $\{\mathcal{T}_k(\epsilon) \subseteq A^k : n \geq 1\}$ satisfy part (a) of the empirical-entropy theorem. Let U_k be the set of all $a_1^k \in \mathcal{T}_k(\epsilon)$ for which $q_k(a_1^k|x_1^n) < 2^{-k(h+2\epsilon)}$, so that for all large enough k and all $n \geq 2^{kh}$,

$$q_k(U_k|x_1^n) \leq |\mathcal{T}_k(\epsilon)| 2^{-k(h+2\epsilon)} \leq 2^{k(h+\epsilon)} 2^{-k(h+2\epsilon)} = 2^{-\epsilon k}. \tag{7}$$

Next let V_k be the set of all a_1^k for which $q_k(a_1^k|x_1^n) > 2^{-k(h-2\epsilon)}$, and note that $|V_k| \leq 2^{k(h-2\epsilon)}$. Part (b) of the empirical-entropy theorem implies that for almost every x there is a $K(x)$ such that

$$q_k(V_k|x_1^n) \leq \epsilon, \quad \text{for } k \geq K(x), n \geq 2^{kh}.$$

This bound, combined with the bound (7) and part (a) of the empirical-entropy theorem, implies that for almost every x, there is a $K_1(x)$ such that

$$q_k(G_k|x_1^n) \geq 1 - 3\epsilon, \quad \text{for } k \geq K_1(x), n \geq 2^{kh},$$

where $G_k = \mathcal{T}_k(\epsilon) - U_k - V_k$.

In summary, the set of a_1^k for which

$$2^{-k(h+2\epsilon)} \leq q_k(a_1^k|x_1^n) \leq 2^{-k(h-2\epsilon)}$$

has $q_k(\cdot|x_1^n)$ measure at least $1 - 3\epsilon$. The same argument that was used to show that entropy-rate is the same as decay-rate entropy, Theorem I.6.9, can now be applied to complete the proof of Theorem II.3.5. □

Remark II.3.6

The entropy-estimation theorem is also true with the overlapping block distribution in place of the nonoverlapping block distribution, see Exercise 2.

II.3.d Universal coding.

A *universal code* for the class of ergodic processes is a faithful code sequence $\{C_n\}$ such that for any ergodic process μ

$$\lim_{n\to\infty} \frac{\ell(x_1^n)}{n} = h, \text{ almost surely,}$$

where h is the entropy of μ. Such a code sequence was constructed in Section II.1.a, see the proof of Theorem II.1.1, and it was also shown that the Lempel-Ziv algorithm gives a universal code. The empirical-entropy theorem provides another way to construct a universal code. The steps of the code are as follows.

Step 1. Partition x_1^n into blocks of length $k = k(n) \sim (1/2)\log_{|A|} n$.

Step 2. Transmit a list $\mathcal{L}$ of these k-blocks in order of decreasing frequency of occurrence in x_1^n.

Step 3. Encode successive k-blocks in x_1^n by giving the index of the block in the code book $\mathcal{L}$.

Step 4. Encode the final block, if k does not divide n, with a per-symbol code.

The number of bits needed to transmit the list $\mathcal{L}$ is short, relative to n, since $k \sim (1/2)\log_{|A|} n$. High frequency blocks appear near the front of the code book and therefore have small indices, so that the empirical-entropy theorem guarantees good performance, since the number of bits needed to transmit an index is of the order of magnitude of the logarithm of the index.

To make this sketch into a rigorous construction, fix $\{k(n)\}$ for which $k(n) \sim (1/2)\log n$. (For simplicity it is assumed that the alphabet is binary.) The code word $C_n(x_1^n)$ is a concatenation of two binary sequences, a header of fixed length $m = k2^k$, where $k = k(n)$, followed by a sequence whose length depends on x_1^n. The header b_1^m is a concatenation of all the members of $\{0, 1\}^k$, subject only to the rule that a_1^k precedes $\tilde{a}_1^k$ whenever

$$q_k(a_1^k|x_1^n) > q_k(\tilde{a}_1^k|x_1^n).$$

The sequence $\{v(j) = b_{jk+1}^{jk+k}\colon\ j = 0, 1, \ldots, 2^k - 1\}$ is called the *code book*.

Suppose $n = tk + r,\ r \in [0, k)$ and $x_1^n = w(1)w(2)\cdots w(t)v$, where each $w(i)$ has length k. For $1 \le i \le t$ and $0 \le j < 2^k$, define the address function by the rule

$$\mathcal{A}(w(i)) = j, \quad \text{if } w(i) = v(j),$$

that is, $w(i)$ is the j-th word in the code book. Let $\mathcal{E}$ be a prefix code on the natural numbers such that the length of the word $\mathcal{E}(j)$ is $\log j + o(\log j)$, that is, an Elias code.

Define b_{m+1}^{m+L} to be the concatenation of the $\mathcal{E}(\mathcal{A}(w(i)))$ in order of increasing i, where L is the sum of the lengths of the $\mathcal{E}(\mathcal{A}(w(i)))$. Finally, define b_{m+L+1}^{m+L+r} to be x_{kq+1}^n. The code $C_n(x_1^n)$ is defined as the concatenation

$$C_n(x_1^n) = b_1^m \cdot b_{m+1}^{m+L} \cdot b_{m+L+1}^{m+L+r}.$$

The code length $\mathcal{L}(x_1^n) = m + L + r$, depends on x_1^n, since L depends on the distribution $q_k(\cdot|x_1^n)$, but the use of the Elias code insures that C_n is a prefix code.

The empirical-entropy theorem will be used to show that for any ergodic μ,

$$\lim_{n\to\infty} \frac{\mathcal{L}(x_1^n)}{n} = h, \text{ a. s.,} \tag{8}$$

where h is the entropy of μ. To establish this, let $\{\mathcal{T}_k(\epsilon) \subseteq A^k\}$ be the sequence given by the empirical-entropy theorem. For each n, let G_k denote the first $2^{k(h+\epsilon)}$ members of the code book, where $k = k(n)$. Note that $|G_k| \geq |\mathcal{T}_k(\epsilon)|$ and, furthermore,

$$q_k(G_k|x_1^n) \geq q_k(\mathcal{T}_k(\epsilon)|x_1^n),$$

since G_k is a set of k-sequences of largest $q_k(\cdot|x_1^n)$ probability whose cardinality is at most $2^{k(h+\epsilon)}$.

The empirical-entropy theorem provides, for almost every x, an integer $K(x)$ such that $q_k(G_k|x_1^n) \geq q_k(\mathcal{T}_k(\epsilon)|x_1^n) \geq 1-\epsilon$ for $k \geq K(x)$. Thus, for $k \geq K(x)$, at least a $(1-\epsilon)$-fraction of the binary addresses $\mathcal{E}(\mathcal{A}(w(i)))$ will refer to members of G_k and thus have lengths bounded above by $k(h+\epsilon)+o(\log n)$. For those $w(i) \notin G_k$, the crude bound $k + o(\log n)$ will do, and hence

$$\mathcal{L}(x_1^n) \leq (1-\epsilon)n(h+\epsilon) + \epsilon n + o(\log n),$$

for $k \geq K(x)$. This proves that

$$\limsup_{n\to\infty} \frac{\mathcal{L}(x_1^n)}{n} \leq h, \text{ a. s.}$$

The reverse inequality follows immediately from the fact that too-good codes do not exist, Theorem II.1.2, but it is instructive to note that it follows from the second part of the empirical-entropy theorem. In fact, let B_k be the first $2^{k(h-\epsilon)}$ sequences in the code book for x_1^n. The empirical-entropy theorem guarantees that eventually almost surely, there are at most ϵt indices i for which $w(i) \in B_k$. These are the only k-blocks that have addresses shorter than $k(h-\epsilon)+o(\log n)$, so that

$$\liminf_{n\to\infty} \frac{\mathcal{L}(x_1^n)}{n} \geq (h-\epsilon)(1-\epsilon), \text{ a. s.}$$

Remark II.3.7

The universal code construction is drawn from [49], which also includes universal coding results for coding in which some distortion is allowed. A second, and in some ways, even simpler algorithm which does not require that the code book be listed in any specific order was obtained in [43], and is described in Exercise 4. Another application of the empirical-entropy theorem will be given in the next chapter, in the context of the problem of estimating the measure μ from observation of a finite sample path, see Section III.3.

II.3.e Exercises

1. Show that the empirical-entropy theorem is true with the overlapping block distribution,

$$p_k(a_1^k|x_1^n) = \frac{|\{i \in [1, n-k+1]\colon\ x_i^{i+k-1} = a_1^k\}|}{n-k+1},$$

in place of the nonoverlapping block distribution $q_k(\cdot|x_1^n)$. (Hint: reduce to the nonoverlapping case for some small shift of the sequence.)

2. Show that the entropy-estimation theorem is true with the overlapping block distribution in place of the nonoverlapping block distribution.

3. Show that the variational distance between $q_k(\cdot|x_1^n)$ and μ_k is asymptotically almost surely lower bounded by $(1-e^{-1})$ for the case when $n = k2^k$ and μ is unbiased coin-tossing. (Hint: if M balls are thrown at random into M boxes, then the expected fraction of empty boxes is asymptotic to $(1-e^{-1})$.)

4. Another simple universal coding procedure is suggested by the empirical-entropy theorem. For each $k < n$ construct a code $C_{k,n}$ as follows. Express x_1^n as the concatenation $w(1)w(2)\cdots w(q)v$, of k-blocks $\{w(i)\}$, plus a possible final block v of length less than k. Make a list in some order of the k-blocks that occur. Transmit k and the list, then code successive $w(i)$ by using a fixed-length code; such a code requires at most $k(1+\log|A|)$ bits per word. Append some coding of the final block v. Call this $C_{k,n}(x_1^n)$ and let $\mathcal{L}_k(x_1^n)$ denote the length of $C_{k,n}(x_1^n)$.

   The final code $C_n(x_1^n)$ transmits the shortest of the codes $\{C_{k,n}(x_1^n)\}$, along with a header to specify the value of k. In other words, let $k_{\min}$ be the first value of $k \in [1, n]$ at which $\mathcal{L}_k(x_1^n)$ achieves its minimum. The code $C_n(x_1^n)$ is the concatenation of $\mathcal{E}(k_{\min})$ and $C_{k_{\min},n}(x_1^n)$.

   Show that $\{C_n\}$ is a universal prefix-code sequence.

Section II.4 Partitions of sample paths.

An interesting connection between entropy and partitions of sample paths into variable-length blocks was established by Ornstein and Weiss, [53]. They show that eventually almost surely, for any partition into distinct words, most of the sample path is covered by words that are not much shorter than $(\log n)/h$, and for any partition into words that have been seen in the past, most of the sample path is covered by words that are not much longer than $(\log n)/h$. Their results were motivated by an attempt to better understand the Lempel-Ziv algorithm, which partitions into distinct words, except possibly for the final word, such that all but the last symbol of each word has been seen before.

As in earlier discussions, a *word* w is a finite sequence of symbols drawn from the alphabet A and the length of w is denoted by $\ell(w)$. A sequence x_1^n is said to be parsed (or partitioned) into the (ordered) set of words $\{w(1), w(2), \ldots, w(t)\}$ if it is the concatenation

$$x_1^n = w(1)w(2)\ldots w(t). \tag{1}$$

If $w(i) \neq w(j)$, for $i \neq j$, then (1) is called a parsing (or partition) into *distinct words.* For example,

$$000110110100 = [000]\,[110]\,[1101]\,[00]$$

is a partition into distinct words, while

$$000110110100 = [00]\,[0110]\,[1101]\,[00]$$

is not.

Partitions into distinct words have the asymptotic property that most of the sample path must be contained in those words that are not too short relative to entropy.

Theorem II.4.1 (The distinct-words theorem.)

Let μ be ergodic with entropy $h > 0$, and let $\epsilon > 0$ be given. For almost every $x \in A^\infty$ there is an $N = N(\epsilon, x)$ such that if $n \geq N$ and $x_1^n = w(1)w(2)\ldots w(t)$ is a partition into distinct words, then

$$\sum_{\ell(w(i))<(h+\epsilon)^{-1}\log n} \ell(w(i)) \leq \epsilon n.$$

The (simple) Lempel-Ziv algorithm discussed in Section II.2 parses a finite sequence into distinct words, except, possibly for the final word, $w(C+1)$. If this final word is not empty, it is the same as a prior word, and hence $\ell(w(C+1)) \leq n/2$. Thus the distinct-words theorem implies that, eventually almost surely, most of the sample path is covered by blocks that are at least as long as $(h+\epsilon)^{-1}\log(n/2)$. But the part covered by shorter words contains only a few words, by the crude bound, Lemma II.2.4, so that

$$\limsup_{n\to\infty} \frac{C(x_1^n)\log n}{n} \leq h, \text{ a. s.} \tag{2}$$

To discuss partitions into words that have been seen in the past, some way to get started is needed. For this purpose, let $\mathcal{F}$ be a fixed finite collection of words, called the *start set*. A partition $x_1^n = w(1)w(2)\ldots w(k)$ is called a partition (parsing) into *repeated words,* if each $w(i)$ is either in $\mathcal{F}$ or has been seen in the past, meaning that there is an index $j \leq \ell(w(1)) + \ldots + \ell(w(i-1))$ such that $w(i) = x_j^{j+\ell(w(i))-1}$.

Partitions into repeated words have the asymptotic property that most of the sample path must be contained in words that are not too long relative to entropy.

Theorem II.4.2 (The repeated-words theorem.)

Let μ be ergodic with entropy $h > 0$, and let $0 < \epsilon < h$ and a start set $\mathcal{F}$ be given. For almost every $x \in A^\infty$ there is an $N = N(\epsilon, x)$ such that if $n \geq N$, and $x_1^n = w(1)w(2)\ldots w(t)$ is a partition into repeated words, then

$$\sum_{\ell(w(i))>(h-\epsilon)^{-1}\log n} \ell(w(i)) \leq \epsilon n.$$

It is not necessary to require exact repetition in the repeated-words theorem. For example, it is sufficient if all but the last symbol appears earlier, see Exercise 1. With this modification, the repeated-words theorem applies to the versions of the Lempel-Ziv algorithm discussed in Section II.2, so that eventually almost surely most of the sample path must be covered by words that are no longer than $(h-\epsilon)^{-1}\log n$, which, in turn, implies that

$$\liminf_{n\to\infty} \frac{C(x_1^n)\log n}{n} \geq h, \text{ a. s.} \tag{3}$$

Thus the distinct-words theorem and the modified repeated-words theorem together provide an alternate proof of the LZ convergence theorem, Theorem II.2.1. Of course, as noted earlier, the lower bound (3) is a consequence of fact that entropy is an upper bound, (2), and the fact that too-good codes do not exist, Theorem II.1.2.

The proofs of both theorems make use of the strong-packing lemma, Lemma II.3.2, this time for parsings into words of variable length.

II.4.a Proof of the distinct-words theorem.

The distinct-words theorem is a consequence of the strong-packing lemma, together with some simple facts about partitions into distinct words. The first observation is that the words must grow in length, because there are at most $|A|^k$ words of length k. The proof of this simple fact is left to the reader.

Lemma II.4.3

Given K and $\delta > 0$, there is an N such that if $n \geq N$ and $x_1^n = w(1)w(2)\ldots w(t)$ is a partition into distinct words, then

$$\sum_{\ell(w(i))<K} \ell(w(i)) \leq \delta n.$$

The second observation is that if the distinct words mostly come from sets that grow in size at a fixed exponential rate α, then words that are too short relative to $(\log n)/\alpha$ cannot cover too much.

Lemma II.4.4

For each k, suppose $G_k \subset A^k$ satisfies $|G_k| \leq 2^{\alpha k}$, where α is a fixed positive number, and let $G = \cup_k G_k$. Given $\epsilon > 0$ there is an N such that if $n \geq N$ and $x_1^n = w(1)w(2)\ldots w(t)$ is a partition into distinct words such that

$$\sum_{w(i)\notin G} \ell(w(i)) \leq \epsilon n,$$

then

$$\sum_{\ell(w(i))\leq(1-\epsilon)(\log n)/\alpha} \ell(w(i)) \leq 2\epsilon n.$$

Proof. First consider the case when *all* the words are required to come from G. The fact that the words are distinct then implies that

$$\sum_{\ell(w(i))\leq(1-\epsilon)(\log n)/\alpha} \ell(w(i)) \leq \sum_{k\leq(1-\epsilon)(\log n)/\alpha} k|G_k|.$$

Since $|G_k| \leq 2^{\alpha k}$, the sum on the right is upper bounded by

$$\left(\frac{2^\alpha}{2^\alpha - 1}\right)\left(\frac{(1-\epsilon)\log n}{\alpha}\right)\exp_2((1-\epsilon)\log n)) = O(n^{1-\epsilon}\log n) = o(n),$$

by an application of the standard bound,

$$\sum_1^m jt^j \leq \left(\frac{t}{t-1}\right) mt^m, \ t > 1.$$

The general case follows since it is enough to estimate the sum over the too-short words that belong to G. This proves the lemma. □

To continue with the proof of the distinct-words theorem, let δ be a positive number to be specified later. The strong-packing lemma, Lemma II.3.2, yields an integer K and $\{\mathcal{T}_k \subset A^k : k \geq 1\}$ such that

$$|\mathcal{T}_k| \leq 2^{k(h+\delta)}, \ k \geq K,$$

and such that eventually almost surely x_1^n is (K, δ)-strongly-packed by $\{\mathcal{T}_k\}$, i. e.,

$$\sum_{w(i)\in\cup_{k=K}^{\infty}\mathcal{T}_k} \ell(w(i)) \geq (1-\delta)n,$$

for *any parsing* $\mathcal{P}$ of x_1^n for which

$$\sum_{\ell(w(i))<K} \ell(w(i)) \leq \frac{\delta n}{2}. \tag{4}$$

Suppose x_1^n is (K, δ)-strongly-packed by $\{\mathcal{T}_k\}$. Suppose also that n is so large that, by Lemma II.4.3, property (4) holds for any parsing of x_1^n into distinct words. Thus, given a parsing $x_1^n = w(1)w(2)\dots w(t)$ into distinct words, strong-packing implies that

$$\sum_{w(i)\in\cup_{k=K}^{\infty}\mathcal{T}_k} \ell(w(i)) \geq (1-\delta)n,$$

and therefore Lemma II.4.4 can be applied with $\alpha = h + \delta$ to yield

$$\sum_{\ell(w(i))\leq(1-\delta)(\log n)/(h+\delta)} \ell(w(i)) \leq 2\delta n,$$

provided only that n is large enough. The distinct-words theorem follows, since δ could have been chosen in advance to be so small that

$$\frac{\log n}{h+\epsilon} \leq \frac{(1-2\delta)\log n}{h+\delta}. \qquad \square$$

II.4.b Proof of the repeated-words theorem.

It will be shown that if the repeated-words theorem is false, then a prefix-code sequence can be constructed which beats entropy by a fixed amount infinitely often on a set of positive measure. The idea is to code each too-long word of a repeated-word parsing by telling where it occurs in the past and how long it is (this is, of course, the basic idea of the version of the Lempel-Ziv code suggested in Remark II.2.8.) If a good code is used on the complement of the too-long words and if the too-long words cover too much, then overall code length will be shorter than entropy allows. As in the proof of the empirical-entropy theorem, the existence of good codes is guaranteed by the strong-packing lemma, this time applied to variable-length parsing.

Throughout this discussion μ will be a fixed ergodic process with positive entropy h and $\epsilon < h$ will be a given positive number. A block of consecutive symbols in x_1^n of length more $(h-\epsilon)^{-1}\log n$ will be said to be *too long*.

The first idea is to merge the words that are between the too-long words. Suppose $x_1^n = w(1)w(2)\dots w(t)$ is some given parsing, for which s of the $w(i)$ are too long. Label these too-long words in increasing order of appearance as

$$V(1), V(2), \dots, V(s).$$

Let u_1 be the concatenation of all the words that precede $V(1)$, for $1 < i < s+1$, let u_i be the concatenation of all the words that come between $V(i-1)$ and $V(i)$, and let

u_{s+1} be the concatenation of all the words that follow $V(s)$. In this way, x_1^n is expressed as the concatenation

$$x_1^n = u_1 V(1) u_2 V(2) \ldots u_s V(s) u_{s+1}.$$

Such a representation is called a *too-long representation* of x_1^n, with the too-long words $\{V(1), \ldots, V(s)\}$ and *fillers* $\{u_1, u_2, \ldots, u_{s+1}\}$.

Let $B(n)$ be the set of all sequences x_1^n for which there is an s and a too-long representation $x_1^n = u_1 V(1) u_2 V(2) \ldots u_s V(s) u_{s+1}$ with the following two properties.

(i) $\sum_j \ell(V(j)) \geq \epsilon n$.

(ii) Each $V(j)$ has been seen in the past.

To say $V(j) = x_{n_j+1}^{n_j+m_j}$ has been seen in the past is to say that there is an index $i \in [0, n_j)$ such that $V(j) = x_{i+1}^{i+m_j}$. Since the start set $\mathcal{F}$ is a fixed finite set, it can be supposed that when n is large enough no too-long word belongs to $\mathcal{F}$, and therefore to prove the repeated-words theorem, it is enough to prove that

$$x_1^n \notin B(n), \quad \text{eventually almost surely.}$$

The idea is to code sequences in $B(n)$ by telling where the too-long words occurred earlier, but to make such a code compress too much a good way to compress the fillers is needed. The strong-packing lemma provides the good codes to be used on the fillers.

Let δ be a positive number to be specified later. An application of the strong-packing lemma, Lemma II.3.2, provides an integer K and for a each $k \geq K$, a set $\mathcal{T}_k \subset A^k$ of cardinality at most $2^{k(h+\delta)}$, such that eventually almost surely x_1^n is (K, δ)-strongly-packed by $\{\mathcal{T}_k\}$. Let $G(n)$ be the set of all x_1^n that are (K, δ)-strongly-packed by $\{\mathcal{T}_k\}$. Since $x_1^n \in G(n)$ eventually almost surely, to complete the proof of the repeated-words theorem it is enough to prove that if K is large enough, then

$$x_1^n \notin B(n) \cap G(n), \quad \text{eventually almost surely.}$$

A code C_n is constructed as follows. Sequences not in $B(n) \cap G(n)$ are coded using some fixed single-letter code on each letter separately. If $x_1^n \in B(n) \cap G(n)$, a too-long representation

$$x_1^n = u_1 V(1) u_2 V(2) \ldots u_s V(s) u_{s+1},$$

is determined for which each too-long $V(j)$ is seen somewhere in its past and the total length of the $\{V(j)\}$ is at least ϵn. The words in the too-long representation are coded sequentially using the following rules.

(a) Each filler $u_j \in \cup_{k=K}^{\infty} \mathcal{T}_k$ is coded by specifying its length and giving its index in the set $\mathcal{T}_k$ to which it belongs.

(b) Each filler $u_j \notin \cup_{k=K}^{\infty} \mathcal{T}_k$ is coded by specifying its length and applying a fixed single-letter code to each letter separately.

(c) Each too-long $V(j)$ is coded by specifying its length and the start position of its earlier occurrence.

An Elias code is used to encode the block lengths, that is, a prefix code $\mathcal{E}$ on the natural numbers such that the length of the code word assign to j is $\log j + o(\log j)$. Two bit headers are appended to each block to specify which of the three types of code is being used. With a one bit header to tell whether or not x_1^n belongs to $B(n) \cap G(n)$, the code becomes a prefix code.

For $x_1^n \in B(n) \cap G(n)$, the principal contribution to total code length $\mathcal{L}(x_1^n) = \mathcal{L}(x_1^n|C_n)$ comes from telling where each $V(j)$ occurs in the past, which requires $\lceil \log n \rceil$ bits for each of the s too-long words, and from specifying the index of each $u_j \in \cup_{k=K}^{\infty} \mathcal{T}_k$, which requires $\ell(u_j)\lceil h + \delta \rceil$ bits per word. This fact is stated in the form needed as the following lemma.

Lemma II.4.5

If $\log K \leq \delta K$, *then*

$$\mathcal{L}(x_1^n) = s \log n + \left(\sum_{u_j \in \cup_{k=K}^{\infty} \mathcal{T}_k} \ell(u_j) \right) (h + \delta) + \delta O(n), \tag{5}$$

uniformly for $x_1^n \in B(n) \cap G(n)$.

The lemma is sufficient to show that $x_1^n \notin B(n) \cap G(n)$, eventually almost surely. The key to this, as well as to the lemma itself, is that the number s of too-long words, while it depends on x_1^n, must satisfy the bound,

$$s \leq \frac{\sum_{i=1}^{s} \ell(V(i))}{\log n} (h - \epsilon), \tag{6}$$

since, by definition, a word is too long if its length is at least $(h - \epsilon)^{-1} \log n$. By assumption, $\sum \ell(V(i) \geq \epsilon n$, so that (5) and (6) yield the code-length bound

$$\mathcal{L}(x_1^n) \leq n(h + \delta - \epsilon(\epsilon + \delta)) + \delta O(n),$$

for $x_1^n \in B(n) \cap G(n)$, and hence if δ is small enough then

$$\mathcal{L}(x_1^n) \leq n(h - \epsilon^2/2), \quad x_1^n \in B(n) \cap G(n),$$

for all sufficiently large n, which, since it is not possible to beat entropy by a fixed amount infinitely often on a set of positive measure, shows that, indeed, $x_1^n \notin B(n) \cap G(n)$, eventually almost surely, completing the proof of the repeated-words theorem. □

Proof of Lemma II.4.5.

It is enough to show that the encoding of the fillers that do not belong to $\cup \mathcal{T}_k$, as well as the encoding of the lengths of all the words, plus the two bit headers needed to tell which code is being used and the extra bits that might be needed to round up $\log n$ and $h + \delta$ to integers, require a total of at most $\delta O(n)$ bits.

For the fillers that do not belong to $\cup \mathcal{T}_k$, first note that there are at most $s + 1$ fillers so the bound

$$s \leq \frac{\sum_{i=1}^{s} \ell(V(i)}{\log n} (h - \epsilon) \leq \frac{n}{\log n} (h - \epsilon),$$

implies that

$$\sum_{\ell(u_j) < K} \ell(u_j) \leq K \left(1 + \frac{n(h - \epsilon)}{\log n} \right),$$

which is $o(n)$ for fixed K. In particular, once n is large enough to make this sum less than $\delta n/2$, then the assumption that $x_1^n \in G(n)$, that is, that x_1^n is (K, δ)-strongly-packed by $\{\mathcal{T}_k\}$, implies that the words longer than K that do not belong to $\cup\mathcal{T}_k$ cover at most a δ-fraction of x_1^n. Since it takes $\ell(u_j)\lceil\log|A|\rceil$ bits to encode a filler $u_j \notin \cup\mathcal{T}_k$, the total number of bits required to encode all such fillers is at most

$$\left(\sum_{u_j\notin\cup_{k=K}^{\infty}\mathcal{T}_k} \ell(u_j)\right)\lceil\log|A|\rceil \leq \delta n\lceil\log|A|\rceil,$$

which is $\delta O(n)$.

Finally, the encoding of the lengths contributes

$$\sum_j \ell(\mathcal{E}(\ell(u_j)) + \sum_j \ell(\mathcal{E}(\ell(V(j)))$$

bits, the dominant terms of which can be expressed as

$$\sum_{\ell(u_j)<K} \log\ell(u_j) + \sum_{\ell(u_j)\geq K} \log\ell(u_j) + \sum_i \log\ell(V(i)).$$

The first sum is at most $(\log K)(s+1)$, which is $o(n)$ for K fixed. Since it can be assumed that the too-long words have length at least K, which can be assumed to be at least e, and since $x^{-1}\log x$ is decreasing for $x \geq e$, the second and third sums together contribute at most

$$\frac{\log K}{K}n$$

to total length, which is δn, since it was assumed that $\log K \leq \delta K$.

Finally, since there are at most $2s+1$ words, the total length contributed by the two bit headers required to tell which type of code is being used plus the possible extra bits needed to round up $\log n$ and $h+\delta$ to integers is upper bounded by $3(2s+1)$ which is $o(n)$ since $s \leq n(h-\epsilon)/(\log n)$.

This completes the proof of Lemma II.4.5. □

Remark II.4.6

The original proof, [53], of the repeated-words theorem parallels the proof of the entropy theorem, using a counting argument to show that the set $B(n) \cap G(n)$ must eventually be smaller than 2^{nh} by a fixed exponential factor. The key bound is that the number of ways to select s disjoint subintervals of $[1, n]$ of length $\ell_j \geq (\log n)/(h-\epsilon)$, is upper bounded by $\exp((h-\epsilon)\sum\ell_j + o(n))$. This is equivalent to showing that the coding of the too-long repeated blocks by telling where they occurred earlier and their lengths requires at most $(h-\epsilon)\sum\ell_j + o(n)$ bits. The coding proof is given here as fits in nicely with the general idea that coding cannot beat entropy in the limit.

II.4.c Exercises

1. Extend the repeated-words theorem to the case when it is only required that all but the last k symbols appeared earlier.

2. Extend the distinct-words and repeated-words theorems to the case where agreement in all but a fixed number of places is required.

3. Carry out the details of the proof that (2) follows from the distinct-words theorem.

4. Carry out the details of the proof that (3) follows from the repeated-words theorem.

5. What can be said about distinct and repeated words in the entropy 0 case?

6. Let μ be the Kolmogorov measure of the binary, equiprobable, i.i.d. process (i. e., unbiased coin-tossing.)

   (a) Show that eventually almost surely, no parsing of x_1^n into repeated words contains a word longer than $4\log n$. (Hint: use Barron's lemma, Lemma II.1.3.)

   (b) Show that eventually almost surely, there are fewer than $n^{1-\epsilon/2}$ words longer than $(1+\epsilon)\log n$ in a parsing of x_1^n into repeated words.

   (c) Show that eventually almost surely, the words longer than $(1+\epsilon)\log n$ in a parsing of x_1^n into repeated words have total length at most $n^{1-\epsilon/2}\log n$.

Section II.5 Entropy and recurrence times.

An interesting connection between entropy and recurrence times for ergodic processes was discovered by Wyner and Ziv, [86], see also the earlier work of Willems, [85]. Wyner and Ziv showed that the logarithm of the waiting time until the first n terms of a sequence x occurs again in x is asymptotic to nh, in probability. This is a sharpening of the fact that the average recurrence time is $1/\mu(x_1^n)$, whose logarithm is asymptotic to nh, almost surely. An almost-sure form of the Wyner-Ziv result was established by Ornstein and Weiss, [53]. These results, along with an application to a prefix-tree problem will be discussed in this section.

The definition of the recurrence-time function is

$$R_n(x) = \min\{m \geq 1 \colon x_{m+1}^{m+n} = x_1^n\}.$$

The theorem to be proved is

Theorem II.5.1 (The recurrence-time theorem.)
For any ergodic process μ with entropy h,

$$\lim_{n\to\infty} \frac{1}{n} \log R_n(x) = h, \quad \textit{almost surely.}$$

Some preliminary results are easy to establish. Define the upper and lower limits,

$$\begin{aligned} \bar{r}(x) &= \limsup_{n\to\infty} \frac{1}{n} \log R_n(x), \\ \underline{r}(x) &= \liminf_{n\to\infty} \frac{1}{n} \log R_n(x). \end{aligned}$$

Since $R_{n-1}(Tx) \leq R_n(x)$, both the upper and lower limits are subinvariant, that is,

$$\bar{r}(Tx) \leq \bar{r}(x), \text{ and } \underline{r}(Tx) \leq \underline{r}(x),$$

hence, both are almost surely invariant. Since the measure is assumed to be ergodic, both upper and lower limits are almost surely constant, and hence, there are constants $\bar{r}$ and $\underline{r}$ such that

$$\bar{r}(x) = \bar{r} \text{ and } \underline{r}(x) = \underline{r}, \text{ almost surely.}$$

Furthermore, $\underline{r} \leq \bar{r}$, so the desired result can be established by showing that $\underline{r} \geq h$ and $\bar{r} \leq h$. The bound $\bar{r} \leq h$ will be proved via a nice argument due to Wyner and Ziv. The bound $\underline{r} \geq h$ was obtained by Ornstein and Weiss by an explicit counting argument similar to the one used to prove the entropy theorem. In the proof given below, a more direct coding construction is used to show that if recurrence happens too soon infinitely often, then there is a prefix-code sequence which beats entropy in the limit, which is impossible, by Theorem II.1.2. Another proof, based on a different coding idea communicated to the author by Wyner, is outlined in Exercise 1.

To establish $\bar{r} \leq h$, fix $\epsilon > 0$ and define

$$D_n = \{x\colon\ R_n(x) > 2^{n(h+\epsilon)}\},\quad n \geq 1.$$

The goal is to show that $x \notin D_n$, eventually almost surely. It is enough to prove this under the additional assumption that x_1^n is "entropy typical." Indeed, if

$$\mathcal{T}_n = \{x\colon\ \mu(x_1^n) \geq 2^{-n(h+\epsilon/2)}\}$$

then $x \in \mathcal{T}_n$, eventually almost surely, so it is enough to prove

$$x \notin D_n \cap \mathcal{T}_n, \text{ eventually almost surely.} \tag{1}$$

To establish this, fix a_1^n and consider only those $x \in D_n$ for which $x_1^n = a_1^n$, that is, the set $D_n(a_1^n) = D_n \cap [a_1^n]$. Note, however, that if $x \in D_n(a_1^n)$, then $T^j x \notin [a_1^n]$, for $j \in [1, 2^{n(h+\epsilon)} - 1]$, by the definition of $D_n(a_1^n)$, and hence the sequence

$$D_n(a_1^n), T^{-1}D_n(a_1^n), \ldots, T^{-2^{n(h+\epsilon)}+1}D_n(a_1^n)$$

must be disjoint. Since these sets all have the same measure, disjointness implies that

$$\mu(D_n(a_1^n)) \leq 2^{-n(h+\epsilon)}. \tag{2}$$

It is now a simple matter to get from (2) to the desired result, (1). Indeed, the cardinality of the projection onto A^n of $D_n \cap \mathcal{T}_n$ is upper bounded by the cardinality of the projection of $\mathcal{T}_n$, which is upper bounded by $2^{n(h+\epsilon/2)}$, so that (2) yields

$$\mu(D_n \cap \mathcal{T}_n) \leq 2^{n(h+\epsilon/2)} 2^{-n(h+\epsilon)} = 2^{-n\epsilon/2}.$$

The Borel-Cantelli principle implies that $x \notin D_n \cap \mathcal{T}_n$, eventually almost surely, so the iterated almost sure principle yields the desired result (1).

The inequality $\underline{r} \geq h$ will be proved by showing that if it is not true, then eventually almost surely, sample paths are mostly covered by long disjoint blocks which are repeated too quickly. A code that compresses more than entropy can then be constructed by specifying where to look for the first later occurrence of such blocks.

To develop the covering idea suppose $\underline{r} < h - \epsilon$, where $\epsilon > 0$. A block x_s^t of consecutive symbols in a sequence x_1^n is said to *recur too soon in* x_1^n if $x_s^t = x_{s+k}^{t+k}$ for some k in the interval

$$[1, 2^{h(t-s+1)}),$$

for which $s+k \leq n$. The least such k is called the *distance from* x_s^t *to its next occurrence in* x_1^n.

Let m be a positive integer and δ be a positive number, both to be specified later. For $n \geq m$, a concatenation

$$x_1^n = u_1 V(1) u_2 V(2) \cdots u_J V(J) u_{J+1},$$

is said to be a (δ, m)*-too-soon-recurrent representation of* x_1^n, if the following hold.

(a) Each $V(j)$ has length at least m and recurs too soon in x_1^n.

(b) The sum of the lengths of the filler words u_j is at most $3\delta n$.

Let $G(n)$ be the set of all x_1^n that have a (δ, m)-too-soon-recurrent representation. It will be shown later that a consequence of the assumption that $\underline{r} < h - \epsilon$ is that

(3) $$x_1^n \in G(n), \text{ eventually almost surely.}$$

Here it will be shown how a prefix n-code C_n can be constructed which, for suitable choice of m and δ, compresses the members of $G(n)$ too well for all large enough n.

The code C_n uses a faithful single-letter code $F \colon A \mapsto \{0, 1\}^d$, where $d \leq 2 + \log |A|$, such that each code word starts with a 0. The code also uses an Elias code $\mathcal{E}$, that is, a prefix code on the natural numbers such that $\ell(\mathcal{E}(j)) = \log j + o(\log j)$.

Sequences not in $G(n)$ are coded by applying F to successive symbols. For $x_1^n \in G(n)$ a (δ, m)-too-soon-recurrent representation

$$x_1^n = u_1 V(1) u_2 V(2) \cdots u_J V(J) u_{J+1}$$

is determined and successive words are coded using the following rules.

(i) Each filler word u_j is coded by applying F to its successive symbols.

(ii) Each $V(j)$ is coded using a prefix 1, followed by $\mathcal{E}(\ell(V(j))$, followed by $\mathcal{E}(k_j)$, where k_j is the distance from $V(j)$ to its next occurrence in x_1^n.

The one bit headers on F and in the encoding of each $V(j)$ are used to specify which code is being used. A one bit header to tell whether or not $x_1^n \in G(n)$ is also used to guarantee that C_n is a prefix n-code.

The key facts here are that $\log k_j \leq \ell(V(j))(h - \epsilon)$, by the definition of "too-soon recurrent," and that the principal contribution to total code length is $\sum_j \log k_j \leq n(h-\epsilon)$, a result stated as the following lemma.

Lemma II.5.2

(4) $$\mathcal{L}(x_1^n) \leq n(h - \epsilon) + n(3d\delta + \alpha_m), \quad x_1^n \in G(n),$$

for $n \geq m$, *where* $\alpha_m \to 0$ *as* $m \to \infty$.

The lemma implies the desired result for by choosing δ small enough and m large enough, the lemma yields $\mathcal{L}(x_1^n) \leq n(h - \epsilon/2)$ on $G(n)$, for all large enough n. But too-good codes do not exist, so that $\mu(G(n))$ must go to 0, contradicting the fact (3) that $x_1^n \in G(n)$, eventually almost surely, a fact that follows from the assumption that $\underline{r} \leq h - \epsilon$. Thus $\underline{r} \geq h$, establishing the recurrence-time theorem. □

Proof of Lemma II.5.2.

The Elias encoding of the distance to the next occurrence of $V(j)$ requires

$$\ell(V(j))(h - \epsilon) + o(\ell(V(j))) \tag{5}$$

bits, by the definition of "too-soon recurrent." Summing the first term over j yields the principal term $n(h - \epsilon)$ in (4).

Since each $V(j)$ is assumed to have length at least m, there are at most n/m such $V(j)$ and hence the sum over j of the second term in (5) is upper bounded by $n\beta_m$, where $\beta_m \to 0$ as $m \to \infty$. The Elias encoding of the length $\ell(V(j))$ requires

$$\log \ell(V(j)) + o(\log \ell(V(j)))$$

bits. Summing on $j \leq n/m$, the second terms contribute at most $n\beta_m$ bits while the total contribution of the first terms is at most

$$\sum_j \log \ell(V(j)) \leq n \frac{\log m}{m},$$

since $(1/x) \log x$ is decreasing for $x \geq e$. The one bit headers on the encoding of each $V(j)$ contribute at most n/m bits to total code length.

In summary, the complete encoding of the too-soon recurring words $V(j)$ requires a total of at most $n(h - \epsilon) + n\alpha_m$ bits, where

$$\alpha_m = 2\beta_m + (1 + \log m)/m \to 0, \text{ as } m \to \infty.$$

Finally, the encoding of each filler u_j requires $d\ell(u_j)$ bits, and the total length of the fillers is at most $3\delta n$, by property (a) of the definition of $G(n)$. Thus the complete encoding of all the fillers requires at most $3d\delta n$ bits. This completes the proof of Lemma II.5.2. □

It remains to be shown that $x_1^n \in G(n)$, eventually almost surely, for fixed m and δ. This is a covering/packing argument. To carry it out, assume $\underline{r} \leq h - \epsilon$ and, for each $n \geq 1$, define

$$B_n = \{x \colon R_n(x) \leq 2^{n(h-\epsilon)}\}.$$

Since $\liminf_n R_n(x) = \underline{r}$ there is an M such that

$$\mu(B) > 1 - \delta, \text{ for } B = \bigcup_{n=m}^{n=M} B_n.$$

The ergodic theorem implies that $(1/N) \sum_1^n \chi_B(T^{i-1}x) > 1 - \delta$, eventually almost surely, and hence the packing lemma implies that for almost every x, there is an $N(x) \geq M/\delta$ such that if $n \geq N(x)$, there is an integer $I = I(n, x)$ and a collection $\{[n_i, m_i] \colon i \leq I\}$ of disjoint subintervals of $[1, n]$, each of length at least m and at most M, for which the following hold.

(a) For each $i \leq I$, $x_{n_i}^{m_i} = x_j^{j+m_i-n_i}$ for some $j \in [n_i+1, n_i + 2^{(m_i-n_i+1)(h-\epsilon)}]$.

(b) $\sum_1^I (m_i - n_i + 1) \geq (1-2\delta)n$.

In addition, it can be supposed that $n \geq 2^{M(h-\epsilon)}/\delta$, so that if J is the largest index i for which $n_i + 2^{M(h-\epsilon)} \leq n$, condition (b) can be replaced by

(c) $\sum_{j\leq J}(m_j - n_j + 1) \geq (1-3\delta)n$.

The sequence x_1^n can then be expressed as a concatenation

$$x_1^n = u_1 V(1) u_2 V(2) \cdots u_J V(J) u_{J+1}, \tag{6}$$

where each $V(j) = x_{n_i}^{m_i}$ recurs too soon, the total length of the $V(j)$ is at least $(1-3\delta)n$, and where each u_i is a (possibly empty) word, that is, $x_1^n \in G(n)$.

This completes the proof that $x_1^n \in G(n)$, eventually almost surely, for fixed m and δ, if it is assumed that $\underline{r} < h - \epsilon$, and thereby finishes the proof of the recurrence-time theorem. □

Remark II.5.3

The recurrence-time theorem suggests a Monte Carlo method for estimating entropy. Let $k \sim \alpha \log n$, where $\alpha < 1$, select indices $\{i_1, \ldots, i_t\}$, independently at random in $[1, n]$, and take $(1/kt)\sum_{j=1}^t \log R_k(T^{i_j}x)$ as an estimate of h. This works reasonably well for suitably nice processes, provided n is large, and can serve, for example, as quick way to test the performance of a coding algorithm.

II.5.a Entropy and prefix trees.

Closely related to the recurrence concept is the prefix tree concept. A sequence $x \in A^\infty$ together with its first $n-1$ shifts, $Tx, T^2x, \ldots, T^{n-1}x$, defines a tree as follows. For each $0 \leq i < n$, let $W(i) = W(i, n, x)$ be the shortest prefix of $T^i x$ which is not a prefix of $T^j x$, for any $j \neq i$, $0 \leq j < n$. Since by definition, the set $\{W(i) \colon i = 0, 1, \ldots, n-1\}$ is prefix-free, it defines a tree, $\mathcal{T}_n(x)$, called *the n-th order prefix tree determined by x*.

For example,

$$x = \underline{00}1010010010\ldots, Tx = \underline{0101}0010010\ldots, T^2x = \underline{101}0010010\ldots,$$

$$T^3x = \underline{0100}10010\ldots, T^4x = \underline{100}10010\ldots$$

where underlining is used to indicate the respective prefixes,

$$W(0) = 00, W(1) = 0101, W(2) = 101, W(3) = 0100, W(4) = 100.$$

The prefix tree $\mathcal{T}_5(x)$ is shown in Figure I.7.9, page 72.

Prefix trees are used to model computer storage algorithms, in which context they are usually called suffix trees. In practice, algorithms for encoding are often formulated in terms of tree structures, as tree searches are easy to program. In particular, as indicated in Remark II.5.7, below, the simplest form of Lempel-Ziv parsing grows a subtree of the prefix tree. This led Grassberger, [16], to propose that use of the full prefix tree might lead to more rapid compression, as it reflects more of the structure of the data. He suggested that if the *depth* of node $W(i)$ is the length $L_i^n(x) = \ell(W(i))$, then average depth should be asymptotic to $(1/h)\log n$, at least in probability. While this is true in the i.i.d. case, and some other cases, see Theorem II.5.5, it is not true in the general ergodic case. What is true in the general case, however, is the following theorem.

Theorem II.5.4 (The prefix-tree theorem.)

If μ is an ergodic process of entropy $h > 0$, then for every $\epsilon > 0$ and almost every x, there is an integer $N = N(x, \epsilon)$ such that for $n \geq N$, all but ϵn of the numbers $L_i^n(x)/\log n$ are within ϵ of $(1/h)$.

Note that the prefix-tree theorem implies a trimmed mean result, namely, for any $\epsilon > 0$, the $(1-\epsilon)$-trimmed mean of the set $\{L_i^n(x)\}$ is almost surely asymptotic to $\log n/h$. The $(1-\epsilon)$-trimmed mean of a set of cardinality M is obtained by deleting the largest $M\epsilon/2$ and smallest $M\epsilon/2$ numbers in the set and taking the mean of what is left.

The $\{L_i^n\}$ are bounded below so average depth $(1/n)\sum L_i^n(x)$ will be asymptotic to $\log n/h$ in those cases where there is a constant C such that

$$\max_i L_i^n(x) \leq C\log n, \quad \text{eventually almost surely.}$$

This is equivalent to the string matching bound $L(x_1^n) = O(\log n)$, eventually almost surely, see Exercise 2, where, as earlier, $L(x_1^n)$ is the length of the longest string that appears twice in x_1^n, since if $L_i^n(x) = k$ then, by definition of $L_i^n(x)$, there is a $j \in [1, n]$, $j \neq i$, such that $x_i^{i+k-2} = x_j^{j+k-2}$.

There is a class of processes, namely, the finite energy processes, for which a simple coding argument gives $L(x_1^n) = O(\log n)$, eventually almost surely. An ergodic process μ has *finite energy* if there is are constants $c < 1$ and K such that

$$\mu(x_{t+1}^{t+L}|x_1^t) \leq Kc^L,$$

for all $t \geq 1$, for all $L \geq 1$, and for all x_1^{t+L} of positive measure.

Theorem II.5.5 (The finite-energy theorem.)

If μ is an ergodic finite energy process with entropy $h > 0$ then there is a constant C such that $L(x_1^n) \leq C\log n$, eventually almost surely, and hence

$$\lim_{n\to\infty} \frac{1}{n\log n}\sum_{i=1}^{n} L_i^n(x) = \frac{1}{h}, \quad \textit{almost surely.}$$

The i.i.d. processes, mixing Markov processes and functions of mixing Markov processes all have finite energy. Another type of finite energy process is obtained by adding i.i.d. noise to a given ergodic process, such as, for example, the binary process defined by

$$X_n = Y_n + Z_n \pmod 2$$

where $\{Y_n\}$ is an arbitrary binary ergodic process and $\{Z_n\}$ is binary i.i.d. and independent of $\{Y_n\}$. (Adding noise is often called "dithering," and is a frequently used technique in image compression.)

The mean length is certainly asymptotically almost surely no smaller than $h\log n$; in general, however, there is no way to keep the longest $W(i)$ from being much larger than $O(\log n)$, which kills the possibility of a general limit result for the mean. One type of counterexample is stated as follows.

Theorem II.5.6

There is a stationary coding F of the unbiased coin-tossing process μ, and an increasing sequence $\{n_k\}$, such that

$$\lim_{k\to\infty} \frac{\sum_{i=0}^{n_k-1} L_i^{n_k}(x)}{n_k \log n_k} = \infty, \text{ almost surely,} \tag{7}$$

with respect to the measure $\nu = \mu \circ F^{-1}$.

The prefix-tree and finite-energy theorems will be proved and the counterexample construction will be described in the following subsections.

Remark II.5.7

The simplest version of Lempel-Ziv parsing $x = w(1)w(2)\cdots$, in which the next word is the shortest new word, grows a sequence of $|A|$-ary labeled trees, where the next word $w(i)$, if it has the form $w(i) = w(j)a$, for some $j < i$ and $a \in A$, is assigned to the node corresponding to a, whose predecessor node was labeled $w(j)$. (See Exercise 3, Section II.2.) The limitations of computer memory mean that at some stage the tree can no longer be allowed to grow. From that point on, several choices are possible, each of which produces a different algorithm. In the simplest of these, mentioned in Remark II.2.10, the tree grows to a certain size then remains fixed. The remainder of the sequence is parsed into words, each of which is one symbol shorter than the shortest new word. Subsequent compression is obtained by giving the successive indices of these words in the fixed tree. Such finite LZ-algorithms do not compress to entropy in the limit, but do achieve something reasonably close to optimal compression for most finite sequences for nice classes of sources; see [87, 88] for some finite version results. Performance can be improved in some cases by designing a finite tree that better reflects the actual structure of the data. Prefix trees may perform better than the LZW tree for they tend to have more long words, which means more compression.

II.5.a.1 Proof of the prefix-tree theorem.

For $\epsilon > 0$, define the sets

$$\begin{aligned} L(x,n) &= \{i \in [0, n-1]\colon\ L_i^n(x) < (1-\epsilon)(\log n)/h\} \\ U(x,n) &= \{i \in [0, n-1]\colon\ L_i^n(x) > (1+\epsilon)(\log n)/h\}. \end{aligned}$$

It is enough to prove the following two results.

$$|L(x,n)| \le \epsilon n, \text{ eventually almost surely.} \tag{8}$$

$$|U(x,n)| \le \epsilon n, \text{ eventually almost surely.} \tag{9}$$

The fact that most of the prefixes cannot be too short, (8), can be viewed as an overlapping form of the distinct-words theorem, Theorem II.4.1, and is a consequence of the fact that, for fixed n, the $W(i, n)$ are distinct. The fact that most of the prefixes cannot be too long, (9), is a consequence of the recurrence-time theorem, Theorem II.5.1, for the longest prefix is only one longer than a word that has appeared twice.

The details of the proof that most of the words cannot be too short will be given first. To control the location of the prefixes an "almost uniform eventual typicality" idea will be needed. For $\delta > 0$ and each k, define

$$\mathcal{T}_k(\delta) = \{x_1^k\colon\ \mu(x_1^k) \ge 2^{-k(h+\delta)}\},$$

so that $|\mathcal{T}_k(\delta)| \leq 2^{k(h+\delta)}$ and $x_1^k \in \mathcal{T}_k(\delta)$, eventually almost surely, by the entropy theorem.

Since $x_1^k \in \mathcal{T}_k(\delta)$, eventually almost surely, there is an M such that the set

$$G(M) = \{x\colon\ x_1^k \in \mathcal{T}_k(\delta), \quad \text{for all } k \geq M\}$$

has measure greater than $1 - \delta$. The ergodic theorem implies that, for almost every x, the shift $T^i x$ belongs to $G(M)$ for all but at most a limiting δ-fraction of the indices i. This result is summarized in the following "almost uniform eventual typicality" form.

Lemma II.5.8

For almost every x there is an integer $N(x)$ such that if $n \geq N(x)$ then for all but at most $2\delta n$ indices $i \in [0, n)$, the i-fold shift $T^i x$ belongs to $G(M)$.

To proceed with the proof that words cannot be too short, fix $\epsilon > 0$. For a given sequence x and integer n, an index $i \leq n$ will be called *good* if $L_i^n(x) \geq M$ and $W(i) = W(i, n, x) \in \cup_{k=M}^{\infty} \mathcal{T}_k(\delta)$. Since the prefixes $W(i)$ are all distinct there are at most $|A|^M$ indices $i \leq n$ such that $L_i^n(x) \leq M$. Thus by making the $N(x)$ of Lemma II.5.8 larger, if necessary, it can be supposed that if $n \geq N(x)$ then the set of non-good indices will have cardinality at most $3\delta n$, which can be assumed to be less than $\epsilon n/2$.

Suppose $k \geq M$ and consider the set of good indices i for which $L_i^n(x) = k$. There are at most $2^{k(h+\delta)}$ such indices because the corresponding $W(i, n, x)$ are distinct members of the collection $\mathcal{T}_k(\delta)$ which has cardinality at most $2^{k(h+\delta)}$. Hence, for any $J > M$, there are at most

$$(\text{constant})\ 2^{J(h+\delta)}$$

good indices i for which $M \leq L_i^n < J$. Thus if $n \geq N(x)$ is sufficiently large and δ is sufficiently small there will be at most $\epsilon n/2$ good indices i for which $L_i^n(x) \leq (1-\epsilon)(\log n)/h$. This, combined with the bound on the number of non-good indices of the preceding paragraph, completes the proof that eventually almost surely most of the prefixes cannot be too short, (8).

Next the upper bound result, (9), will be established by using the fact that, except for the final letter, each $W(i, n, x)$ appears at least twice. If a too-long block appears twice, then it has too-short return-time. The recurrence-time theorem shows that this can only happen rarely.

Both a forward and a backward recurrence-time theorem will be needed. The corresponding recurrence-time functions are respectively defined by

$$\begin{aligned} F_k(x) &= \min\left\{m \geq 1\colon\ x_{m+1}^{m+k} = x_1^k\right\} \\ B_k(x) &= \min\left\{m \geq 1\colon\ x_{-m-k+1}^{-m} = x_{-k+1}^0\right\} \end{aligned}$$

The recurrence-time theorem directly yields the forward result,

$$\lim_{k\to\infty} \frac{1}{k} \log F_k(x) = h, \ \text{a.s.,}$$

and when applied to the reversed process it yields the backward result,

$$\lim_{k\to\infty} \frac{1}{k} \log B_k(x) = h, \ \text{a.s.,}$$

since the reversed process is ergodic and has the same entropy as the original process. (See Exercise 7, Section I.6 and Exercise 11, Section I.7.) These limit results imply that there is a K, and two sets F and B, each of measure at most $\epsilon n/4$, such that if $k \geq K$, then

$$\log F_k(x) \geq kh(1+\epsilon/4)^{-1}, \quad x \notin F,$$

and

$$\log B_k(x) \geq kh(1+\epsilon/4)^{-1}, \quad x \notin B.$$

The ergodic theorem implies that for almost every x there is an integer $N(x)$ such that if $n \geq N(x)$ then $T^i x \in F$, for at most $\epsilon n/3$ indices $i \in [0, n-1]$, and $T^{-i}x \in B$, for at most $\epsilon n/3$ indices $i \in [0, n-1]$. Fix such an x and assume $n \geq N(x)$.

Let k be the least integer that exceeds $(1+\epsilon/2)\log n/h$. By making n larger, if necessary, it can be supposed that both of the following hold.

(a) $K \leq (1+\epsilon/2)\log n/h$.

(b) $k+1 \leq (1+\epsilon)\log n/h$.

Assume $i \in U(x,n)$, that is, $L_i^n > (1+\epsilon)\log n/h$. Condition (b) implies that $L_i^n - 1 \geq k$, and hence, by the definition of $W(i,n,x)$ as the shortest prefix of $T^i x$ that is not a prefix of any other $T^j x$, for $j < n$, there is an index $j \in [0,n)$, such that $j \neq i$ and $x_i^{i+k-1} = x_j^{j+k-1}$. The two cases $i < j$ and $i > j$ will be discussed separately.

Case 1. $i < j$.

In this case, the k-block starting at i recurs in the future within the next n steps, that is, $F_k(T^i x) \leq n$, so that

$$\frac{1}{k}\log F_k(T^i x) \leq \frac{\log n}{k} < h(1+\epsilon/4)^{-1},$$

which means that $T^i x \in F$.

Case 2. $j < i$.

If $i+k \leq n$, this means that $B_k(T^{i+k}x) \leq n$, which means that $T^{-(i+k)}x \in B$.

In summary, if $i \in U_k(x,n)$ then $T^i \in F$, $T^{-(i+k)}x \in B$, or $i+k \leq n$. Thus, if n also satisfies $k \leq \epsilon n/3$, then $|U(x,n)| \leq \epsilon n$. This completes the proof of the prefix-tree theorem. □

II.5.a.2 Proof of the finite-energy theorem.

The idea of the proof is suggested by earlier code constructions, namely, code the second occurrence of the longest repeated block by telling its length and where it was seen earlier, and use an optimal code on the rest of the sequence. An upper bound on the length of the longest repeated block is obtained by comparing this code with an optimal code. Here "optimal code" means Shannon code. The key to making this method yield the desired bound is Barron's almost-sure code-length bound, Lemma 5, which asserts that no prefix code can asymptotically beat the Shannon code by more than $2\log n$.

A Shannon code for a measure μ is a prefix code such that $\mathcal{L}(w) = \lceil -\log \mu(w) \rceil$, for each w, see Remark I.7.13. A version of Barron's bound sufficient for the current result asserts that for any prefix-code sequence $\{C_n\}$,

$$\ell(C_n(x_1^n)) + \log \mu(x_1^n) \geq -2\log n, \quad \text{eventually almost surely.} \tag{10}$$

The details of the coding idea can be filled in as follows. Suppose a block of length L occurs at position s, and then again at position $t+1 > s$ in the sequence x_1^n. The block x_1^t is encoded by specifying its length t, then using a Shannon code with respect to the given measure μ. The block x_{t+1}^{t+L} is encoded by specifying s and L. The final block x_{t+L+1}^n is encoded using a Shannon code with respect to the conditional measure $\mu(\cdot\,|x_1^{t+L})$.

The prefix property is guaranteed by encoding the lengths with the Elias code $\mathcal{E}$ on the natural numbers, that is, a prefix code for which $\ell(\mathcal{E}(n)) = \log n + o(\log n)$. Since t, s, and L must be encoded, total code length satisfies

$$3\log n + \log\frac{1}{\mu(x_1^t)} + \log\frac{1}{\mu(x_{t+L+1}^n|x_1^{t+L})}, \tag{11}$$

where the extra $o(\log n)$ terms required by the Elias codes and the extra bits needed to roundup the logarithms to integers are ignored, as they are asymptotically negligible.

To compare code length with the length of the Shannon code of x_1^n with respect to μ, the probability $\mu(x_1^n)$ is factored to produce

$$\log\mu(x_1^n) = \log\mu(x_1^t) + \log\mu(x_{t+1}^{t+L}|x_1^t) + \log\mu(x_{t+L+1}^n|x_1^{t+L}).$$

This is now added to the code length (11) to yield

$$3\log n + \log\mu(x_{t+1}^{t+L}|x_1^t).$$

Thus Barron's lemma gives

$$3\log n + \log\mu(x_{t+1}^{t+L}|x_1^t) \geq -2\log n,$$

to which an application of the finite energy assumption $\mu(x_{t+1}^{t+L}|x_1^t) \leq Kc^L$, yields

$$\limsup_{n\to\infty}\frac{L(x_1^n)}{\log n} \leq \frac{5}{-\log c}, \quad \text{almost surely.}$$

This completes the proof of the finite-energy theorem, since $c < 1$. □

II.5.a.3 A counterexample to the mean conjecture.

Let μ denote the measure on two-sided binary sequences $\{0,1\}^Z$, given by unbiased coin-tossing, that is, the shift-invariant measure defined by requiring that $\mu(a_1^n) = 2^{-n}$, for each n and each $a_1^n \in A^n$. It will be shown that given $\epsilon > 0$, there is a measurable shift invariant function $F\colon \{0,1\}^Z \mapsto \{0,1\}^Z$ and an increasing sequence $\{n_k\}$ of positive integers with the following properties.

(a) If $y = F(x)$ then, with probability greater than $1 - 2^{-k}$, there is an i in the range $0 < i < n_k - n_k^{3/4}$, such that $y_{i+j} = 0,\ 0 \leq j \leq n_k^{3/4}$.

(b) $\mu(\{x\colon x_0 \neq (F(x)_0\}) \leq \epsilon$.

The process $\nu = \mu\circ F^{-1}$ will then satisfy the conditions of Theorem II.5.6, for property (a) guarantees that eventually almost surely $L_i^{n_k}(x) \geq n_k^{5/8}$ for at least $n_k^{5/8}$ indices $i < n_k$, so that $\sum_{i\leq n_k} L_i^n(x) \geq n_k^{5/4}$ and hence, (7) holds.

Property (b) is not really needed, but it is added to emphasize that such a ν can be produced by making an arbitrarily small limiting density of changes in the sample paths of the i.i.d. process μ, hence, in particular, the entropy of ν can be as close to $\log 2$ as desired.

The coding F can be constructed by only a minor modification of the coding used in the string-matching example, discussed in Section I.8.c, as an illustration of the block-to-stationary coding method. In that example, blocks of 0's of length $2n_k^{1/2}$ were created. Replacing $2n_k^{1/2}$ by $n_k^{3/4}$ will produce the coding F needed here.

Remark II.5.9

The prefix tree theorem was proved in [72], which also contains the counterexample construction given here. The coding proof of the finite-energy theorem is new; after its discovery it was learned that the theorem was proved using a different method in [31]. The asymptotics of the length of the longest and shortest word in the Grassberger tree are discussed in [82] for processes satisfying memory decay conditions.

II.5.b Exercises.

1. Another proof, due to Wyner, that $\underline{r} \geq h$ is based on the following coding idea. A function $f\colon A_{-\infty}^{n} \mapsto B^*$ will be said to be *conditionally invertible* if $x_{-\infty}^{n} = y_{-\infty}^{n}$, whenever $f(x_{-\infty}^{n}) = f(y_{-\infty}^{n})$ and $x_{-\infty}^{0} = y_{-\infty}^{0}$.

   (a) Show that $\liminf_n \ell(f(x_{-\infty}^{n}))/n \geq h$, almost surely, for any ergodic measure μ of entropy h. (Hint: let $B_n = \{x_{-\infty}^{n}\colon \mu(x_1^n | x_{-\infty}^{0}) \geq 2^{-n(H-\epsilon)}$ and $C_n = \{x_{-\infty}^{n}\colon \ell(f(x_{-\infty}^{n})) \leq 2^{n(H-2\epsilon)}\}$, where H is the entropy-rate of μ and show that $\sum \mu(B_n \cap C_n | x_{-\infty}^{0})$ is finite.)

   (b) Deduce that the entropy of a process and its reversed process are the same.

   (c) Define $f(x_{-\infty}^{n})$ to be the minimum $m \geq n$ for which $x_{-m+1}^{-m+n} = x_1^n$ and apply the preceding result.

   (d) The preceding argument establishes the recurrence result for the reversed process. Show that this implies the result for μ.

2. Show that if $L(x_1^n) = O(\log n)$, eventually almost surely, then $\max_i L_i^n(x) = O(\log n)$, eventually almost surely. (Hint: $L(x_1^{2n}) \leq K \log n$ implies $\max L_i^n(x) \leq K \log n$.)

3. Show that dithering an ergodic process produces a finite-energy process.

Chapter III

Entropy for restricted classes.

Section III.1 Rates of convergence.

Many of the processes studied in classical probability theory, such as i.i.d. processes, Markov processes, and finite-state processes have exponential rates of convergence for frequencies of all orders and for entropy, but some standard processes, such as renewal processes, do not have exponential rates. In the general ergodic case, no uniform rate is possible, that is, given any convergence rate for frequencies or for entropy, there is an ergodic process that does not satisfy the rate. These results will be developed in this section.

In this section μ denotes a stationary, ergodic process with alphabet A, with μ_k denoting the projection onto A^k defined by $\mu_k(a_1^k) = \mu(\{x\colon x_1^k = a_1^k\})$. The k-th order empirical distribution, or k-type, is the measure $p_k = p_k(\cdot\,|x_1^n)$ on A^k defined by

$$p_k(a_1^k|x_1^n) = \frac{\left|\{i \in [1, n-k+1]\colon x_i^{i+k-1} = a_1^n\}\right|}{n-k+1}, \quad a_1^k \in A^k.$$

The distance between μ_k and p_k will be measured using the distributional (variational) distance, defined by

$$|\mu_k - p_k| = |\mu_k - p_k(\cdot\,|x_1^n)| = \sum_{a_1^k} |\mu_k(a_1^k) - p_k(a_1^k|x_1^n)|.$$

The ergodic theorem implies that if k and ϵ are fixed then

$$\mu_n(\{x_1^n\colon |\mu_k - p_k(\cdot\,|x_1^n)| \geq \epsilon\}) \to 0,$$

as $n \to \infty$. Likewise, the entropy theorem implies that

$$\mu_n(\{x_1^n\colon 2^{-n(h+\epsilon)} \leq \mu_n(x_1^n) \leq 2^{-n(h-\epsilon)}\} \to 1,$$

as $n \to \infty$. The rate of convergence problem is to determine the rates at which these convergences takes place.

A mapping $r_k\colon R_+ \times Z_+ \mapsto R_+$, where R_+ denotes the positive reals and Z_+ the positive integers, is called a *rate function for frequencies* for a process μ, if for each fixed k and $\epsilon > 0$,

$$\mu_n(\{x_1^n\colon |\mu_k - p_k(\cdot\,|x_1^n)| \geq \epsilon\}) \leq r_k(\epsilon, n),$$

and $r_k(\epsilon, n) \to 0$ as $n \to \infty$. An ergodic process is said to have *exponential rates for frequencies* if there is a rate function such that for each $\epsilon > 0$ and k, $(-1/n)\log r_k(\epsilon, n)$ is bounded away from 0. (The determination of the best value of the exponent is known as "large deviations" theory; while ideas from that theory will be used, it will not be the focus of interest here.)

A mapping $r\colon R_+ \times Z_+ \mapsto R_+$ is called a *rate function for entropy* for a process μ of entropy h, if for each $\epsilon > 0$,

$$\mu_n(\{x_1^n\colon 2^{-n(h+\epsilon)} \leq \mu_n(x_1^n) \leq 2^{-n(h-\epsilon)}\}) \geq 1 - r(\epsilon, n),$$

and $r(\epsilon, n) \to 0$ as $n \to \infty$. An ergodic process has *exponential rates for entropy* if it has a rate function for entropy such that for each $\epsilon > 0$, $(-1/n)\log r(\epsilon, n)$ is bounded away from 0.

Exponential rates of convergence for i.i.d processes will be established in the next subsection, then extended to Markov and other processes that have appropriate asymptotic independence properties. Examples of processes without exponential rates and even more general rates will be discussed in the final subsection.

Remark III.1.1

A unrelated concept with a similar name, called speed of convergence, is concerned with what happens in the ergodic theorem when $(1/n)\sum_1^n f(T^i x)$ is replaced by $(1/a_n)\sum_1^n f(T^i x)$, for some unbounded, nondecreasing sequence $\{a_n\}$. See [32] for a discussion of this topic.

III.1.a Exponential rates for i.i.d. processes.

In this section the following theorem will be proved.

Theorem III.1.2 (Exponential rates for i.i.d. processes)

If μ is i.i.d., then it has exponential rates for frequencies and for entropy.

The theorem is proved by establishing the first-order case. The overlapping-block problem is then reduced to the nonoverlapping-block problem, which is treated by applying the first-order result to the larger alphabet of blocks. Exponential rates for entropy follow immediately from the first-order frequency theorem, for the fact that μ is product measure implies that $\mu(x_1^n)$ is a continuous function of $p_1(\cdot\,|x_1^n)$.

The first-order theorem is a consequence of the bound given in the following lemma, which is, in turn, a combination of a large deviations bound due to Hoeffding and Sanov and an inequality of Pinsker, [22, 62, 58].

Lemma III.1.3 (First-order rate bound.)

There is a positive constant c such that

$$\mu\left(\{x_1^n\colon |p_1(\cdot\,|x_1^n) - \mu_1| \geq \epsilon\}\right) \leq (n+1)^{|A|} 2^{-nc\epsilon^2}, \tag{1}$$

for any n, for any finite set A, and for any i.i.d. process μ with alphabet A.

Proof. A direct calculation, using the fact that μ is a product measure, produces

$$\mu(x_1^n) = \prod_{a \in A} (\mu(a))^{np_1(a|x_1^n)} = 2^{-n(H(p_1)+D(p_1\|\mu_1))} \tag{2}$$

where

$$H(p_1) = H(p_1(\cdot\,|x_1^n)) = -\sum_a p_1(a|x_1^n) \log p_1(a|x_1^n),$$

is the entropy of the empirical 1-block distribution, and

$$D(p_1 \| \mu_1) = \sum_a p_1(a|x_1^n) \log \frac{p_1(a|x_1^n)}{\mu_1(a)}$$

is the divergence of $p_1(\cdot\,|x_1^n)$ relative to μ_1. The desired result will follow from an application of the theory of type classes discussed in I.6.d. The type class of x_1^n is the set $\mathcal{T}(x_1^n) = \{y_1^n\colon\ p_1(\cdot\,|y_1^n) = p_1(\cdot\,|x_1^n)\}$. The two key facts, Theorem I.6.13 and Theorem I.6.12, are

(a) $|\mathcal{T}(x_1^n)| \leq 2^{nH(p_1)}$.

(b) There are at most $(n+1)^{|A|}$ type classes.

Fact (a) in conjunction with the product formula (2) produces the bound,

$$\mu\left(\mathcal{T}(x_1^n)\right) \leq 2^{-nD(p_1\|\mu_1)}. \tag{3}$$

The bad set $B(n,\epsilon) = \left\{x_1^n\colon\ |p_1(\cdot\,|x_1^n) - \mu_1| \geq \epsilon\right\}$ can be partitioned into disjoint sets of the form $B(n,\epsilon) \cap \mathcal{T}(x_1^n)$, and hence the bound (3) on the measure of the type class, together with fact (b), produces the bound

$$\mu(B(n,\epsilon)) \leq (n+1)^{|A|} 2^{-nD_*},$$

where

$$D_* = \min\left\{D(p_1\|\mu_1)\colon\ |p_1(\cdot\,|x_1^n) - \mu_1| \geq \epsilon\right\}.$$

Since $D(P\|Q) \geq |P-Q|^2/(2\ln 2)$ is always true, see Pinsker's inequality, Exercise 6, Section I.6, it follows that

$$D_* \geq \frac{1}{2\ln 2}|p_1(\cdot\,|x_1^n) - \mu_1|^2,$$

which completes the proof of Lemma III.1.3. □

Lemma III.1.3 gives the desired exponential-rate theorem for first-order frequencies, since

$$(n+1)^{|A|} 2^{-nc\epsilon^2} = \exp(-n(c\epsilon^2 + \delta_n))$$

where $\delta_n \to 0$ as $n \to \infty$. The extension to k-th order frequencies is carried out by reducing the overlapping-block problem to k separate nonoverlapping-block problems. To assist in this task some notation and terminology will be developed.

For $n \geq 2k$ define integers $t = t(n,k)$ and $r \in [0,k)$ such that $n = tk + k + r$. For each $s \in [0, k-1]$, the sequence x_1^n can be expressed as the concatenation

$$x_1^n = w(0)w(1)\cdots w(t)w(t+1), \tag{4}$$

where the first word $w(0)$ has length $s < k$, the next t words, $w(1),\ldots,w(t)$, all have length k, and the final word $w(t+1)$ has length $n - tk - s$. This is called *the s-shifted, k-block parsing of x_1^n*. The sequences x_1^n and y_1^n are said to be *(s,k)-equivalent* if their

s-shifted, k-block parsings have the same (ordered) set $\{w(1), \ldots, w(t)\}$ of k-blocks, that is, $x_{s+1}^{s+tk} = y_{s+1}^{s+tk}$. The (s, k)-equivalence class of x_1^n is denoted by $\mathcal{S}_k(x_1^n, s)$.

The *s-shifted (nonoverlapping) empirical k-block distribution* $p_k^s = p_k^s(\cdot\,|x_1^n)$ is the distribution on A^k defined by

$$p_k^s(a_1^k) = p_k^s(a_1^k|x_1^n) = \frac{|\{j \in [1, t]\colon\ w(j) = a_1^k\}|}{t},$$

where x_1^n is given by (4). It is, of course, constant on the (s, k)-equivalence class $\mathcal{S}_k(x_1^n, s)$. The overlapping-block measure p_k is (almost) an average of the measures p_k^s, where "almost" is needed to account for end effects, a result summarized as the following lemma.

Lemma III.1.4 (Overlapping to nonoverlapping lemma.)

Given $\epsilon > 0$ there is a $\gamma > 0$ such that if $k/n < \gamma$ and $|p_k(\cdot\,|x_1^n) - \mu_k| \geq \epsilon$ then there is an $s \in [0, k-1]$ such that $|p_k^s(\cdot\,|x_1^n) - \mu_k| \geq \epsilon/2$.

Proof. Left to the reader.

If k is fixed and n is large enough the lemma yields the containment relation,

$$\left\{x_1^n\colon\ |p_k(\cdot\,|x_1^n) - \mu_k| \geq \epsilon\right\} \subseteq \bigcup_{s=0}^{k-1} \left\{x_1^n\colon\ |p_k^s(\cdot\,|x_1^n) - \mu_k| \geq \epsilon/2\right\}. \tag{5}$$

Fix such a k and n, and fix $s \in [0, k-1]$. The fact that μ is a product measure implies

$$\mu_n(\mathcal{S}_k(x_1^n, s)) \leq \prod_{j=1}^{t} \mu_k(w(j)), \tag{6}$$

where $\mathcal{S}_k(x_1^n, s)$ is the (s, k)-equivalence class of x_1^n, so that if B denotes A^k and ν denotes the product measure on B^t defined by the formula

$$\nu(w_1^t) = \prod_{j=1}^{t} \mu(w_j),\ \ w_j \in B.$$

then

$$\mu\left(\{x_1^n\colon\ |p_k^s(\cdot\,|x_1^n) - \mu_k| \geq \epsilon/2\}\right) \leq \nu\left(\left\{w_1^t\colon\ |p_1(\cdot\,|w_1^t) - \nu_1| \geq \epsilon/2\right\}\right). \tag{7}$$

The latter, however, is upper bounded by

$$(t+1)^{|A|^k} 2^{-tc\epsilon^2/4}, \tag{8}$$

by the first-order result, Lemma III.1.3, applied to the measure ν and super-alphabet $B = |A|^k$. The containment relation, (5), then provides the desired k-block bound

$$\mu\left(\left\{x_1^n\colon\ |p_k(\cdot\,|x_1^n) - \mu_k| \geq \epsilon\right\}\right) \leq k(t+1)^{|A|^k} 2^{-tc\epsilon^2/4}. \tag{9}$$

If k and $\epsilon > 0$ are fixed, the logarithm of the right-hand side is asymptotic to $-tc\epsilon^2/4n \sim -c\epsilon^2/4k$, hence the bound decays exponentially in n. This completes the proof of Theorem III.1.2. □

III.1.b The Markov and related cases.

Exponential rates for Markov chains will be established in this section.

Theorem III.1.5 (Exponential rates for Markov sources.)
If μ is an ergodic Markov source, then it has exponential rates for frequencies and for entropy.

The entropy part follows immediately from the second-order frequency result, since $\mu(x_1^n)$ is a continuous function of $p_1(\cdot\,|x_1^n)$ and $p_2(\cdot\,|x_1^n)$. The theorem for frequencies is proved in stages. First it will be shown that aperiodic Markov chains satisfy a strong mixing condition called ψ-mixing. Then it will be shown that ψ-mixing processes have exponential rates for both frequencies and entropy. Finally, it will be shown that periodic, irreducible Markov chains have exponential rates.

A process is *ψ-mixing* if there is a nonincreasing sequence $\{\psi(g)\}$ such that $\psi(g) \to 1$ as $g \to \infty$, and such that

$$\mu(uvw) \leq \psi(g)\mu(u)\mu(w),\ u, w \in A^*, v \in A^g.$$

The ψ-mixing concept is stronger than ordinary mixing, which allows the length of the gap to depend on the length of the past and future.

An i.i.d. process is ψ-mixing, with $\psi(g) \equiv 1$. It requires only a bit more effort to obtain the mixing Markov result.

Lemma III.1.6
An aperiodic Markov chain is ψ-mixing.

Proof. Let M be the transition matrix and $\pi(\cdot)$ the stationary vector for μ. Fix a gap g, let a be the final symbol of u, and let b be the first symbol of w. The formula (10), Section I.2, for computing Markov probabilities yields

$$\sum_{\ell(v)=g} \mu(uvw) = \mu(u)\frac{M_{ab}^g}{\pi(b)}\mu(w).$$

The right-hand side is upper bounded by $\psi(g)\mu(u)\mu(w)$, where

$$\psi(g) = \max_a \max_b \frac{M_{ab}^g}{\pi(b)}.$$

The function $\psi(g) \to 1$, as $g \to \infty$, since the aperiodicity assumption implies that $M_{ab}^g \to \pi(b)$. This proves the lemma. □

The next task is to prove

Theorem III.1.7
If μ is ψ-mixing then it has exponential rates for frequencies of all orders.

Proof. A simple modification of the i.i.d. proof, so as to allow gaps between blocks, is all that is needed to establish this lemma. Fix positive integers k and g. For $n \geq 2k+g$, define $t = t(n, k, g)$ such that

(10) $$n = t(k+g) + (k+g) + r,\ 0 \leq r < k+g.$$

For each $s \in [0, k+g-1]$, the sequence x_1^n can be expressed as the concatenation

$$x_1^n = w(0)g(1)w(1)g(2)w(2)\dots g(t)w(t)w(t+1), \tag{11}$$

where $w(0)$ has length $s < k+g$, the $g(j)$ have length g and alternate with the k-blocks $w(j)$, for $1 \le j \le t$, and the final block $w(t+1)$ has length $n - t(k+g) - s < 2(k+g)$. This is called the *the s-shifted, k-block parsing of x_1^n, with gap g*. The sequences x_1^n and y_1^n are said to be *(s, k, g)-equivalent* if their s-shifted, g-gapped, k-block parsings have the same (ordered) set $\{w(1), \dots, w(t)\}$ of k-blocks. The (s, k, g)-equivalence class of x_1^n is denoted by $\mathcal{S}_k(x_1^n, s, g)$.

The *s-shifted (nonoverlapping) empirical k-block distribution, with gap g*, is the distribution on A^k defined by

$$p_{k,g}^s(a_1^k) = p_{k,g}^s(a_1^k|x_1^n) = \frac{|\{j \in [1, t]\colon\ w(j) = a_1^k\}|}{t}.$$

It is, of course, constant on the (s, k, g)-equivalence class $\mathcal{S}_k(x_1^n, s, g)$. The overlapping-block measure p_k is (almost) an average of the measures $p_{k,g}^s$, where "almost" is now needed to account both for end effects and for the gaps. If k is large relative to gap length and n is large relative to k, this is no problem, however, and Lemma III.1.4 easily extends to the following.

Lemma III.1.8 (Overlapping to nonoverlapping with gaps.)
Given $\epsilon > 0$ and g, there is a $\gamma > 0$ and a $K > 0$, such that if $k/n < \gamma$, if $k > K$, and if $|p_k(\cdot\,|x_1^n) - \mu_k| \ge \epsilon$, then there is an $s \in [0, k+g-1]$ such that $|p_{k,g}^s(\cdot\,|x_1^n) - \mu_k| \ge \epsilon/2$.

Completion of proof of Theorem III.1.7.
The i.i.d. proof adapts to the ψ-mixing case, with the product measure bound (6) replaced by

$$\mu_n(\mathcal{S}_k(x_1^n, s, g)) \le [\psi(g)]^t \prod_{j=1}^{t} \mu_k(w(j)), \tag{12}$$

where the $w(j)$ are given by (11), the s-shifted, k-block parsing of x_1^n, with gap g. The upper bound (8) on $\mu\left(\{x_1^n\colon\ |p_k^s(\cdot\,|x_1^n) - \mu_k| \ge \epsilon/2\}\right)$ is replaced by the upper bound

$$[\psi(g)]^t (t+1)^{|A|^k} 2^{-tc\epsilon^2/4} \tag{13}$$

and the final upper bound (9) on the probability $\mu\left(\left\{x_1^n\colon\ |p_k(\cdot\,|x_1^n) - \mu_k| \ge \epsilon\right\}\right)$ is replaced by the bound

$$(k+g)[\psi(g)]^t\ (t+1)^{|A|^k} 2^{-tc\epsilon^2/4}, \tag{14}$$

assuming, of course, that k is enough larger than g and n enough larger than k to guarantee that Lemma III.1.8 holds.

The proof of Theorem III.1.7 is completed by using the ψ-mixing property to choose g so large that $\psi(g) \le 2^{c\epsilon^2/8}$. The logarithm of the bound in (14) is then asymptotically at most $-tc\epsilon^2/8n \sim -c\epsilon^2/8(k+g)$, hence the bound decays exponentially, provided only that k is large enough. Since it is enough to prove exponential rates for k-th order frequencies for large k, this completes the proof of Theorem III.1.7, and establishes the exponential rates theorem for aperiodic Markov chains. □

The following lemma removes the aperiodicity requirement in the Markov case.

Lemma III.1.9

An ergodic Markov chain has exponential rates for frequencies of all orders.

Proof. The only new case is the periodic case. Let μ be an ergodic Markov chain with period $d > 1$, and partition A into the (periodic) classes, $C_1, C_2, \ldots, C_d$, such that

$$\text{Prob}(X_{n+1} \in C_{s\oplus 1} | X_n \in C_s) = 1, \ 1 \le s \le d,$$

where $\oplus$ denotes addition mod d. Define the function $c: A \mapsto [1, d]$ by putting $c(a) = s$, if $a \in C_s$. Also let $\mu^{(s)}$ denote the measure μ conditioned on $X_1 \in C_s$.

Let g be a gap length, which can be assumed to be divisible by d and small relative to k. It can also be assumed that k is divisible by d, for k can always be increased or decreased by no more than d to achieve this, which has no effect on the asymptotics. For $s \in [0, k+g-1]$, the nonoverlapping block measure $p^s_{k,g} = p^s_{k,g}(\cdot\, | x_1^n)$ satisfies the condition

$$p^s_{k,g}(a_1^k) = 0, \ \text{unless } c(a_1) = c(x_s).$$

Since μ_k is an average of the $\mu_k^{(s)}$, it follows that if k/g is large enough then

$$\{x_1^n\colon |p_k(\cdot\, | x_1^n) - \mu_k| > \epsilon\} \subseteq \bigcup_{s=0}^{k+g-1} \{x_1^n\colon |p^s_{k,g}(\cdot\, | x_1^n) - \mu_k^{(c(x_s))}| > \epsilon/2\}.$$

The measure $\mu^{(s)}$ is, however, an aperiodic Markov measure with state space

$$C(s) \times C(s \oplus 1) \times \cdots \times C(s \oplus d - 1),$$

so the previous theory applies to each set $\{x_1^n\colon |p^s_{k,g} - \mu_k^{(c(x_s))}| > \epsilon/2\}$ separately. Thus, as before,

$$\mu(\{x_1^n\colon |p_k - \mu_k| \ge \epsilon\})$$

decays exponentially as $n \to \infty$, establishing the lemma, and thereby completing the proof of the exponential-rates theorem for Markov chains, Theorem III.1.5. □

III.1.c Counterexamples.

Examples of renewal processes without exponential rates for frequencies and entropy are not hard to construct; see Exercise 4. A simple cutting and stacking procedure which produces counterexamples for general rate functions will be presented, in part, because it is conceptually quite simple, and, in part, because it illustrates several useful cutting and stacking ideas.

The following theorem summarizes the goal.

Theorem III.1.10

Let $n \mapsto r(n)$ be a positive decreasing function with limit 0. There is a binary ergodic process μ and an integer N such that

$$\mu_n\left(\{x_1^n\colon |p_1(\cdot\, | x_1^n) - \mu_1| \ge 1/2\}\right) \ge r(n), \ n \ge N.$$

Counterexamples for a given rate would be easy to construct if ergodicity were not required. For example, if μ were the average of two processes, $\mu = (\nu + \widetilde{\nu})/2$, where ν is concentrated on the sequence of all 1's and $\widetilde{\nu}$ is concentrated on the sequence of all 0's, then for any n, $p_1(1|x_1^n)$ is either 1 or 0, according to whether x_1^n consists of 1's or 0's, while $\mu_1(1) = 1/2$, so that

$$\mu_n\left(\{x_1^n\colon |p_1(\cdot\,|x_1^n) - \mu_1| \geq 1/2\}\right) = 1,\ n \geq 1.$$

Ergodic counterexamples are constructed by making μ look like one process on part of the space and something else on the other part of the space, mixing the first part into the second part so slowly that the rate of convergence is as large as desired. The cutting and stacking method was designed, in part, to implement both of these goals, that is, to make a process look like another process on part of the space, and to mix one part slowly into another.

As an example of what it means to "look like one process on part of the space and something else on the other part of the space," suppose μ is defined by a (complete) sequence of column structures. Suppose some column $\mathcal{C}$ at some stage has all of its levels labeled '1'. If x is a point in some level of $\mathcal{C}$ below the top m levels, then it is certain that $x_i = 1$, for $1 \leq i \leq m$, so that, $p_1(1|x_1^m) = 1$. In particular, if $\mathcal{C}$ has height L and measure β, then the chance that a randomly selected x lies in the first $L - m$ levels of $\mathcal{C}$ is $(1 - m/L)\beta$, and hence

$$\mu_n\left(\{x_1^m\colon p_1(1|x_1^m) = 1\}\right) \geq \left(1 - \frac{m}{L}\right)\beta, \tag{15}$$

no matter what the complement of $\mathcal{C}$ looks like. Furthermore, if exactly half the measure of $\mathcal{S}(1)$ is concentrated on intervals labeled '1', then

$$\mu_n\left(\{x_1^m\colon |p_1(\cdot\,|x_1^m) - \mu_1| \geq 1/2\}\right) \geq \left(1 - \frac{m}{L}\right)\beta,$$

so that if $1/2 > \beta > r(m)$ and L is large enough then

$$\mu_n\left(\{x_1^m\colon |p_1(\cdot\,|x_1^m) - \mu_1| \geq 1/2\}\right) > r(m). \tag{16}$$

The mixing of "one part slowly into another" is accomplished by cutting $\mathcal{C}$ into two columns, say $\mathcal{C}_1$ and $\mathcal{C}_2$. Mixing is done by applying repeated independent cutting and stacking to the union of the second column $\mathcal{C}_2$ with the complement of $\mathcal{C}$. The bound (16) can be achieved with m replaced by $m + 1$ by making sure that the first column $\mathcal{C}_1$ has measure slightly more than $r(m + 1)$, for it can then be cut and stacked into a much longer column, long enough to guarantee that (16) will indeed hold with m replaced by $m + 1$. As long as enough independent cutting and stacking is done at each stage and all the mass is moved in the limit to the second part, the final process will be ergodic and will satisfy (16) for all m for which $r(m) < 1/2$.

The details of the above outline will now be given. Without loss of generality, it can be supposed that $r(m) < 1/2$, for all m. Let $m \to \beta(m)$ be a nonincreasing function with limit 0, such that $\beta(1) = 1/2$ and $r(m) < \beta(m) < 1/2$, for all $m > 1$.

The desired (complete) sequence $\{\mathcal{S}(m)\}$ will now be defined. The construction will be described inductively in terms of two auxiliary unbounded sequences, $\{\ell_m\}$ and $\{r_m\}$, of positive integers which will be specified later. Each $\mathcal{S}(m)$ will contain a column

$\mathcal{C}(m)$, of height ℓ_m and measure $\beta(m)$, all of whose entries are labeled '1', together with a column structure $\mathcal{R}(m)$, disjoint from $\mathcal{C}(m)$.

To get started, let $\mathcal{C}(1)$ consist of one interval of length 1/2, labeled '1', and let $\mathcal{R}(1)$ consist of one interval of length 1/2, labeled '0'. This guarantees that at all subsequent stages the total measure of the intervals of $\mathcal{S}(m)$ that are labeled with a '1' is 1/2, and the total measure of the intervals of $\mathcal{S}(m)$ that are labeled with a '0' is 1/2, and hence the final process must satisfy $\mu_1(1) = \mu_1(0) = 1/2$.

For $m > 1$, $\mathcal{S}(m)$ is constructed as follows. First $\mathcal{C}(m-1)$ is cut into two columns, $\mathcal{C}(m-1, 1)$ and $\mathcal{C}(m-1, 2)$, where $\mathcal{C}(m-1, 1)$ has measure $\beta(m)$ and $\mathcal{C}(m-1, 2)$ has measure $\beta(m-1) - \beta(m)$. The column $\mathcal{C}(m-1, 1)$ is cut into ℓ_m/ℓ_{m-1} columns of equal width, which are then stacked to obtain $\mathcal{C}(m)$. The new remainder $\mathcal{R}(m)$ is obtained by applying r_m-fold independent cutting and stacking to $\mathcal{C}(m-1, 2) \cup \mathcal{R}(m-1)$.

Since $r_m \to \infty$, and since $\beta(m) \to 0$, the total width of $\mathcal{S}(m)$ goes to 0, so the sequence $\{\mathcal{S}(m)\}$ is complete and hence defines a process μ. All that remains to be shown is that for suitable choice of ℓ_m and r_m the final process μ is ergodic and satisfies the desired condition

$$\mu_n\left(\{x_1^m\colon |p_1(\cdot\,|x_1^m) - \mu_1| \geq 1/2\}\right) > r(m), \quad m \geq 1. \tag{17}$$

The condition (17) is guaranteed by choosing ℓ_m so large that $(1 - m/\ell_m)\beta(m) > r(m)$, for this is all that is needed to make sure that

$$\mu_n\left(\{x_1^m\colon |p_1(\cdot\,|x_1^m) - \mu_1| \geq 1/2\}\right) \geq \left(1 - \frac{m}{\ell_m}\right)\beta(m),$$

holds. Once this holds, then it can be combined with the fact that 1's and 0's are equally likely and the assumption that $1/2 > \beta(m) > r(m)$, to guarantee that (17) holds.

The sequence $\{r_m\}$ is chosen, sequentially, so that $\mathcal{C}(m-1, 2) \cup \mathcal{R}(m-1)$ and $\mathcal{R}(m)$ are $(1 - 2^{-m})$-independent. This is possible by Theorem I.10.11. Since the measure of $\mathcal{R}(m)$ goes to 1, this guarantees that the final process is ergodic, by Theorem I.10.7. This completes the proof of Theorem III.1.10. □

III.1.d Exercises.

1. Suppose μ has exponential rates for frequencies.

   (a) Choose k such that $\mathcal{T}_k = \{x_1^k\colon \mu(x_1^k) \geq 2^{-k(h+\delta)}\}$ has measure close to 1. Show that the measure of the set of n-sequences that are not almost strongly-packed by $\mathcal{T}_k$ goes to 0 exponentially fast.

   (b) Show that the measure of $\{x_1^n\colon \mu(x_1^n) < 2^{-n(h+\epsilon)}\}$ goes to 0 exponentially fast. (Hint: as in the proof of the entropy theorem it is highly unlikely that a sequence can be mostly packed by $\mathcal{T}_k$ and have too-small measure.)

   (c) Show that a ψ-mixing process has exponential rates for entropy.

2. Let ν be a finite stationary coding of μ and assume μ has exponential rates for frequencies and entropy.

   (a) Show that ν has exponential rates for frequencies.

(b) Show that ν has exponential rates for entropy. (Hint: bound the number of n-sequences that can have good k-block frequencies, conditional k-type too much larger than $2^{-k(h(\mu)-h(\nu)}$ and measure too much larger than $2^{-nh(\nu)}$.)

3. Use cutting and stacking to construct a process with exponential rates for entropy but not for frequencies.

4. Show that there exists a renewal process that does not have exponential rates for frequencies of order 1. (Hint: make sure that the expected recurrence-time series converges slowly.)

5. Exercise 4 shows that renewal processes are not ψ-mixing. Establish this by directly constructing a renewal process that is not ψ-mixing.

Section III.2 Entropy and joint distributions.

The kth-order joint distribution for an ergodic finite-alphabet process can be estimated from a sample path of length n by sliding a window of length k along the sample path and counting frequencies of k-blocks. If k is fixed the procedure is almost surely consistent, that is, the resulting empirical k-block distribution almost surely converges to the true distribution of k-blocks as $n \to \infty$, a fact guaranteed by the ergodic theorem. The consistency of such estimates is important when using training sequences, that is, finite sample paths, to design engineering systems. The empirical k-block distribution for a training sequence is used as the basis for design, after which the system is run on other, independently drawn sample paths. There are some situations, such as data compression, where it is good to make the block length as long as possible. Thus it would be desirable to have consistency results for the case when the block length function $k = k(n)$ grows as rapidly as possible, as a function of sample path length n. This is the problem addressed in this section.

As before, let $p_k(\cdot\,|x_1^n)$ denote the empirical distribution of overlapping k-blocks in the sequence x_1^n. A nondecreasing sequence $\{k(n)\}$ will be said to be *admissible* for the ergodic measure μ if

$$\lim_{n\to\infty} \left|p_{k(n)}(\cdot\,|x_1^n) - \mu_{k(n)}\right| = 0, \quad \text{a. s.},$$

where $|\cdot|$ denotes variational, that is, distributional, distance. The definition of *admissible in probability* is obtained by replacing almost-sure convergence by convergence in probability. Every ergodic process has an admissible sequence such that $\lim_n k(n) = \infty$, by the ergodic theorem. It is also can be shown that for any sequence $k(n) \to \infty$ there is an ergodic measure for which $\{k(n)\}$ is not admissible, see Exercise 2.

The problem addressed here is whether it is possible to make a universal choice of $\{k(n)\}$ for "nice" classes of processes, such as the i.i.d. processes or Markov chains. Here is where entropy enters the picture, for if k is large then, with high probability, the probability of a k-block will be roughly 2^{-kh}. In particular, if

$$k(n) \geq (1+\epsilon)(\log n)/h,$$

or, equivalently, $k(n) \geq (\log n)/(h-\epsilon)$, there is no hope that the empirical k-block distribution will be close to the true distribution, see Theorem III.2.1, below. Consistent estimation also may not be possible for the choice $k(n) \sim (\log n)/h$. For example, in the unbiased coin-tossing case when $h = 1$, the choice $k(n) \sim \log n$ is not admissible, for it is easy to see that, with high probability, an approximate $(1 - e^{-1})$ fraction of the k-blocks will fail to appear in a given sample path of length n, see Exercise 3. Thus the case of most interest is when $k(n) \sim (\log n)/(h + \epsilon)$.

The principle results are the following.

Theorem III.2.1 (The nonadmissibility theorem.)
If μ is ergodic with positive entropy h, if $0 < \epsilon < h$, and if $k(n) \geq (\log n)/(h - \epsilon)$ then $\{k(n)\}$ is not admissible in probability for μ.

Theorem III.2.2 (The positive admissibility theorem.)
If μ is i.i.d., Markov, or ψ-mixing, and $k(n) \leq (\log n)/(h + \epsilon)$ then $\{k(n)\}$ is admissible for μ.

Theorem III.2.3 (Weak Bernoulli admissibility.)
If μ is weak Bernoulli and $k(n) \leq (\log n)/(h + \epsilon)$ then $\{k(n)\}$ is admissible for μ.

The ψ-mixing concept was introduced in the preceding section. The weak Bernoulli property, which will be defined carefully later, requires that past and future blocks be almost independent if separated by a long enough gap, independent of the length of past and future blocks.

The nonadmissibility theorem is a simple consequence of the fact that no more than $2^{k(n)(h-\epsilon)}$ sequences of length k can occur in a sequence of length $n \leq 2^{k(n)(h-\epsilon)}$, and hence there is no hope of seeing even a fraction of the full distribution. The positive admissibility result for the i.i.d. and ψ-mixing cases is a consequence of a slight strengthening of the exponential rate bounds of the preceding section. The weak Bernoulli result follows from an even more careful look at what is really used for the ψ-mixing proof.

Remark III.2.4

A first motivation for the problem discussed in this section was the training sequence problem described in the opening paragraph. A second motivation was the desire to obtain a more classical version of the positive results of Ornstein and Weiss, who used the $\bar{d}$-distance rather than the variational distance, [52]. They showed, in particular, that if the process is a stationary coding of an i.i.d. process then the $\bar{d}$-distance between the empirical $k(n)$-block distribution and the true $k(n)$-block distribution goes to 0, almost surely, provided $k(n) \sim (\log n)/h$. The $\bar{d}$-distance is upper bounded by half the variational distance, and hence the results described here are a sharpening of the Ornstein-Weiss theorem for the case when $k \leq (\log n)/(h + \epsilon)$ and the process satisfies strong enough forms of asymptotic independence. The Ornstein-Weiss results will be discussed in more detail in Section III.3.

Remark III.2.5

A third motivation for the problem discussed here was a waiting-time result obtained by Wyner and Ziv, [86]. They showed that if $W_n(x, y)$ is the waiting time until the first n terms of x appear in the sequence y then, for ergodic Markov chains, $(1/n) \log W_n(x, y)$ converges in probability to h, provided x and y are independently chosen sample paths. The admissibility results of obtained here can be used to prove stronger versions of their theorem. These applications, along with various related results and counterexamples, are presented in Section III.5.

III.2.a Proofs of admissibility and nonadmissibility.

First the nonadmissibility theorem will be proved. Let μ be an ergodic process with entropy h and suppose $k = k(n) \geq (\log n)/(h - \epsilon)$, where $0 < \epsilon < h$. Define the *empirical universe of k-blocks* to be the set

$$\mathcal{U}_k(x_1^n) = \left\{a_1^k \colon x_i^{i+k-1} = a_1^k, \text{ for some } i \in [1, n-k+1]\right\}.$$

Since $p_k(a_1^k|x_1^n) = 0$, whenever $a_1^k \notin \mathcal{U}_k(x_1^n)$, it follows that

$$\left|p_k(\cdot\,|x_1^n) - \mu_k\right| \geq \mu_k\left((\mathcal{U}_k(x_1^n))^c\right). \tag{1}$$

Let

$$T_k(\epsilon/2) = \left\{x_1^k \colon \mu(x_1^k) \leq 2^{-k(h-\epsilon/2)}\right\}.$$

The assumption $n \leq 2^{k(h-\epsilon)}$ implies that $|\mathcal{U}_k(x_1^n)| \leq 2^{k(h-\epsilon)}$, and hence

$$\mu_k\left(\mathcal{U}_k(x_1^n) \cap T_k(\epsilon/2)\right) \leq 2^{k(h-\epsilon)}2^{-k(h-\epsilon/2)} = 2^{-k\epsilon/2},$$

since each member of $T_k(\epsilon/2)$ has measure at most $2^{-k(h-\epsilon/2)}$. The entropy theorem guarantees that $\mu_k(T_k(\epsilon/2)) \to 1$, so that $\mu_k\left((\mathcal{U}_k(x_1^n))^c\right)$ also goes to 1, and hence $\left|p_k(\cdot\,|x_1^n) - \mu_k\right| \to 1$ for *every* $x \in A^\infty$, by the distance lower bound, (1). This implies the nonadmissibility theorem. □

Next the positive admissibility theorem will be established. First note that, without loss of generality, it can be supposed that $\{k(n)\}$ is unbounded, since any bounded sequence is admissible for any ergodic process, by the ergodic theorem.

The rate bound,

$$\mu\left(\left\{x_1^n \colon |p_k(\cdot\,|x_1^n) - \mu_k| \geq \epsilon\right\}\right) \leq k(t+1)^{|A|^k}2^{-t\epsilon^2/4}, \tag{2}$$

where $t \sim n/k$, see the bound (9) of Section III.1, is enough to prove the positive admissibility theorem for unbiased coin-tossing, for, as the reader can show, the error is summable in n, if $n \geq |A|^{k(1+\epsilon)}$. The key to such results for other processes is a similar bound which holds when the full alphabet is replaced by a subset of large measure.

Lemma III.2.6 (Extended first-order bound.)
There is a positive constant C such that for any $\epsilon > 0$, for any finite set A, for any $n > 0$, for any i.i.d. process μ with finite alphabet A, and for any $B \subset A$ such that $\mu(B) > 1 - \epsilon$ and $|B| \geq 2$, the following holds.

$$\mu(\{x_1^n \colon |p_1(\cdot\,|x_1^n) - \mu_1| > 5\epsilon\} \leq 2(n+1)^{|B|}2^{-nC\epsilon^2}.$$

Proof. Define

$$y_n = y_n(x) = \begin{cases} 0 & \text{if } x_n \in B \\ 1 & \text{otherwise,} \end{cases}$$

so that $\{y_n\}$ is a binary i.i.d. process with Prob $(y_n = 1) < \epsilon$. Let

$$C_n = \left\{x_1^n \colon \sum_1^n y_i > 2\epsilon n\right\}$$

and apply the bound (2) to obtain

$$\mu(C_n) \leq (n+1)^2 2^{-cn4\epsilon^2}. \tag{3}$$

The idea now is to partition A^n according to the location of the indices i for which $x_i \notin B$. For each $m \leq n$ and $1 \leq i_1 < i_2 < \cdots < i_m \leq n$, let $B(i_1, \ldots, i_m)$ denote the set of all x_1^n for which $x_i \in B$ if and only if $i \notin \{i_1, i_2, \ldots, i_m\}$. The sets $B(i_1, \ldots, i_m)$ are disjoint for different $\{i_1, i_2, \ldots, i_m\}$ and have union A^n. Furthermore, the sets $\{B(i_1, \ldots, i_m)\colon m > 2\epsilon n\}$ have union C_n so that (3) yields the bound

$$\sum_{\{i_1,\ldots,i_m\}: m > 2\epsilon n} \mu(B(i_1, \ldots, i_m)) \leq (n+1)^2 2^{-cn4\epsilon^2}. \tag{4}$$

Fix a set $\{i_1, \ldots, i_m\}$ with $m \leq 2\epsilon n$, put $s = n - m$, and for $x_1^n \in B(i_1, \ldots, i_m)$ define $\bar{x} = \bar{x}(x_1^n)$ to be the sequence of length s obtained by deleting $x_{i_1}, x_{i_2}, \ldots, x_{i_m}$ from x_1^n. Also let $\mu_{1,B}$ be the conditional distribution on B defined by μ_1 and let μ_B be the corresponding product measure on B^s defined by $\mu_{1,B}$. The assumption that $m \leq 2\epsilon n$ the implies that the probability

$$\mu\left(\{x_1^n \in B(i_1, \ldots, i_m)\colon |p_1(\cdot\,|x_1^n) - \mu_1| > 5\epsilon\}\right)$$

is upper bounded by

$$\mu(B(i_1, \ldots, i_m))\mu_B\left(\{\bar{x}_1^s \in B^s\colon |p_1(\cdot\,|\bar{x}_1^s) - \mu_1| > 3\epsilon\}\right),$$

which is, in turn, upper bounded by

$$\mu(B(i_1, \ldots, i_m))\mu_B\left(\{\bar{x}_1^s \in B^s\colon |p_1(\cdot\,|\bar{x}_1^s) - \mu_{1,B}| > \epsilon\}\right), \tag{5}$$

since

$$|\mu_1 - \mu_{1,B}| = \sum_{b \in B}\left[\frac{\mu(b)}{\mu(B)} - \mu(b)\right] + \sum_{b \notin B} \mu(b) = 2(1 - \mu(B)) \leq 2\epsilon.$$

The exponential bound (2) can now be applied with $k = 1$ to upper bound (5) by

$$\mu(B(i_1, \ldots, i_m))(s+1)^{|B|} 2^{-sc\epsilon^2} \leq \mu(B(i_1, \ldots, i_m))(n+1)^{|B|} 2^{-n(1-2\epsilon)c\epsilon^2}.$$

The sum of the $\mu(B(i_1, \ldots, i_m))$ for which $m \leq 2\epsilon n$ cannot exceed 1, and hence, combined with (4) and the assumption that $|B| \geq 2$, the lemma follows. □

The positive admissibility result will be proved first for the i.i.d. case, then extended to the ψ-mixing and periodic Markov cases. Assume μ is i.i.d. with entropy h, and assume $k(n) \leq (\log n)/(h + \epsilon)$. Let δ be a given positive number. It will be shown that

$$\sum_n \mu\left(\{x_1^n\colon |p_{k(n)}(\cdot\,|x_1^n) - \mu_{k(n)}| \geq \delta\}\right) < \infty, \tag{6}$$

which immediately implies the desired result, by the Borel-Cantelli lemma.

The proof of (6) starts with the overlapping to nonoverlapping relation derived in the preceding section, see (5), page 168,

$$\left\{x_1^n\colon |p_k(\cdot\,|x_1^n) - \mu_k| \geq \delta\right\} \subseteq \bigcup_{s=0}^{k-1}\left\{x_1^n\colon |p_k^s(\cdot\,|x_1^n) - \mu_k| \geq \delta/2\right\}, \tag{7}$$

which is valid for k/n smaller than some fixed $\gamma > 0$, where $n = tk+k+r$, $0 \le r < k$, and $p_k^s(\cdot\,|x_1^n)$ denotes the s-shifted nonoverlapping empirical k-block distribution, for $s \in [0, k-1]$. As in the preceding section, see (7) on page 168,

$$\mu\Big(\{x_1^n\colon\ |p_k^s(\cdot\,|x_1^n) - \mu_k| \ge \delta/2\}\Big) \le \nu\Big(\{w_1^t\colon\ |p_1(\cdot\,|w_1^t) - \nu_1| \ge \delta/2\}\Big), \tag{8}$$

where each $w_j \in \widetilde{A} = A^k$, and ν denotes the product measure on $\widetilde{A}^t$ induced by μ.

The idea now is to upper bound the right-hand side in (8) by applying the extended first-order bound, Lemma III.2.6, to the super-alphabet A^k, with B replaced by a suitable set of entropy-typical sequences. Since the number of typical sequences can be controlled, as k grows, the polynomial factor in that lemma takes the form, $(1+n/k)^{2^{k(h+\epsilon')}}$, which, though no longer a polynomial in n is still dominated by the exponential factor. The rigorous proof is given in the following paragraphs.

The set

$$\mathcal{T}_k = \{x_1^k\colon\ \mu(x_1^k) \ge 2^{-k(h+\epsilon/2)}\}$$

has cardinality at most $2^{k(h+\epsilon/2)}$, and, furthermore, by the entropy theorem, there is a K such that $\nu_1(\mathcal{T}_k) = \mu_k(\mathcal{T}_k) > 1 - \delta/10$, for $k \ge K$. The extended first-order bound, Lemma III.2.6, with A replaced by $\widetilde{A} = A^k$ and B replaced by $\mathcal{T}_k$, therefore yields the bound

$$\nu\Big(\{w_1^t\colon\ |p_1(\cdot\,|w_1^t) - \nu_1| \ge \delta/2\}\Big) \le 2(t+1)^{2^{k(h+\epsilon/2)}} 2^{-tC\delta^2/100},$$

which, combined with (8) and (7), produces the bound

$$\mu\Big(\{x_1^n\colon\ |p_k(\cdot\,|x_1^n) - \mu_k| > \delta\}\Big) \le 2k(t+1)^{2^{k(h+\epsilon/2)}} 2^{-tC\delta^2/100}, \tag{9}$$

valid for all $k \ge K$ and $n \ge k/\gamma$. Note that k can be replaced by $k(n)$ in this bound, since it was assumed that $k(n) \to \infty$, as $n \to \infty$.

The bound (9) is actually summable in n. To show this put

$$\alpha = C\delta^2/100,\quad \beta = \frac{h+\epsilon/2}{h+\epsilon}.$$

The right-hand side of (9) is then upper bounded by

$$2\frac{\log n}{h+\epsilon}(t+1)^{n^\beta} 2^{-\alpha t}, \tag{10}$$

since $k(n) \le (\log n)/(h+\epsilon)$, by assumption. This is, indeed, summable in n, since $t \sim n/k(n)$ and $\beta < 1$. This establishes that the sum in (6) is finite, thereby completing the proof of the admissibility theorem for i.i.d. processes. □

The preceding argument applied to the ψ-mixing case yields the bound

$$\mu\Big(\{x_1^n\colon\ |p_k(\cdot\,|x_1^n) - \mu_k| \ge \delta\}\Big) \le 2(k+g)[\psi(g)]^t\ (t+1)^{2^{k(h+\epsilon/2)}} 2^{-tC\delta^2/100}, \tag{11}$$

valid for all g, all $k \ge K = K(g)$, and all $n \ge k/\gamma$, where γ is a positive constant which depends only on g. To obtain the desired summability result it is only necessary to choose

g so large that $\psi(g) < 2^{C\delta^2/200}$. With $\alpha = C\delta^2/200$ and $\beta = (h+\epsilon/2)/(h+\epsilon)$, and assuming that $k(n) \to \infty$, subject only to the requirement that $k(n) \le (\log n)/(h+\epsilon)$, the right-hand side in (11) is, in turn, upper bounded by

$$\frac{\log n}{h+\epsilon} 2(t+1)^{n^\beta} 2^{-\alpha t},$$

which is again summable in n, since since $t \sim n/k(n)$ and $\beta < 1$. This establishes the positive admissibility theorem for the ψ-mixing case, hence for the aperiodic Markov case. The extension to the periodic Markov case is obtained by a similar modification of the exponential-rates result for periodic chains, see Lemma III.1.9. □

III.2.b The weak Bernoulli case.

A process has the weak Bernoulli property if past and future become almost independent when separated by enough, where the variational distance is used as the measure of approximation. To make this precise, let $\mu_n(\cdot|x_{-g-m}^{-g})$ denote the conditional measure on n-steps into the future, conditioned on the past $\{x_j,\ -g-m \le j \le -g\}$, where g and m are nonnegative integers, that is, the measure defined for $a_1^n \in A^n$ by

$$\mu_n(a_1^n|x_{-g-m}^{-g}) = \frac{\text{Prob}(X_1^n = a_1^n \text{ and } X_{-g-m}^{-g} = x_{-g-m}^{-g})}{\text{Prob}(X_{-g-m}^{-g} = x_{-g-m}^{-g})}.$$

Also let $E_{X_{-g-m}^{-g}}(f)$ denote conditional expectation of a function f with respect to the random vector X_{-g-m}^{-g}.

A stationary process $\{X_i\}$ is *weak Bernoulli (WB)* or *absolutely regular*, if given $\epsilon > 0$ there is a gap $g \ge 0$ such that

$$E_{X_{-g-m}^{-g}}\left(|\mu_n(\cdot|X_{-g-m}^{-g}) - \mu_n|\right) \le \epsilon, \tag{12}$$

for all $n \ge 1$ and all $m \ge 0$. An equivalent form is obtained by letting $m \to \infty$ and using the martingale theorem to obtain the condition

$$E_{X_{-\infty}^{-g}}\left(|\mu_n(\cdot|X_{-\infty}^{-g}) - \mu_n|\right) \le \epsilon,\quad n \ge 1. \tag{13}$$

The weak Bernoulli property is much stronger than mixing, which only requires that for each $m \ge 0$ and $n \ge 1$, there is a gap g for which (12) holds. It is much weaker than ψ-mixing, however, which requires that $\mu_n(a_1^n|x_{-g-m}^{-g}) \le (1+\epsilon)\mu_n(a_1^n)$, uniformly in $m \ge 0$, $n \ge 1$, a_1^n, and x_{-g-m}^{-g}, provided only that g be large enough.

The key to the ψ-mixing result was the fact that the measure on shifted, nonoverlapping blocks could be upper bounded by the product measure on the blocks, with only a small exponential error. The weak Bernoulli property leads to a similar bound, at least for a large fraction of shifts, provided a small fraction of blocks are omitted and conditioning on the past is allowed, and this is enough to obtain the admissibility theorem.

As before, the s-shifted, k-block parsing of x_1^n, with gap g is the expression

$$x_1^n = w(0)g(1)w(1)g(2)w(2)\dots g(t)w(t)w(t+1), \tag{14}$$

where $w(0)$ has length s, the g-blocks $g(j)$ alternate with the k-blocks $w(j)$, for $1 \leq j \leq t$, and the final block $w(t+1)$ has length $n - t(k+g) - s < 2(k+g)$. Note that the k-block $w(j)$ starts at index $s + (j-1)(k+g) + g + 1$ and ends at index $s + j(k+g)$.

An index $j \in [1, t]$ is called a *(γ, s, k, g)-splitting index* for $x \in A^Z$ if

$$\mu(w(j) \mid x_{-\infty}^{s+(j-1)(k+g)}) < (1+\gamma)\mu(w(j)).$$

Here, and later, $\mu(\cdot \mid x_{-\infty}^i)$ denotes the conditional measure on the infinite past, defined as the limit of $\mu(\cdot \mid x_{-m}^i)$ as $m \to \infty$, which exists almost surely by the martingale theorem. The set of all $x \in A^Z$ for which j is a (γ, s, k, g)-splitting index will be denoted by $B_j(\gamma, s, k, g)$, or by B_j if s, γ, k, and g are understood. Note that B_j is measurable with respect to the coordinates $i \leq s + j(k+g)$.

Lemma III.2.7

Fix (γ, s, k, g) and fix a finite set J of positive integers. Then for any assignment $\{w(j): j \in J\}$ of k-blocks

$$\mu\left(\bigcap_{j \in J}([w(j)] \cap B_j)\right) \leq (1+\gamma)^{|J|} \prod_{j \in J} \mu(w(j)).$$

Proof. Put $j_m = \max\{j : j \in J\}$ and condition on $B^* = \bigcap_{j \in J - \{j_m\}} \left([w(j)] \cap B_j\right)$ to obtain

$$\mu\left(\bigcap_{j \in J}([w(j)] \cap B_j)\right) = \mu\left([w(j_m)] \cap B_{j_m} \,\middle|\, B^*\right) \mu(B^*). \tag{15}$$

The first factor is an average of the measures $\mu\left([w(j_m)] \cap B_{j_m} \mid x_{-\infty}^{s+(j_m-1)(k+g)}\right)$, each of which satisfies

$$\mu\left([w(j_m)] \cap B_{j_m} \mid x_{-\infty}^{s+(j_m-1)(k+g)}\right) \leq (1+\gamma)\mu(w(j_m)),$$

by the definition of B_{j_m}. Thus (15) yields

$$\mu\left(\bigcap_{j \in J}([w(j)] \cap B_j)\right) \leq (1+\gamma) \cdot \mu(w(j_m)) \cdot \mu\left(\bigcap_{j \in J - \{j_m\}}([w(j)] \cap B_j)\right),$$

and Lemma III.2.7 follows by induction. □

The weak Bernoulli property guarantees the almost-sure existence of a large density of splitting indices for most shifts.

Lemma III.2.8 (The weak Bernoulli splitting-set lemma.)

If μ is weak Bernoulli and $0 < \gamma < 1/2$, then there is a gap $g = g(\gamma)$, there are integers $k(\gamma)$ and $t(\gamma)$, and there is a sequence of measurable sets $\{G_n(\gamma)\}$, such that the following hold.

(a) *$x \in G_n(\gamma)$, eventually almost surely.*

(b) *If $k \geq k(\gamma)$, if $t \geq t(\gamma)$, and if $(t+1)(k+g) \leq n < (t+2)(k+g)$, then for $x \in G_n(\gamma)$ there are at least $(1-\gamma)(k+g)$ values of $s \in [0, k+g-1]$ for each of which there are at least $(1-\gamma)t$ indices j in the interval $[1, t]$ that are (γ, s, k, g)-splitting indices for x.*

Proof. By the weak Bernoulli property there is a gap $g = g(\gamma)$ so large that for any k

$$\int \mu(x_1^k | x_{-\infty}^{-g}) \left| 1 - \frac{\mu(x_1^k)}{\mu(x_1^k | x_{-\infty}^{-g})} \right| d\mu(x) < \frac{\gamma^4}{4}.$$

Fix such a g and for each k define

$$f_k(x) = \frac{\mu(x_1^k)}{\mu(x_1^k | x_{-\infty}^{-g})}.$$

Let Σ_k be the σ-algebra determined by the random variables, $\{X_i : i \leq -g\} \cup \{X_i : 1 \leq i \leq k\}$. Direct calculation shows that each f_k has expected value 1 and that $\{f_k\}$ is a martingale with respect to the increasing sequence $\{\Sigma_k\}$, see Exercise 7b. Thus f_k converges almost surely to some f. Fatou's lemma implies that

$$\int \mid 1 - f(x) \mid d\mu(x) \leq \frac{\gamma^4}{4},$$

so there is an M such that if

$$C_M = \left\{ x : \mid 1 - f_k(x) \mid \leq \frac{\gamma^2}{2}, \ \forall k \geq M \right\},$$

then $\mu(C_M) > 1 - (\gamma^2/2)$. The ergodic theorem implies that

$$\lim_{N \to \infty} \frac{1}{N} \sum_{i=0}^{N-1} \mathcal{K}_{C_M}(T^i x) > 1 - \frac{\gamma^2}{2} \text{ a.s.},$$

where $\mathcal{K}_{C_M}$ denotes the indicator function of C_M, so that if

$$G_n(\gamma) = \left\{ x : \frac{1}{n} \sum_{i=0}^{n-1} \mathcal{K}_{C_M}(T^i x) > 1 - \frac{\gamma^2}{2} \right\},$$

then $x \in G_n(\gamma)$, eventually almost surely.

Put $k(\gamma) = M$ and let $t(\gamma)$ be any integer larger than $2/\gamma^2$. Fix $k \geq M$, $t \geq t(\gamma)$, and $(t+1)(k+g) \leq n < (t+2)(k+g)$, and fix an $x \in G_n(\gamma)$. The definition of $G_n(\gamma)$ and the assumption $t \geq 2/\gamma^2$ imply that

$$\frac{1}{t(k+g)} \sum_{i=0}^{t(k+g)-1} \mathcal{K}_{C_M}(T^i x) = \frac{1}{k+g} \sum_{s=0}^{k+g-1} \frac{1}{t} \sum_{j=1}^{t} \mathcal{K}_{C_M}(T^{s+(j-1)(k+g)} x) > 1 - \gamma^2,$$

so there is a subset $S = S(x) \subseteq [0, k+g-1]$ of cardinality at least $(1-\gamma)(k+g)$ such that for $x \in G_n(\gamma)$ and $s \in S(x)$

$$\frac{1}{t} \sum_{j=1}^{t} \mathcal{K}_{C_M}(T^{s+(j-1)(k+g)} x) > 1 - \gamma.$$

In particular, if $s \in S(x)$, then $T^{s+(j-1)(k+g)}x \in C_M$, for at least $(1-\gamma)t$ indices $j \in [1, t]$. But if $T^{r+(j-1)(k+g)}x \in C_M$ then

$$\mu\big(w(j)|x_{-\infty}^{s+(j-1)(k+g)-g}\big) < (1+\gamma)\mu\big(w(j)\big),$$

which implies that j is a (γ, s, k, g)-splitting index for x.

In summary, for $x \in G_n(\gamma)$ and $s \in S(x)$ there are at least $(1-\gamma)t$ indices j in the interval $[1, t]$ which are (γ, s, k, g)-splitting indices for x. Since $|S(x)| \geq (1-\delta)(k+g)$, this completes the proof of Lemma III.2.8. □

For $J \subseteq [1, t]$ define

$$p_{k,g}^{s,J}(a_1^k|x_1^n) = \frac{|\{j \in J : w(j) = a_1^k\}|}{|J|}, \quad a_1^k \in A^k,$$

that is, the empirical distribution of k-blocks obtained by looking only at those k-blocks $w(j)$ for which $j \in J$.

The overlapping-block measure p_k is (almost) an average of the measures $p_{k,g}^{s,J}$, provided the sets J are large fractions of $[1, t]$, where "almost" is now needed to account for end effects, for gaps, and for the $j \notin J$. If k is large relative to gap length and n is large relative to k, this is no problem, however, and Lemma III.1.8 easily extends to the following sharper form in which the conclusion holds for a positive fraction of shifts.

Lemma III.2.9

Given $\delta > 0$ there is a positive $\gamma < 1/2$ such that for any g there is a $K = K(g, \gamma)$ such that if $k \geq K$, if $k/n < \gamma$, and if $|p_k(\cdot|x_1^n) - \mu_k| \geq \delta$ then $|p_{k,g}^{s,J}(\cdot|x_1^n) - \mu_k| \geq \delta/4$ for at least $2\gamma(k+g)$ indices $s \in [0, k+g-1]$, for any subset $J \subset [1, t]$ of cardinality at least $(1-\gamma)t$.

Proof of weak Bernoulli admissibility, Theorem III.2.3.

Assume μ is weak Bernoulli of entropy h, and $k(n) \leq (\log n)/(h+\epsilon)$, $n \geq 1$. Fix $\delta > 0$, choose a positive $\gamma < 1/2$, then choose integers $g = g(\gamma)$, $k(\gamma)$ and $t(\gamma)$, and measurable sets $G_n = G_n(\gamma)$, $n \geq 1$, so that conditions (a) and (b) of Lemma III.2.8 hold.

Fix $t \geq t(\gamma)$ and $(t+1)(k+g) \leq n < (t+2)(k+g)$, where $k(\gamma) \leq k \leq (\log n)/(h+\epsilon)$. For $s \in [0, k+g-1]$ and $J \subseteq [1, t]$, let $D_n(s, J)$ be the set of those sequences x for which every $j \in J$ is a (γ, s, k, g)-splitting index, and note that

$$\bigcap_{j \in J} B_j = D_n(s, J),$$

so that Lemma III.2.7 and the fact that $|J| \leq t$ yield

$$\mu\left(\bigcap_{j \in J}[w(j)] \cap D_n(s, J)\right) \leq (1+\gamma)^t \prod_{j \in J} \mu\left(w(j)\right). \tag{16}$$

If $x \in G_n(\gamma)$ then Lemma III.2.8 implies that there are $(1-\gamma)(k+g)$ indices $s \in [0, k+g-1]$ for each of which there are at least $(1-\gamma)t$ indices j in the interval $[1, t]$ that are (γ, s, k, g)-splitting indices for x.

On the other hand, γ can be assumed to be so small and t so large that Lemma III.2.9 assures that if $\left|p_k(\cdot|x_1^n) - \mu_k\right| \geq \delta$ then $\left|p_{k,g}^{s,J}(\cdot|x_1^n) - \mu_k\right| \geq \delta/4$ for at least $2\gamma(k+g)$ indices s, for any subset $J \subset [1, t]$ of cardinality at least $(1-\gamma)t$. Thus if γ is sufficiently small, and $k \geq k(\gamma)$ and $t \geq t(\gamma)$ are sufficiently large, then for any $x \in G_n(\gamma)$ there exists *at least one* $s \in [0, k+g-1]$ and *at least one* $J \subseteq [1, t]$ of cardinality at least $(1-\gamma)t$, for which $x \in D_n(s, J)$ and $\left|p_{k,g}^{s,J}(\cdot|x_1^n) - \mu_k\right| \geq \delta/4$. In particular, the set

$$\left\{x: |p_k - \mu_k| \geq \delta\right\} \cap G_n(\gamma) \tag{17}$$

is contained in the set

$$\bigcup_{s=0}^{k+g-1} \bigcup_{\substack{J \subseteq [1,t] \\ |J| \geq (1-\gamma)t}} \left(\left\{x: |p_{k,g}^{s,J}(\cdot|x_1^n) - \mu_k| \geq \delta/4\right\} \bigcap D_n(s, J)\right).$$

The proof of weak Bernoulli admissibility can now be completed very much as the proof for the ψ-mixing case. Using the argument of that proof, the measure of the set (17) is upper bounded by

$$2 \cdot 2^{-2t\gamma \log \gamma}(1+\gamma)^t [k(n)+g](t+1)^{2^{k(n)(h+\epsilon/2)}} 2^{-t(1-\gamma)C\delta^2/400}, \tag{18}$$

for t sufficiently large. This bound is the counterpart of (11), with an extra factor, $2^{-2t\gamma \log \gamma}$, to bound the number of subsets $J \subseteq [1, t]$ of cardinality at least $(1-\gamma)t$. If γ is small enough, then, as in the ψ-mixing case, the bound (18) is summable in n. Since $x \in G_n(\gamma)$, eventually almost surely, this establishes Theorem III.2.3. □

Remark III.2.10

The admissibility results are based on joint work with Marton, [38]. The weak Bernoulli concept was introduced by Friedman and Ornstein, [13], as part of their proof that aperiodic Markov chains are isomorphic to i.i.d. processes in the sense of ergodic theory. Aperiodic Markov chains are weak Bernoulli because they are ψ-mixing. Aperiodic renewal and regenerative processes, which need not be ψ-mixing are, however, weak Bernoulli, see Exercises 2 and 3, in Section IV.3. It is shown in Chapter 4 that weak Bernoulli processes are stationary codings of i.i.d. processes.

III.2.c Exercises.

1. Show that there exists an ergodic measure μ with positive entropy h such that $k(n) \sim (\log n)/(h+\epsilon)$ is admissible for μ, yet μ does not have exponential rates of convergence for frequencies.

2. Show that for any nondecreasing, unbounded sequence $\{k(n)\}$ there is an ergodic process μ such that $\{k(n)\}$ is not admissible for μ.

3. Show that $k(n) = \lfloor \log n \rfloor$ is not admissible for unbiased coin-tossing.

4. Show that if μ is i.i.d., if $k(n) \leq (\log n)/(h+\epsilon)$, and if $B_n = \{x_1^n\colon |p_{k(n)}(\cdot|x_1^n) - \mu_{k(n)}| \geq \delta\}$, then $\mu(B_n)$ goes to 0 exponentially fast.

5. Assume $t = \lfloor n/k \rfloor$ and $x_1^n = w(1)\cdots w(t)r$, where each $w(i)$ has length k. Let $\mathcal{R}(x_1^n, k, \epsilon)$ be the set of all kt-sequences that can be obtained by changing an ϵ-fraction of the $w(i)$'s and permuting their order. Show that if $k(n) \sim (\log n)/(h+\epsilon)$, and μ is i.i.d., then $\mu(\mathcal{R}(x_1^n, k(n), \epsilon)) \to 1$, as $n \to \infty$, almost surely.

6. Show that the preceding result holds for weak Bernoulli processes.

7. Let $\{Y_j : j \geq 1\}$ be a sequence of finite-valued random variables, let $\Sigma(m)$ be the σ-algebra determined by $Y_1, \ldots, Y_m$, and let Σ be the smallest complete σ-algebra containing $\cup_m \Sigma(m)$. For each x let $P_m(x)$ be the atom of $\Sigma(m)$ that contains x.

   (a) Show that

$$E(g|\Sigma)(x) = \lim_{m\to\infty} \frac{1}{\mu(P_m(x))} \int_{P_m(x)} g \, d\mu, \text{ a. s.,}$$

   for any integrable function g. (Hint: use the martingale theorem.)

   (b) Show that the sequence f_k defined in the proof of Lemma III.2.8 is indeed a martingale. (Hint: use the previous result with $Y_j = (x_1^k, x_{-j}^{-g})$ to evaluate $E(f_{k+1}|\Sigma_k)$.)

Section III.3 The $\bar{d}$-admissibility problem.

The $\bar{d}$-metric forms of the admissibility results of the preceding section are also of interest. A sequence $\{k(n)\}$ is called *$\bar{d}$-admissible* for the ergodic process μ if

$$\lim_n \bar{d}_n(p_{k(n)}(\cdot\,|x_1^n),\ \mu_{k(n)}) = 0,\ a.s. \tag{1}$$

The definition of $\bar{d}$-admissible in probability is obtained by replacing almost-sure convergence by convergence in probability. The earlier concept of admissible will now be called *variationally admissible.* Since the bound

$$\bar{d}_n(\mu, \nu) \leq \frac{1}{2} \sum_{a_1^n} |\mu(a_1^n) - \nu(a_1^n)|, \tag{2}$$

always holds a variationally-admissible sequence is also $\bar{d}$-admissible.

The principal positive result is

Theorem III.3.1 (The $\bar{d}$-admissibility theorem.)
If $k(n) \leq (\log n)/h$ then $\{k(n)\}$ is $\bar{d}$-admissible for any process of entropy h which is a stationary coding of an i.i.d. process.

Stated in a somewhat different form, the theorem was first proved by Ornstein and Weiss, [52]. A similar result holds with the nonoverlapping-block empirical distribution in place of the overlapping-block empirical distribution, though only the latter will be discussed here. Note that in the $\bar{d}$-metric case, the admissibility theorem only requires that $k(n) \leq (\log n)/h$, while in the variational case the condition $k(n) \leq (\log n)/(h+\epsilon)$ was needed.

The $\bar{d}$-admissibility theorem has a fairly simple proof, based on the empirical entropy theorem, Theorem II.3.1, and a deep property, called the finitely determined property, which holds for stationary codings of i.i.d. processes, and which will be proved in Section IV.2. A proof of the $\bar{d}$-admissibility theorem, assuming this finitely determined property, is given in Section III.3.a.

The negative results are two-fold. One is the simple fact that if n is too short relative to k, then a small neighborhood of the empirical universe of k-blocks in an n-sequence is too small to allow admissibility. The other is a much deeper result asserting in a very strong way that no sequence can be admissible for every ergodic process.

Theorem III.3.2 (The $\bar{d}$-nonadmissibility theorem.)
If $k(n) \geq (\log n)/(h-\epsilon)$, then $\{k(n)\}$ is not admissible for any ergodic process of entropy h.

Theorem III.3.3 (The strong-nonadmissibility theorem.)
For any nondecreasing unbounded sequence $\{k(n)\}$ and $0 < \alpha < 1/2$, there is an ergodic process μ such that

$$\liminf_{n\to\infty} \bar{d}_{k(n)}(p_{k(n)}(\cdot\,|x_1^n),\ \mu_{k(n)}) \geq \alpha,\ \textit{almost surely.}$$

The $\bar{d}$-nonadmissibility theorem is a consequence of the fact that no more than $2^{k(n)(h-\epsilon)}$ sequences of length $k(n)$ can occur in a sequence of length $n \leq 2^{k(n)(h-\epsilon)}$, so that at most $2^{k(n)(h-\epsilon/2)}$ sequences can be within δ of one of them, provided δ is small enough. It will be proved in the next subsection. A weaker, subsequence version of strong nonadmissibility was first given in [52], then extended to the form given here in [50]. These results will be discussed in a later subsection.

III.3.a Admissibility and nonadmissibility proofs.

To establish the $\bar{d}$-nonadmissibility theorem, define the (empirical) universe of k-blocks of x_1^n,

$$\mathcal{U}_k(x_1^n) = \left\{a_1^k : x_i^{i+k-1} = a_1^k,\ \text{for some } i \in [0, n-k+1]\right\},$$

and its δ-blowup

$$[\mathcal{U}_k(x_1^n)]_\delta = \{a_1^k:\ d_k(a_1^k, \mathcal{U}_k(x_1^n)) \leq \delta\}.$$

If $k \geq (\log n)/(h-\epsilon)$ then $|\mathcal{U}_k(x_1^n)| \leq 2^{k(h-\epsilon)}$, for any x_1^n. If δ is small enough then for all k,

$$|[\mathcal{U}_k(x_1^n)]_\delta| \leq 2^{k(h-\epsilon/2)},$$

by the blowup-bound, Lemma I.7.5, and hence, intersecting with the entropy-typical set, $\mathcal{T}_k = \left\{x_1^k:\ \mu(x_1^k) \leq 2^{-k(h-\epsilon/4)}\right\}$, produces

$$\mu_k\left([\mathcal{U}_k(x_1^n)]_\delta \cap \mathcal{T}_k\right) \leq 2^{k(h-\epsilon/2)}2^{-k(h-\epsilon/4)} \leq 2^{-k\epsilon/4},$$

The entropy theorem implies that $\mu(\mathcal{T}_k) \to 1$, so that $\mu([\mathcal{U}_k(x_1^n)]_\delta) \leq \delta$, if k is large enough. But if this holds then

$$\bar{d}_k(\mu_k, p_k(\cdot\,|x_1^n)) \geq \delta^2,$$

for otherwise Lemma I.9.14(a) gives $\mu_k([\mathcal{U}_k(x_1^n)]_\delta) \geq 1-2\delta$, since $p_k(\mathcal{U}_k(x_1^n)|x_1^n) = 1$. This completes the proof of the $\bar{d}$-nonadmissibility theorem, Theorem III.3.2. □

The finitely determined property asserts that a process has the finitely determined property if any process close enough in joint distribution and entropy must also be $\bar{d}$-close. An equivalent finite form, see Theorem IV.2.2, which is more suitable for use here and in Section III.4, is expressed in terms of the averaging of finite distributions. If ν is a measure on A^n, and $m \leq n$, *the ν-distribution of m-blocks in n-blocks* is the measure $\phi_m = \phi(m, \nu)$ on A^m defined by

$$\phi_m(a_1^m) = \frac{1}{n-m+1}\sum_{i=0}^{n-m}\sum_{u\in A^i}\sum_{v\in A^{n-m-i}} \nu(ua_1^m v),$$

that is, the average of the ν-probability of a_1^m over all except the final $m-1$ starting positions in sequences of length n. An alternative expression is

$$\phi_m(a_1^m) = \sum_{x_1^n} p_m(a_1^m|x_1^n)\nu(x_1^n) = E_\nu(p_m(a_1^m|X_1^n)),$$

which makes it clear that $\phi(m, \nu_n) = \nu_m$ for stationary ν.

A stationary process μ is *finitely determined* if for any $\epsilon > 0$ there is a $\delta > 0$ and positive integers m and K such that if $k \geq K$ then any measure ν on A^k which satisfies the two conditions

(i) $|\mu_m - \phi(m, \nu)| < \delta$,

(ii) $|H(\mu_k) - H(\nu)| < k\delta$,

must also satisfy $\bar{d}_k(\mu_k, \nu_k) < \epsilon$.

The important fact needed for the current discussion is that *a stationary coding of an i.i.d. process is finitely determined*, see Section IV.2.

The $\bar{d}$-admissibility theorem is proved by taking $\nu = p_k(\cdot\,|x_1^n)$, and noting that the sequence $\{k(n)\}$ can be assumed to be unbounded, since the bounded result is true for any ergodic process, by the ergodic theorem. Thus, by the definition of finitely determined, it is sufficient to prove that if $\{k(n)\}$ is unbounded and $k(n) \leq (1/h)\log n$, then, almost surely as $n \to \infty$, the following hold for any ergodic process μ of entropy h.

(a) $|\phi(m, p_{k(n)}(\cdot\,|x_1^n)) - \mu_m| \to 0$, for each m.

(b) $(1/k(n))H(p_{k(n)}(\cdot\,|x_1^n)) \to h$.

Convergence of entropy, fact (b), is really just the empirical-entropy theorem, Theorem II.3.1, for it asserts that $(1/k)H(p_k(\cdot\,|x_1^n)) \to h = H(\mu)$, almost surely, as k and n go to infinity, provided only that $k \leq (1/h)\log n$.

To establish convergence in m-block distribution, fact (a), note that the averaged distribution $\phi(m, p_k(\cdot\,|x_1^n))$ is almost the same as $p_m(\cdot\,|x_1^{n-m})$, the only difference being that the former gives less than full weight to those m-blocks that start in the initial $k-1$ places or end in the final $k-1$ places x_1^n, a negligible effect if n is large enough. The desired result then follows since $p_m(a_1^m|x_1^n) \to \mu(a_1^m)$, almost surely, by the ergodic theorem.

This completes the proof of the $\bar{d}$-admissibility theorem, modulo, of course, the proof that stationary codings of i.i.d. processes are finitely determined. □

III.3.b Strong-nonadmissibility examples.

Throughout this section $\{k(n)\}$ denotes a fixed, nondecreasing, unbounded sequence of positive integers. Associated with the window function $k(\cdot)$ is the path-length function, $n(k) = \max\{n\colon k(n) \le k\}$. Let

$$B_k(\alpha) = \left\{x_1^{n(k)}\colon \bar{d}_k(p_k(\cdot\,|x_1^{n(k)}), \mu_k) < \alpha\right\}.$$

It is enough to show that for any $0 < \alpha < 1/2$ there is an ergodic measure μ for which

$$\lim_{K\to\infty} \mu\left(\bigcup_{k\ge K} B_k(\alpha)\right) = 0 \tag{3}$$

The construction is easy if ergodicity is not an issue, and $k(n) \le \log n$, for one can take μ to be the average of a large number, $\{\mu^{(j)}\colon j \le J\}$, of $\{k(n)\}$-$\bar{d}$-admissible ergodic processes which are mutually α' apart in $\bar{d}$, where $\alpha'(1-1/J) \ge \alpha$. A given x is typical for at most one component $\mu^{(j)}$, in which case $p_k(\cdot\,|x_1^{n(k)})$ is mostly concentrated on $\mu^{(j)}$-typical sequences, and hence

$$\liminf_{k\to\infty} \bar{d}_k(p_k(\cdot\,|x_1^{n(k)}), \mu_k) \ge \alpha'(1-1/J),$$

almost surely, see Exercise 1.

This simple nonergodic example suggests the basic idea: build an ergodic process whose n-length sample paths, when viewed through a window of length $k(n)$, look as though they are drawn from a large number of mutually far-apart ergodic processes. This must happen for every value of n, yet at the same time, merging of these far-apart processes must be taking place in order to obtain a final ergodic process. The trick is to do the merging itself in different ways on different parts of the space, keeping the almost the same separation at the same time as the merging is happening. The tool for managing all this is cutting and stacking, which is well-suited to the tasks of both merging and separation.

Before beginning the construction, some discussion of the relation between d_k-far-apart sequences and $\bar{d}_k$-far-apart measures is in order. The easiest way to make measures $\bar{d}$-far-apart is to make supports disjoint, even when blown-up by α. Two sets $C, D \subseteq A^k$ are said to be *α-separated*, if $C \cap [D]_\alpha = \emptyset$, where, as before,

$$[D]_\alpha = \{x_1^k\colon d_k(x_1^k, y_1^k) \le \alpha, \text{ for some } y_1^k \in D\}$$

is the α-blowup of D. Thus, α-separated means that at least αk changes must be made in a member of C to produce a member of D.

The support $\sigma(\mu)$ of a measure μ on A^k is the set of all x_1^k for which $\mu(x_1^k) > 0$. Two measures μ and ν on A^k are said to be *α-separated* if their supports are α-separated. A simple, but important fact, is

(4) If μ and ν are α-separated, then $\bar{d}_k(\mu, \nu) \ge \alpha$.

Indeed, if λ is any joining of two α-separated measures, μ and ν, then

$$E_\lambda(d_k(x_1^k, y_1^k)) \ge \alpha\lambda(\sigma(\mu)\times\sigma(\nu)) = \alpha.$$

The result (4) extends to averages of separated families in the following form.

Lemma III.3.4

If $\mu = (1/J)\sum_j \nu_j$ is the average of a family $\{\nu_j\colon\ j \le J\}$ of pairwise α-separated measures, then $\bar{d}(\mu, \nu) \ge \alpha(1 - 1/J)$, for any measure ν whose support is entirely contained in the support of one of the ν_j.

Proof. The α-separation condition implies that the α-blowup of the support of ν_j does not meet the support of ν_i, for $i \ne j$, and hence

$$\mu([\sigma(\nu_j)]_\alpha) = \frac{1}{J}\sum_{i=1}^{J} \nu_i([\sigma(\nu_j)]_\alpha) = \nu_j([\sigma(\nu_j)]_\alpha) = \frac{1}{J}.$$

Thus, if the support of ν is contained in the support of ν_j, then $\mu([\sigma(\nu)]_\alpha) \le 1/J$, and hence

$$E_\lambda(d_k(x_1^k, y_1^k)) \ge \alpha \cdot \nu(\sigma(\nu)) \cdot [1 - \mu([\sigma(\nu)]_\alpha)] \ge \alpha(1 - 1/J),$$

for any joining λ of μ and ν, from which the lemma follows. □

With these simple preliminary ideas in mind, the basic constructions can begin. A weaker liminf result will be described in the next subsection, after which a sketch of the modifications needed to produce the stronger result (3) will be given.

III.3.b.1 A limit inferior result.

Here it will be shown, in detail, how to produce an ergodic μ and an increasing sequence $\{k_m\}$ such that

$$\lim_{m\to\infty} \mu(B_{k_m}(\alpha)) = 0. \tag{5}$$

The construction, based on [52], illustrates most of the ideas involved and is a bit easier to understand than the one used to obtain the full limit result, (3).

To achieve (5), a complete sequence $\{\mathcal{S}(m)\}$ of column structures and an increasing sequence $\{k_m\}$ will be constructed such that a randomly chosen point x which does not lie in the top $n(k_m)$-levels of any column of $\mathcal{S}(m)$ must satisfy

$$\bar{d}_{k_m}(p_{k_m}(\cdot\ |x_1^{n(k_m)}), \mu_{k_m}) \ge \alpha_m, \tag{6}$$

where x_i is the label of the interval containing $T^{i-1}x$, for $1 \le i \le n(k_m)$. The sequence α_m will decrease to 0 and the total measure of the top $n(k_m)$-levels of $\mathcal{S}(m)$ will be summable in m, guaranteeing that (5) holds. Furthermore, ergodicity will be guaranteed by making sure that $\mathcal{S}(m)$ and $\mathcal{S}(m+1)$ are sufficiently independent.

Lemma III.3.4 suggests a strategy for making (6) hold, namely, take $\mathcal{S}(m)$ to be the union of a disjoint collection of column structures $\{\mathcal{S}_j(m)\colon\ j \le J_m\}$ such that any k_m-block which appears in the name of any column of $\mathcal{S}_j(m)$ must be at least α_m apart from any k_m-block which appears in the name of any column of $\mathcal{S}_i(m)$ for $i \ne j$. If all the columns in $\mathcal{S}(m)$ have the same width and height $\ell(m)$, and if the cardinality of $\mathcal{S}_j(m)$ is constant in j, then, conditioned on being $n(k_m)$-levels below the top, the measure is the average of the conditional measures on the separate $\mathcal{S}_j(m)$. Thus if $\ell(m)/n(k_m)$ is summable in m, then Lemma III.3.4 guarantees that the $\bar{d}$-property (6) will indeed hold.

Some terminology will be helpful. A column structure $\mathcal{S}$ is said to be *uniform* if its columns have the same height $\ell(\mathcal{S})$ and width. The k-block universe $\mathcal{U}(\mathcal{S})$ of a column structure $\mathcal{S}$ is the set of all a_1^k that appear as a block of consecutive symbols in the name of any column in $\mathcal{S}$. Disjoint column structures $\mathcal{S}$ and $\mathcal{S}'$ are said to be (α, k)-*separated* if their k-block universes are α-separated.

To summarize the discussion up to this point, the goal (6) can be achieved with an ergodic measure for a given $\alpha \in (0, 1/2)$ by constructing a complete sequence $\{\mathcal{S}(m)\}$ of column structures and an increasing sequence $\{k_m\}$ with the following properties.

(A1) $\mathcal{S}(m)$ is uniform with height $\ell(m) \geq 2^m n(k_m)$.

(A2) $\mathcal{S}(m)$ and $\mathcal{S}(m+1)$ are 2^{-m}-independent.

(A3) $\mathcal{S}(m)$ is a union of a disjoint family $\{\mathcal{S}_j(m)\colon j \leq J_m\}$ of pairwise (α_m, k_m)-separated column structures, for which the cardinality of $\mathcal{S}_j(m)$ is constant in j, and for which $\alpha_m(1 - 1/J_m)$ decreases to α, as $m \to \infty$.

Since all the columns of $\mathcal{S}(m)$ will have the same width and height, and there is no loss in assuming that distinct columns have the same name, the simpler concatenation language will be used in place of cutting and stacking language. For this reason $\mathcal{S}(m)$ will be taken to be a subset of $A^{\ell(m)}$. Conversion back to cutting and stacking language is achieved by replacing $\mathcal{S}(m)$ by its columnar representation with all columns equally likely, and concatenations of sets by independent cutting and stacking of appropriate copies of the corresponding column structures.

There are many ways to force the initial stage $\mathcal{S}(1)$ to have property (A3). The real problem is how to go from stage m to stage $m+1$ so that separation holds, yet asymptotic independence is guaranteed. This is, of course, the simultaneous merging and separation problem. The following four sequences suggest a way to do this.

$$(7)\qquad \begin{aligned} a &= 01010\ldots0101\ldots && 01 &&= (01)^{64} \\ b &= 00001111000011110\ldots && 01111 &&= (0^4 1^4)^{16} \\ c &= 000000000000000001\ldots\ldots && 11 &&= (0^{16} 1^{16})^4 \\ d &= 000\ldots && \ldots 11 &&= (0^{64} 1^{64})^1. \end{aligned}$$

These sequences are created by concatenating the two symbols, 0 and 1, using a rapidly increasing period from one sequence to the next. Each sequence has the same frequency of occurrence of 0 and 1, so first-order frequencies are good, yet if a block x_1^{32} of 32 consecutive symbols is drawn from one of them and a block y_1^{32} is drawn from another, then $d_{32}(x_1^{32}, y_1^{32}) \geq 1/2$.

The construction (7) suggests a way to merge while keeping separation, namely, concatenate blocks according to a rule that specifies to which set the m-th block in the concatenation belongs. If one starts with enough far-apart sets, then by using cyclical rules with rapidly growing periods many different sets can be produced that are almost as far apart.

The idea of merging rule is formalized as follows. For M divisible by J, an (M, J)-*merging rule* is a function

$$\phi\colon \{1, 2, \ldots, M\} \mapsto \{1, 2, \ldots, J\},$$

whose level sets are constant, that is,

$$|\phi^{-1}(j)| = M/J, \quad j \leq J.$$

An (M, J)-merging rule ϕ, when applied to a collection $\{\mathcal{S}_j\colon j \in J\}$ of disjoint subsets of A^ℓ produces the subset $\mathcal{S}(\phi) = \mathcal{S}(\phi, \{\mathcal{S}_j\})$ of $A^{M\ell}$, formed by concatenating the sets in the order specified by ϕ, that is, the direct product

$$\mathcal{S}(\phi) = \prod_{m=1}^{M} \mathcal{S}_{\phi(m)} = \left\{ w(1)w(2)\cdots w(M)\colon\ w(m) \in \mathcal{S}_{\phi(m)},\ m \in [1, M] \right\}.$$

In other words, $\mathcal{S}(\phi)$ is the set of all concatenations $w(1)\cdots w(M)$ that can be formed by selecting $w(m)$ from the $\phi(m)$-th member of $\{\mathcal{S}_j\}$, for each m. The set $\mathcal{S}(\phi)$ is called the *ϕ-merging* of $\{\mathcal{S}_j\colon j \in J\}$. In general, a *merging* of the collection $\{\mathcal{S}_j\colon j \in J\}$ is just an (M, J)-merging of the collection for some M.

The two important properties of this merging idea are that each factor $\mathcal{S}_j$ appears exactly M/J times, and, given that a block comes from $\mathcal{S}_j$, it is equally likely to be any member of $\mathcal{S}_j$. In cutting and stacking language, each $\mathcal{S}_j$ is cut into exactly M/J copies and these are independently cut and stacked in the order specified by ϕ. Use of this type of merging at each stage will insure asymptotic independence.

Cyclical merging rules are defined as follows. Assume p divides M/J and J divides p. The merging rule ϕ defined by the two conditions

$$\text{(i)}\quad \phi(m) = j,\quad (j-1)\frac{p}{J} < m \le j\frac{p}{J},\quad 1 \le j \le J,$$
$$\text{(ii)}\quad \phi(m + np) = \phi(m),\quad m \in [1, p],\quad 0 \le n < M/p,$$

is called the *cyclic rule with period p*.

The desired "cyclical rules with rapidly growing periods" are obtained as follows. Let $\exp_b(\cdot)$ denote the exponential function with base b. Given J and J^*, let

$$M = \exp_J(\exp_2(J^* - 1)) = J^{(2^{J^*-1})},$$

and for each $t \in [1, J^*]$, let ϕ_t be the cyclic (M, J)-merging rule with period $p_t = \exp_J(\exp_2(t-1))$. The family $\{\phi_t\colon t \le J^*\}$ is called the *canonical family of cyclical merging rules defined by J and J^*.*

When applied to a collection $\{\mathcal{S}_j\colon j \in J\}$ of disjoint subsets of A^ℓ, the canonical family $\{\phi_t\}$ produces the collection $\{\mathcal{S}(\phi_t)\}$ of disjoint subsets of $A^{M\ell}$. The key to the construction is that if J is large enough, then the new collection will be almost as well separated as the old collection. In proving this it is somewhat easier to use a stronger infinite concatenation form of the separation idea. The *full k-block universe* of $\mathcal{S} \subseteq A^\ell$ is the set

$$\mathcal{U}_k(\mathcal{S}^\infty) = \{a_1^k\colon a_1^k = x_i^{i+k-1},\ \text{for some } x \in \mathcal{S}^\infty \text{ and some } i \ge 1\}.$$

Two subsets $\mathcal{S}, \mathcal{S}' \subset A^\ell$ are *(α, K)-strongly-separated* if their full k-block universes are α-separated for any $k \ge K$.

The following lemma, stated in a form suitable for iteration, is the key to producing an ergodic measure with the desired property (5).

Lemma III.3.5 (The merging/separation lemma.)

Given $0 < \alpha^* < \alpha < 1/2$, there is a $J_0 = J_0(\alpha, \alpha^*)$ such that if $J \ge J_0$ and $\{\mathcal{S}_j\colon j \le J\}$ is a collection of pairwise (α, K)-strongly-separated subsets of A^ℓ of the same cardinality, then for any J^* there is a K^* and an ℓ^*, and a collection $\{\mathcal{S}_t^*\colon t \le J^*\}$ of subsets of A^{ℓ^*} of equal cardinality, such that

(a) *$\{\mathcal{S}_t^*: t \leq J^*\}$ is pairwise (α^*, K^*)-strongly-separated.*

(b) *Each $\mathcal{S}_t^*$ is a merging of $\{\mathcal{S}_j: j \in J\}$.*

Proof. Without loss of generality it can be supposed that $K \leq \ell$, for otherwise $\{\mathcal{S}_j: j \leq J\}$ can be replaced by $\{\mathcal{S}_j^N: j \in J\}$ for $N \geq K/\ell$, without losing (α, K)-strong-separation, since $(\mathcal{S}_j^N)^\infty = \mathcal{S}_j^\infty$, for any j, and, furthermore, any merging of $\{\mathcal{S}_j^N\}$ is also a merging of $\{\mathcal{S}_j\}$.

Let $\{\phi_t: t \leq J^*\}$ be the canonical family of cyclical merging rules defined by J and J^*, and for each t, let $\mathcal{S}_t^* = \mathcal{S}(\phi_t)$. Part (b) is certainly true, so it only remains to show that if J is large enough, then K^* can be chosen so that property (a) holds.

Suppose $x \in (\mathcal{S}_t^*)^\infty$ and $y \in (\mathcal{S}_s^*)^\infty$ where $s > t$. The condition $s > t$ implies that there are integers n and m such that m is divisible by nJ and such that $x = b(1)b(2)\cdots$, where each $b(i)$ is a concatenation of the form

$$w(1)w(2)\cdots w(J), \quad w(j) \in \mathcal{S}_j^n, \quad 1 \leq j \leq J, \tag{8}$$

while $y = c(1)c(2)\cdots$, where each $c(i)$ has the form

$$v(1)v(2)\cdots v(J), \quad v(j) \in \mathcal{S}_j^m, \quad 1 \leq j \leq J. \tag{9}$$

Furthermore, each block $v(j)$ is at least as long as each concatenation (8), so that if $y_u^{u+(J-1)n\ell+1}$ is a subblock of such a $v(j)$, and

$$x_v^{v+(J-1)n\ell+1} = w(t+1)w(t+2)\cdots w(J)w(1)\cdots w(t)$$

where $w(k) \in \mathcal{S}_k^n$, for each k, then there can be at most one index k which is equal to j, and hence

$$d_{(J-1)n\ell}(x_v^{v+(J-1)n\ell+1}, y_u^{u+(J-1)n\ell+1}) \geq \alpha\left(1 - \frac{1}{J-1}\right), \tag{10}$$

by the definition of (α, K)-strongly-separated and the assumption that $K \leq \ell$. Since all but a limiting $(1/J)$-fraction of y is covered by such $y_u^{u+(J-1)n\ell+1}$, and since J^* is finite, the collection $\{\mathcal{S}_t^*: t \leq J^*\}$ is indeed pairwise (α^*, K^*)-strongly-separated, for all large enough K^*, provided J is large enough.

This completes the proof of Lemma III.3.5. □

The construction of the desired sequence $\{\mathcal{S}(m)\}$ is now carried out by induction. In concatenation language, this may be described as follows. Fix a decreasing sequence $\{\alpha_m\}$ such that $\alpha < \alpha_m < 1/2$, and, for each m, choose $J_m > J_0(\alpha_m, \alpha_{m+1})$ from Lemma III.3.5. Suppose $\mathcal{S}(m) \subset A^{\ell(m)}$ and k_m have been determined so that the following hold.

(B1) $\mathcal{S}(m)$ is a disjoint union of a collection $\{\mathcal{S}_j(m): j \leq J_m\}$ of pairwise (α_m, k_m)-strongly-separated sets of the same cardinality.

(B2) Each $\mathcal{S}_j(m)$ is a merging of $\{\mathcal{S}_j(m-1): j \leq J_{m-1}\}$.

(B3) $2^n n(k_m) \leq \ell(m)$.

Define k_{m+1} to be the K^* of Lemma III.3.5 for the family $\{\mathcal{S}_j(m): j \leq J_m\}$ with $\alpha_m = \alpha$ and $\alpha_{m+1} = \alpha^*$. Let $\{\mathcal{S}_t^*: t \leq J_{m+1}\}$ be the collection given by Lemma III.3.5 for $J^* = J_{m+1}$. By replacing $\mathcal{S}_t^*$ by $(\mathcal{S}_t^*)^N$ for some suitably large N, if necessary, it can be supposed that $2^{m+1}n(k_{m+1}) \leq \ell(\mathcal{S}_t^*)$. Put $\mathcal{S}_j(m+1) = \mathcal{S}_j^*$, $j \leq J_{m+1}$. The union $\mathcal{S}(m+1) = \cup_j \mathcal{S}_j(m+1)$ then has properties (B1), (B2), and (B3), for m replaced by $m+1$.

Since $\ell(m) \to \infty$, by (B3), the sequence $\{\mathcal{S}(m)\}$ defines a stationary A-valued measure μ by the formula

$$\mu(a_1^k) = \lim_{m\to\infty} \frac{1}{|\mathcal{S}(m)|} \sum_{x_1^{\ell(m)} \in \mathcal{S}(m)} p_k(a_1^k | x_1^{\ell(m)}), \tag{11}$$

the analogue of Theorem I.10.4 in this case where all columns have the same height and width. Furthermore, the measure μ is ergodic since the definition of merging implies that if M is large relative to m, then each member of $\mathcal{S}(m)$ appears in most members of $\mathcal{S}(M)$ with frequency almost equal to $1/|\mathcal{S}(m)|$. Since (6) clearly holds, and $\sum_m n(k_m)/\ell(m) < \infty$, the measure of the bad set $\mu(B_{k_m}(\alpha))$ is summable in m, establishing the desired goal (5). □

III.3.b.2 The general case.

To obtain the stronger property (3) it is necessary to control what happens in the interval $k_m \leq k \leq k_{m+1}$. This is accomplished by doing the merging in separate intermediate steps, on each of which only a small fraction of the space is merged. At each intermediate step, prior separation properties are retained on the unmerged part of the space until the somewhat smaller separation for longer blocks is obtained. Only a brief sketch of the ideas will be given here; for the complete proof see [50].

The new construction goes from $\mathcal{S}(m)$ to $\mathcal{S}(m+1)$ through a sequence of J_{m+1} intermediate steps

$$\mathcal{S}(m, 0) \to \mathcal{S}(m, 1) \to \ldots \to \mathcal{S}(m, J_{m+1}),$$

merging only a $(1/J_{m+1})$-fraction of the space at each step. All columns at any substage have the same height, but two widths are possible, one on the part waiting to be merged at subsequent steps, and the other on the part already merged, and hence cutting and stacking language, rather than simple concatenation language, is needed. The only merging idea used is that of *cyclical merging of* $\{\mathcal{V}_j: j \leq J\}$ *with period* J, which is just the independent cutting and stacking in the order given, that is, $\mathcal{V}_1 * \mathcal{V}_2 * \cdots * \mathcal{V}_J$.

As in the earlier construction, the structure $\mathcal{S}(m) = \mathcal{S}(m, 0)$ is a union of pairwise (α_m, k_m)-strongly separated substructures $\{\mathcal{S}_j(m, 0): j \leq J_m\}$ all of whose columns have the same width and the same height $\ell(m) = \ell(m, 0)$. In the first substage, each $\mathcal{S}_j(m, 0)$ is cut into two copies, $\mathcal{S}_j'$ and $\mathcal{S}_j''$, where

(i) $\lambda(\mathcal{S}_j') = (1 - 1/J_{m+1})\lambda(\mathcal{S}_j(m, 0))$.

(ii) $\lambda(\mathcal{S}_j'') = (1/J_{m+1})\lambda(\mathcal{S}_j(m, 0))$.

Cyclical merging with period J_{m+1} is applied to the collection $\{\mathcal{S}_j''\}$ to produce a new column structure $R_1(m, 1)$. Meanwhile, J_{m+1}-fold independent cutting and stacking is applied separately to each S'_j to obtain a new column structure $\mathcal{L}_j(m, 1)$. Note, by the way, that the columns of $R_1(m, 1)$ have the same height as those of $L_j(m, 1)$, but are only $1/J_{m+1}$ as wide.

The collection $\{\mathcal{L}_j(m, 1)\colon j \leq J\}$ is pairwise (α_m, k_m)-strongly-separated, since neither cutting into copies nor independent cutting and stacking applied to separate $\mathcal{L}_j(m, 1)$ affects this property. An argument similar to the one used to prove the merging/separation lemma, Lemma III.3.5, can be used to show that if $\ell(m)$ is long enough and J_m large enough, then for a suitable choice of $\alpha_m > \beta(m, 1) > \alpha_{m+1}$, there is a $k(m, 1) \geq k_m$ such that each unmerged part $\mathcal{L}_j(m, 1)$ is $(\beta(m, 1), k(m, 1))$-strongly separated from the merged part $\mathcal{R}_1(m, 1)$. In particular, since strong-separation for any value of k implies strong-separation for all larger values, the following holds.

(a) The collection $\mathcal{S}(m, 1) = \{\mathcal{L}_j(m, t)\colon j \leq J\} \cup \{\mathcal{R}_1(m, t))\}$ is $(\beta(m, 1), k(m, 1))$-strongly-separated.

Each of the separate substructures can be extended by applying independent cutting and stacking to it, which makes each substructure longer without changing any of the separation properties. Thus if $M(m, 1)$ is large enough and each $\mathcal{L}_j(m, 1)$ is replaced by its $M(m, 1)$-fold independent cutting and stacking, and $\mathcal{R}(m, 1)$ is replaced by its $M(m, 1)$-fold independent cutting and stacking, then $\{\mathcal{L}_j(m, 1)\colon j \leq J\}$ remains pairwise (α_m, k_m)-strongly-separated, and (a) continues to hold. But, if $M(m, 1)$ is chosen large enough, then it can be assumed that $\ell(m, 1) \geq n(k(m, 1))J_{m+1}$, which, in turn, implies the following.

(b) If x and y are picked at random in $\mathcal{S}(m, 1)$, then for any k in the range $k_m \leq k \leq k(m, 1)$, the probability that $\bar{d}_k(p_k(\cdot|x_1^{n(k)}), p_k(\cdot|y_1^{n(k)})) \geq \alpha$ is at least $(1 - 2/J_{m+1})^2(1 - 1/J_m^2)$.

This is because, for the stated range of k-values, the k-block universes of $x_1^{n(k)}$ and $y_1^{n(k)}$ are at least α-apart, if they lie at least $n(k(m, 1))$-levels below the top and in different $\mathcal{L}_j(m, 1)$, an event of probability at least $(1-2/J_{m+1})^2(1-1/J_m^2)$, since $\mathcal{R}(m, 1)$ has measure $1/J_{m+1}$ and the top $n(k(m, 1))$-levels have measure at most $n(k(m, 1))/\ell(m, 1) \leq 1/J_{m+1}$, and since the probability that both lie in in the same $\mathcal{L}_j(m, 1)$ is upper bounded by $1/J_m^2$.

The merging-of-a-fraction idea can now be applied to $\mathcal{S}(m, 1)$, merging a copy of $\{\mathcal{L}_j(m, 1)\colon j \leq J\}$ of measure $1/J_{m+1}$, then applying enough independent cutting and stacking to it and to the separate pieces to achieve almost the same separation for the separate structures, while making each structure so long that (b) holds for $k(m, 1) \leq k \leq k(m, 2)$. After J_{m+1} iterations, the unmerged part has disappeared, and the whole construction can be applied anew to $\mathcal{S}(m + 1) = \mathcal{S}(m, J_{m+1})$, producing in the end an ergodic process μ for which

$$\lim_{k\to\infty} \mu(B_k(\alpha)) = 0.$$

This is, of course, weaker than the desired almost-sure result, but a more careful look shows that if $k(m, t) \leq k \leq k(m, t + 1)$, then k-blocks drawn from the unmerged part or previously merged parts must be far-apart from a large fraction of k-blocks drawn from the part merged at stage $t + 1$, and this is enough to obtain

$$\sum_m \mu\Bigg(\bigcup_{k_m \leq k \leq k_{m+1}} B_k \Bigg) < \infty, \tag{12}$$

for suitable choice of the sequence $\{J_m\}$, which is, in turn, enough to guarantee the desired result (3). □

Remark III.3.6

It is important to note here that enough separate independent cutting and stacking can always be done at each stage to make $\ell(m)$ grow arbitrarily rapidly, relative to k_m. This fact will be applied in III.5.e to construct waiting-time counterexamples.

Remark III.3.7

By starting with a well-separated collection of n-sequences of cardinality more than $2^{n(h+\delta)}$, one can obtain a final process of positive entropy h. In particular, this shows the existence of an ergodic process for which the sequence $k(n) = \lceil (\log n)/h \rceil$ is not $\bar{d}$-admissible.

III.3.c Exercises.

1. Let $\mu = (1/J)\sum_j \nu_j$, where each v_j is ergodic, $\bar{d}_k(p_k(\cdot|x_1^{n(k)}), (\nu_j)_k) \to 0$, almost surely, and $\bar{d}(\nu_i, \nu_j) \geq \alpha$, for $i \neq j$. Show that

$$\liminf_{k\to\infty} \bar{d}_k(p_k(\cdot\ |x_1^{n(k)}), \mu_k) \geq \alpha(1 - 1/J),$$

almost surely. (Hint: use Exercise 11, Section I.9, and Lemma III.3.4.)

2. Show that if $\widehat{\nu}$ is the concatenated-block process defined by a measure ν on A^n, then $|\phi(k, \nu) - \widehat{\nu}_k| \leq 2(k-1)/n$.

Section III.4 Blowing-up properties.

An interesting property closely connected to entropy ideas, called the blowing-up property, has recently been shown to hold for i.i.d. processes, for aperiodic Markov sources, and for other processes of interest, including a large family called the finitary processes. Informally, a stationary process has the blowing-up property if sets of n-sequences that are not exponentially too small in probability have a large blowup. Processes with the blowing-up property are characterized as those processes that have exponential rates of convergence for frequencies and entropy and are stationary codings of i.i.d. processes. A slightly weaker concept, called the almost blowing-up property is, in fact, equivalent to being a stationary coding of an i.i.d. process. The blowing-up property and related ideas will be introduced in this section. A full discussion of the connections between blowing-up properties and stationary codings of i.i.d. processes is delayed to Chapter 4.

If $C \subseteq A^n$ then $[C]_\epsilon$ denotes the ϵ-neighborhood (or ϵ-blowup) of C, that is

$$[C]_\epsilon = \left\{b_1^n\colon\ d_n(a_1^n, b_1^n) \leq \epsilon,\ \text{ for some } a_1^n \in C\right\}.$$

An ergodic process μ has the *blowing-up property (BUP)* if given $\epsilon > 0$ there is a $\delta > 0$ and an N such that if $n \geq N$ then $\mu([C]_\epsilon) > 1 - \epsilon$, for any subset $C \subseteq A^n$ for which $\mu(C) \geq 2^{-k\delta}$.

The following theorem characterizes those processes with the blowing-up property.

Theorem III.4.1 (Blowing-up property characterization.)
A stationary process has the blowing-up property if and only if it is a stationary coding of an i.i.d. process and has exponential rates of convergence for frequencies and entropy.

Note, in particular, that the theorem asserts that an i.i.d. process has the blowing-up property. The proof of this fact as well as most of the proof of the theorem will be delayed to Chapter 4. The fact that processes with the blowing-up property have exponential rates will be established later in the section.

A particular kind of stationary coding, called finitary coding, does preserve the blowing-up property. A stationary coding $F: A^Z \mapsto B^Z$ is said to be *finitary, relative to an ergodic process* μ, if there is a nonnegative, almost surely finite integer-valued measurable function $w(x)$, called the *window function*, such that, for almost every $x \in A^Z$, the time-zero encoder f satisfies the following condition.

$$\text{If } \tilde{x} \in A^Z \text{ and } x_{-w(x)}^{w(x)} = \tilde{x}_{-w(x)}^{w(x)} \text{ then } f(x) = f(\tilde{x}).$$

A finitary coding of an i.i.d. process is called a *finitary process.* It is known that aperiodic Markov chains and finite-state processes, m-dependent processes, and some renewal processes are finitary, [28, 79]. Later in this section the following will be proved.

Theorem III.4.2 (The finitary-coding theorem.)
Finitary coding preserves the blowing-up property.

In particular, since i.i.d. processes have the blowing-up property it follows that finitary processes have the blowing-up property.

Not every stationary coding of an i.i.d. process has the blowing-up property. The difficulty is that only sets of sequences that are mostly both frequency and entropy typical can possibly have a large blowup, and, without exponential rates there can be sets that are not exponentially too small yet fail to contain any typical sequences. Once these are suitably removed, however, then a blowing-up property will hold for an arbitrary stationary coding of an i.i.d. process.

A set $B \subseteq A^k$ has the *(δ, ϵ)-blowing-up property* if $\mu([C]_\epsilon) > 1 - \epsilon$, for any subset $C \subseteq B$ for which $\mu(C) \geq 2^{-k\delta}$. An ergodic process μ has the *almost blowing-up property (ABUP)* if for each k there is a set $B_k \subseteq A^k$ such that the following hold.

(i) $x_1^k \in B_k$, eventually almost surely.

(ii) For any $\epsilon > 0$ there is a $\delta > 0$ and a K such that B_k has the (δ, ϵ)-blowing-up property for $k \geq K$.

Theorem III.4.3 (Almost blowing-up characterization.)
A stationary process has the almost blowing-up property if and only if it is a stationary coding of an i.i.d. process.

By borrowing one concept from Chapter 4, it will be shown later in this section that a stationary coding of an i.i.d. process has the almost blowing-up property, a result that will be used in the waiting-time discussion in the next section. A proof that a process with the almost blowing-up property is a stationary coding of an i.i.d. process will be given in Section IV.3.c.

III.4.a Blowing-up implies exponential rates.

Suppose μ has the blowing-up property. First it will be shown that μ has exponential rates for frequencies. The idea of the proof is that if the set of sequences with bad frequencies does not have exponentially small measure then it can be blown up by a small amount to get a set of large measure. If the amount of blowup is small enough, however, then frequencies won't change much and hence the blowup would produce a set of large measure all of whose members have bad frequencies, contradicting the ergodic theorem.

To fill in the details let $p_k(\cdot\ |x_1^n)$ denote the empirical distribution of overlapping k blocks, and define

$$B(n,k,\epsilon) = \{x_1^n\colon\ |p_k(\cdot\ |x_1^n) - \mu_k| \ge \epsilon\}.$$

Note that if $\gamma = \epsilon/(2k+1)$ then

$$d_n(x_1^n, y_1^n) \le \gamma \implies |p_k(\cdot\ |x_1^n) - p_k(\cdot\ |y_1^n)| \le \epsilon/2,$$

so that, in particular,

$$[B(n,k,\epsilon)]_\gamma \subseteq B(n,k,\epsilon/2). \tag{1}$$

The blowing-up property provides $\delta > 0$ and N so that if $n \ge N$, $C \subseteq A^n$, and $\mu(C) \ge 2^{-\delta n}$ then $\mu([C]_\gamma) \ge 1-\epsilon$. If, however, $n \ge N$ and $\mu_n(B(n,k,\epsilon)) \ge 2^{-\delta n}$ then $\mu_n([B(n,k,\epsilon)]_\gamma)$ would have to be at least $1-\epsilon$, and hence (1) would force the measure $\mu_n(B(n,k,\epsilon/2))$ to be at least $1-\epsilon$. But this cannot be true for all large n since the ergodic theorem guarantees that $\lim_n \mu_n(B(n,k,\epsilon/2)) = 0$ for each fixed k and ϵ. Thus $\mu_n(B(n,k,\epsilon)) \le 2^{-\delta n}$, for all sufficiently large n.

Next it will be shown that blowing-up implies exponential rates for entropy. One part of this is easy, for there cannot be too many sequences whose measure is too large, hence a small blowup of such a set cannot possibly produce enough sequences to cover a large fraction of the measure. To fill in the details of this part of the argument, let

$$B^*(n,\epsilon) = \{x_1^n\colon\ \mu_n(x_1^n) \ge 2^{-n(h-\epsilon)}\},$$

so that $|B^*(n,\epsilon)| \le 2^{n(h-\epsilon)}$, and hence there is an $\alpha > 0$ such that

$$|[B^*(n,\epsilon)]_\alpha| \le 2^{n(h-\epsilon/2)},$$

for all n. Intersecting with the set $\mathcal{T}_n = \{x_1^n\colon \mu(x_1^n) \le 2^{-n(h-\epsilon/4)}\}$ gives

$$\mu_n([B^*(n,\epsilon)]_\alpha \cap \mathcal{T}_n) \le 2^{n(h-\epsilon/2)} 2^{-n(h-\epsilon/4)} \le 2^{-n\epsilon/4},$$

so $\mu_n([B^*(n,\epsilon)]_\alpha) \to 0$, since $\mu_n(\mathcal{T}_n) \to 1$, by the entropy theorem.

The blowing-up property provides δ and N so that if $n \ge N$, $C \subseteq A^n$, and $\mu(C) \ge 2^{-\delta n}$ then $\mu([C]_\alpha) \ge 1-\alpha$. Since $\mu_n([B^*(n,\epsilon)]_\alpha) \to 0$ it follows that $\mu(B^*(n,\epsilon))$ must be less than $2^{-\delta n}$, for all n sufficiently large.

An exponential bound for the measure of the set

$$B_*(n,\epsilon) = \{x_1^n\colon\ \mu_n(x_1^n) \le 2^{-n(h+\epsilon)}\}$$

of sequences of too-small probability is a bit trickier to obtain, but depends only on the existence of exponential rates for frequencies, since this implies that if n is sufficiently large then, except for a set of exponentially small probability, most of x_1^n will be covered by k-blocks whose measure is about 2^{-kh}. As in the proof of the entropy theorem, this gives an exponential bound on the number of such x_1^n, which in turn means that it is exponentially very unlikely that such an x_1^n can have probability much smaller than 2^{-nh}.

To fill in the details of the preceding argument apply the entropy theorem to choose k so large that

$$\mu_k(B_*(k, \epsilon/4)) < \alpha,$$

where α will be specified in a moment. For $n \geq k$ let

$$G_n = \{x_1^n \colon\ p_k(B_*(k, \epsilon/4)\ |x_1^n) < 2\alpha\},$$

that is, the set of n-sequences that are $(1-2\alpha)$-covered by the complement of $B_*(k, \epsilon/4)$. Since the complement of $B_*(k, \epsilon/4)$ has cardinality at most $2^{k(h+\epsilon/4)}$, the built-up set bound, Lemma I.7.6, implies that there is an $\alpha > 0$ and an N such that if $n \geq N$, then $|G_n| \leq 2^{n(h+\epsilon/2)}$, and hence

$$\mu_n(G_n \cap B_*(n, \epsilon)) \leq |G_n| 2^{-n(h+\epsilon)} \leq 2^{-n\epsilon/2},\ n \geq N.$$

Since exponential rates have already been established for frequencies, there is a $\delta > 0$ and an $N_1 \geq N$ such that $\mu_n(G_n) \geq 1 - 2^{-\delta n}$, if $n \geq N_1$. Thus

$$\mu_n(B_*(n, \epsilon)) \leq 2^{-\delta n} + 2^{-n\epsilon/2},\quad n \geq N_1,$$

which gives the desired exponential bound. □

III.4.b Finitary coding and blowing-up.

The finitary-coding theorem will now be established. Fix $\epsilon > 0$ and a process μ with the blowing-up property. Let ν be a finitary coding of μ, with full encoder $F \colon B^Z \mapsto A^Z$, time-zero encoder $f \colon B^Z \mapsto A$, and window function $w(x)$, $x \in B^Z$. Thus $f(x)$ depends only on the values $x_{-w(x)}^{w(x)}$, where $w(x)$ is almost surely finite, and hence there is a k such that if G_k is the set of all x_{-k}^k such that $w(x) \leq k$, then $\mu(G_k) \geq 1 - \epsilon^2$. In particular, the set

$$T = \{x_{-k+1}^{n+k} \colon\ p_{2k+1}(G_k | x_{-k+1}^{n+k}) \geq 1 - \epsilon\},$$

satisfies $\mu(T) \geq 1 - \epsilon$ by the Markov inequality.

Suppose $C \subseteq A^n$ satisfies $\nu_n(C) \geq 2^{-n\delta}$, where $\delta > 0$. Let D be the projection of $F^{-1}C$ onto B_{-k+1}^{n+k}, and note that $\mu(D) \geq 2^{-n\delta}$. Thus, since k is fixed and μ has the blowing-up property, δ can be assumed to be so small and n so large that if $\gamma = \epsilon/(2k+1)$ then $\mu([D]_\gamma) \geq 1 - \epsilon$.

Now consider a sequence $x_{-k+1}^{n+k} \in [D]_\gamma \cap T$. For such a sequence $(1 - \epsilon)n$ of its $(2k + 1)$-blocks belong to G_k, and, moreover, there is a sequence $\bar{x}_{-k+1}^{n+k} \in D$ such that

$$d_{n+2k}(x_{-k+1}^{n+k}, \bar{x}_{-k+1}^{n+k}) < \gamma = \frac{\epsilon}{2k+1}.$$

Thus fewer than $(2k + 1)\gamma n \leq \epsilon n$ of the $(2k + 1)$-blocks in x_{-k+1}^{n+k} can differ from the corresponding block in $\bar{x}_{-k+1}^{n+k}$. In particular, there is at least a $(1 - 2\epsilon)$-fraction of $(2k + 1)$-blocks in x_{-k+1}^{n+k} that belong to G_k and, at the same time agree entirely with the corresponding block in $\bar{x}_{-k+1}^{n+k}$.

Choose y and $\bar{y} \in D$ such that $y_{-k+1}^{n+k} = x_{-k+1}^{n+k}$, and $\bar{y}_{-k+1}^{n+k} = \bar{x}_{-k+1}^{n+k}$, and put $z = F(y)$, $\bar{z} = F(\bar{y})$. The sequence $\bar{z}_1^n$ belongs to C and $d_n(z_1^n, \bar{z}_1^n) \le 2\epsilon$, so that $z_1^n \in [C]_{2\epsilon}$. This proves that

$$\nu_n([C]_{2\epsilon}) \ge \mu([D]_\gamma \cap T) \ge 1 - 2\epsilon,$$

which completes the proof of the finitary-coding theorem. □

III.4.c Almost blowing-up and stationary coding.

In this section it is shown that stationary codings of i.i.d. processes have the almost blowing-up property (ABUP). Towards this end a simple connection between the blowup of a set and the $\bar{d}_n$-distance concept will be needed. Let $\mu_n(\cdot\,|C)$ denote the conditional measure defined by the set $C \subseteq A^n$. Also let $d_n(x_1^n, C)$ denote the distance from x_1^n to C, that is, the minimum of $d_n(x_1^n, y_1^n)$, $y_1^n \in C$.

Lemma III.4.4
If $\bar{d}_n(\mu, \mu_n(\cdot\,|C)) < \epsilon^2$ then $\mu([C]_\epsilon) > 1 - \epsilon$.

Proof. The relation

$$\sum_{y_1^n \in C} \lambda(x_1^n, y_1^n) d_n(x_1^n, y_1^n) \ge p(x_1^n) d_n(x_1^n, C). \tag{2}$$

holds for any joining λ of μ_n and $\mu_n(\cdot\,|C)$, and hence $E_\mu(d_n(x_1^n, C)) \le \bar{d}_n(\mu, \mu_n(\cdot\,|C))$. Thus if $\bar{d}_n(\mu, \mu_n(\cdot\,|C)) < \epsilon^2$, then, by the Markov inequality, the set of x_1^n such that $d_n(x_1^n, C) \ge \epsilon$ has μ measure less than ϵ, that is, $\mu([C]_\epsilon) > 1 - \epsilon$. This proves the lemma. □

The proof that stationary codings of i.i.d. process have the almost blowing-up property makes use of the fact that such coded processes have the finitely determined property, a property introduced to prove the $\bar{d}$-admissibility theorem in Section III.3, and which will be discussed in detail in Section IV.2. An ergodic process μ is finitely determined (FD) if given $\epsilon > 0$, there is a $\delta > 0$ and positive integers k and N such that if $n \ge N$ then any measure ν on A^n which satisfies the two conditions

(a) $|\mu_k - \phi(k, \nu)| < \delta$,

(b) $|H(\mu_n) - H(\nu)| < n\delta$,

must also satisfy $\bar{d}_n(\mu, \nu) < \epsilon$, where $\phi(k, \nu)$ is the measure on A^k obtained by averaging the ν-probability of k-blocks over all starting positions in sequences of length n, that is,

$$\phi(a_1^k) = \sum_{x_1^n} p_k(a_1^k|x_1^n)\nu(x_1^n) = E_\nu(p_k(a_1^k|X_1^n)). \tag{3}$$

This simply a translation of the definition of Section III.3.a into the notation used here.

Theorem III.4.5 (FD implies ABUP.)
A finitely determined process has the almost blowing-up property.

Proof. Fix a finitely determined process μ. The basic idea is that if $C \subseteq A^n$ consists of sequences that have good k-block frequencies and probabilities roughly equal to 2^{-nh}, and $\mu(C)$ is not exponentially too small, then $\mu_n(\cdot\ |C)$ will have (averaged) k-block probabilities close to μ_k and entropy close to $H(\mu_n)$. The finitely determined property then implies that $\mu_n(\cdot\ |C)$ and μ_n are $\bar{d}$-close, which, by Lemma III.4.4, implies that C has a large blowup.

The first step in the rigorous proof is to remove the sequences that are not suitably frequency and entropy typical. A sequence x_1^n will be called (k, α)-frequency-typical if $|p_k(\cdot\ |x_1^n) - \mu_k| < \alpha$, where, as usual, $p_k(\cdot\ |x_1^n)$ is the empirical distribution of overlapping k-blocks in x_1^n. A sequence x_1^n will be called entropy-typical, relative to α, if

$$2^{-H(\mu_n)-n\alpha} \le \mu(x_1^n) \le 2^{-H(\mu_n)+n\alpha}.$$

Note that in this setting n-th order entropy, $H(\mu_n)$, is used, rather than the usual nh in the definition of entropy-typical, which is all right since $H(\mu_n)/n \to h$.

Let $B_n(k, \alpha)$ be the set of n-sequences that are both entropy-typical, relative to α, and (j, α)-frequency-typical for all $j \le k$. If k and α are fixed, then $x_1^n \in B_n(k, \alpha)$, eventually almost surely, so there is a nondecreasing, unbounded sequence $\{k(n)\}$, and a nonincreasing sequence $\{\alpha(n)\}$ with limit 0, both of which depend on μ, such that $x_1^n \in B_n(k(n), \alpha(n))$, eventually almost surely. Put $B_n = B_n(k(n), \alpha(n))$. It will be shown that for any $\epsilon > 0$, there is a $\delta > 0$ such that B_n eventually has the (δ, ϵ)-blowing-up property.

Given $\epsilon > 0$, the finitely determined property provides $\delta > 0$ and k such that if n is large enough then any measure ν on A^n which satisfies $|\mu_k - \phi(k, \nu)| < \delta$ and $|H(\mu_n) - H(\nu)| < n\delta$, must also satisfy $\bar{d}_n(\mu, \nu) < \epsilon^2$.

Suppose $C \subseteq B_n$ and $\mu(C) \ge 2^{-n\delta}$. The definition of B_n and formula (3) together imply that if $k(n) \ge k$, then the conditional measure $\mu_n(\cdot\ |C)$ satisfies the following.

(i) $|\phi(k, \mu_n(\cdot\ |C)) - \mu_k| < \alpha(n)$.

(ii) $H(\mu_n) - n\alpha(n) - n\delta \le H(\mu_n(\cdot\ |C)) \le H(\mu_n) + n\alpha(n)$.

Thus if n is large enough then $\bar{d}_n(\mu, \mu_n(\cdot\ |C)) < \epsilon^2$, for any $C \subseteq B_n$ for which $\mu(C) \ge 2^{-n\delta}$ which, by Lemma III.4.4, implies that $\mu([C]_\epsilon) > 1 - \epsilon$. This completes the proof that finitely determined implies almost blowing-up. □

Remark III.4.6

The results in this section are mostly drawn from joint work with Marton, [37, 39]. For references to earlier papers on blowing-up ideas the reader is referred to [37] and to Section 1.5 in the Csiszár-Körner book, [7], which also includes applications to some problems in multi-user information theory.

III.4.d Exercises.

1. Show that (2) is indeed true.

2. Show that a process with the almost blowing-up property must be ergodic.

3. Show that a process with the almost blowing-up property must be mixing.

4. Assume that i.i.d. processes have the blowing-up property.

   (a) Show that an m-dependent process has the blowing-up property.

   (b) Show that a ψ-mixing process has the blowing-up property.

5. Assume that aperiodic renewal processes are stationary codings of i.i.d. processes. Show that some of them do not have the blowing-up property.

6. Show that condition **(i)** in the definition of ABUP can be replaced by the condition that $\mu(B_n) \to 1$.

7. Show that a coding is finitary relative to μ if and only if for each b the set $f^{-1}(b)$ is a countable union of cylinder sets together with a null set.

Section III.5 The waiting-time problem.

A connection between entropy and recurrence times, first noted by Wyner and Ziv, [86], was shown to hold for any ergodic process in Section Section II.5. Wyner and Ziv also established a positive connection between a waiting-time concept and entropy, at least for certain classes of ergodic processes. They showed that if $W_k(x, y)$ is the waiting time until the first k terms of x appear in an independently chosen y, then $(1/k)\log W_k(x, y)$ converges in probability to h, for irreducible Markov chains. This result was extended to somewhat larger classes of processes, including the weak Bernoulli class in [44, 76]. The surprise here, of course, is the positive result, for the well-known waiting-time paradox, [11, pp. 10ff.], suggests that waiting times are generally longer than recurrence times.

An almost sure version of the Wyner-Ziv result will be established here for the class of weak Bernoulli processes by using the joint-distribution estimation theory of III.2.b. In addition, an approximate-match version will be shown to hold for the class of stationary codings of i.i.d. processes, by using the $\bar{d}$-admissibility theorem, Theorem III.3.1, in conjunction with the almost blowing-up property discussed in the preceding section. Counterexamples to extensions of these results to the general ergodic case will also be discussed. The counterexamples show that waiting time ideas, unlike recurrence ideas, cannot be extended to the general ergodic case, thus further corroborating the general folklore that waiting times and recurrence times are quite different concepts.

The *waiting-time function* $W_k(x, y)$ is defined for $x, y \in A^\infty$ by

$$W_k(x, y) = \min\{m \geq 1\colon\ y_m^{m+k-1} = x_1^k\}.$$

The *approximate-match waiting-time function* $W_k(x, y, \delta)$ is defined by

$$W_k(x, y, \delta) = \min\{m \geq 1\colon\ d_k(x_1^k, y_m^{m+k-1}) \leq \delta\}.$$

Two positive theorems will be proved.

Theorem III.5.1 (The exact-match theorem.)
If μ is weak Bernoulli with entropy h, then

$$\lim_{k\to\infty} \frac{1}{k} \log W_k(x, y) = h,$$

almost surely with respect to the product measure $\mu \times \mu$.

Theorem III.5.2 (The approximate-match theorem.)
If μ is a stationary coding of an i.i.d. process, then

$$\limsup_{k\to\infty} \frac{1}{k} \log W_k(x, y, \delta) \leq h, \quad \forall\, \delta > 0,$$

and

$$\lim_{\delta\to 0} \liminf_{k\to\infty} \frac{1}{k} \log W_k(x, y, \delta) \geq h,$$

almost surely with respect to the product measure $\mu \times \mu$.

The easy part of both results is the lower bound, for there are exponentially too few k-blocks in the first $2^{k(h-\epsilon)}$ terms of y, hence it is very unlikely that a typical x_1^k can be among them or even close to one of them.

The proofs of the two upper bound results will have a parallel structure, though the parallelism is not complete. In each proof two sequences of measurable sets, $\{B_k \subseteq A^k\}$ and $\{G_n\}$, are constructed for which the following hold.

(a) $x_1^k \in B_k$, eventually almost surely.

(b) $y \in G_n$, eventually almost surely.

For the weak Bernoulli case, it is necessary to allow the set G_n to depend on the infinite past, hence G_n is taken to be a subset of the space A^Z of doubly infinite sequences. For the approximate-match result, the set G_n can be taken to depend only on the coordinates from 1 to n, hence is taken to be a subset of A^n.

In the weak Bernoulli case, the sequences $\{B_k\}$ and $\{G_n\}$ have the additional property that if $k \leq (\log n)/(h+\epsilon)$ then for any fixed $x_1^k \in B_k$, the probability that $y \in G_n$ does not contain x_1^k anywhere in its first n places is almost exponentially decaying in n. In the approximate-match proof, $\{B_k\}$ and $\{G_n\}$ have the property that if $n \geq 2^{k(h+\epsilon)}$, then for any fixed $y_1^n \in G_n$ the probability that $x_1^k \in B_k$ is not within δ of some k-block in y_1^n will be exponentially small in k. In both cases an application the Borel-Cantelli lemma establishes the desired result.

In the weak Bernoulli case, B_k consists of the entropy-typical sequences, so that property (a) holds by the entropy theorem. The set G_n consists of those $y \in A^Z$ for which y_1^n can be split into k-length blocks separated by a fixed length gap, such that a large fraction of these k-blocks are almost independent of each other. In other words, the G_n are the sets given by the weak Bernoulli splitting-set lemma, Lemma III.2.8.

In the approximate-match case, G_n consists of those y_1^n whose set of k-blocks has a large δ-neighborhood, for $k \sim (\log n)/(h+\epsilon)$. The $\bar{d}$-admissibility theorem, Theorem III.3.1, guarantees that property (b) holds. The set $B_k \subseteq A^k$ is chosen to have the property that any subset of it that is not exponentially too small in measure must have a large δ-blowup. In other words, the B_k are the sets given by the almost blowing-up property of stationary codings of i.i.d. processes, see Theorem III.4.3.

The lower bound and upper bound results are established in the next three subsections. In the final two subsections a stationary coding of an i.i.d. process will be constructed for which the exact-match version of the approximate-match theorem is false, and an ergodic process will be constructed for which the approximate-match theorem is false.

III.5.a Lower bound proofs.

The following notation will be useful here and later subsections. The (empirical) universe of k-blocks of y_1^n is the set

$$\mathcal{U}_k(y_1^n) = \left\{a_1^k \colon y_i^{i+k-1} = a_1^k, \text{ for some } i \in [0, n-k+1]\right\}.$$

Its δ-blowup is

$$[\mathcal{U}_k(y_1^n)]_\delta = \{x_1^k \colon d_k(x_1^k, \mathcal{U}_k(y_1^n)) \leq \delta\}.$$

In the following discussion, the set of entropy-typical k-sequences (with respect to $\alpha > 0$) is taken to be

$$\mathcal{T}_k(\alpha) = \left\{x_1^k \colon 2^{-k(h+\alpha)} \leq \mu(x_1^k) \leq 2^{-k(h-\alpha)}\right\}.$$

Theorem III.5.3

Let μ be an ergodic process with entropy $h > 0$ and let $\epsilon > 0$ be given. If $k(n) \geq (\log n)/(h-\epsilon)$ then for every $y \in A^\infty$ and almost every x, there is an $N = N(x, y)$ such that $x_1^{k(n)} \notin \mathcal{U}_{k(n)}(y_1^n)$, $n \geq N$.

Proof. If $k \geq (\log n)/(h-\epsilon)$ then $n \leq 2^{k(h-\epsilon)}$, and hence $|\mathcal{U}_k(y_1^n)| \leq 2^{k(h-\epsilon)}$, $y_1^n \in A^n$, so that intersecting with the entropy-typical set $\mathcal{T}_k(\epsilon/2)$ gives

$$\mu(\mathcal{U}_k(y_1^n) \cap \mathcal{T}_k(\epsilon/2)) \leq 2^{k(h-\epsilon)} 2^{-k(h-\epsilon/2)} \leq 2^{-k\epsilon/2}.$$

Since $x_1^k \in \mathcal{T}_k(\epsilon/2)$, eventually almost surely, an application of the Borel-Cantelli lemma yields the theorem. □

The theorem immediately implies that

$$\liminf_{k\to\infty} \frac{1}{k} \log W_k(x, y) \geq h,$$

almost surely with respect to the product measure $\mu \times \mu$, for any ergodic measure μ with entropy h.

Theorem III.5.4

Let μ be an ergodic process with entropy $h > 0$ and let $\epsilon > 0$ be given. There is a $\delta > 0$ such that if $k(n) \geq (\log n)/(h-\epsilon)$ then for every $y \in A^\infty$ and almost every x, there is an $N = N(x, y)$ such that $x_1^{k(n)} \notin [\mathcal{U}_{k(n)}(y_1^n)]_\delta$, $n \geq N$.

Proof. If $k \geq (\log n)/(h-\epsilon)$ then $n \leq 2^{k(h-\epsilon)}$, and hence $|\mathcal{U}_k(y_1^n)| \leq 2^{k(h-\epsilon)}$. If δ is small enough then for all k, $|[\mathcal{U}_k(y_1^n)]_\delta| \leq 2^{k(h-3\epsilon/4)}$, by the blowup-bound lemma, Lemma I.7.5. Intersecting with the entropy-typical set $\mathcal{T}_k(\epsilon/2)$ gives

$$\mu\left([\mathcal{U}_k(y_1^n)]_\delta \cap \mathcal{T}_k(\epsilon/2)\right) \leq 2^{k(h-3\epsilon/4)} 2^{-k(h-\epsilon/2)} \leq 2^{-k\epsilon/4},$$

which establishes the theorem. □

The theorem immediately implies that

$$\lim_{\delta\to 0} \liminf_{k\to\infty} \frac{1}{k} \log W_k(x, y, \delta) \geq h,$$

almost surely with respect to the product measure $\mu \times \mu$, for any ergodic measure μ with entropy h.

III.5.b Proof of the exact-match waiting-time theorem.

Fix a weak Bernoulli measure μ on A^Z with entropy h, and fix $\epsilon > 0$. Let $n(k) = n(k, \epsilon)$ denote the least integer n such that $k \leq (\log n)/(h + \epsilon)$.

The sets $\{B_k\}$ are just the entropy-typical sets, that is,

$$B_k = \mathcal{T}_k(\alpha) = \{x_1^k \colon 2^{-k(h+\alpha)} \leq \mu(x_1^k) \leq 2^{-k(h-\alpha)}\},$$

where α is a positive number to be specified later. The entropy theorem implies that $x_1^k \in B_1^k$, eventually almost surely.

The sets $G_n \subset A^Z$ are defined in terms of the splitting-index concept introduced in III.2.b. The basic terminology and results needed here will be restated in the form they will be used. The s-shifted, k-block parsing of y_1^n, with gap g is

$$y_1^n = w(0)g(1)w(1)g(2)w(2)\dots g(t)w(t)w(t+1),$$

where $w(0)$ has length $s < k+g$, the $g(j)$ have length g and alternate with the k-blocks $w(j)$, for $1 \leq j \leq t$, and the final block $w(t+1)$ has length $n - t(k+g) - s < 2(k+g)$. An index $j \in [1, t]$ is a (γ, s, k, g)-splitting index for $y \in A^Z$ if

$$\mu(w(j) \mid y_{-\infty}^{s+(j-1)(k+g)}) < (1+\gamma)\mu(w(j)).$$

If $B_{s,j}$ is the set of all $y \in A^Z$ that have j as a (γ, s, k, g)-splitting index, and $J \subseteq [1, t]$, then

$$\mu\left(\bigcap_{j\in J}([w(j)] \cap B_{s,j})\right) \leq (1+\gamma)^{|J|} \prod_{j \in J} \mu(w(j)). \tag{1}$$

The sets $\{G_n\}$ are given by Lemma III.2.8 and depend on a positive number γ, to be specified later, and a gap g, which depends on γ. Property (a) and a slightly weaker version of property (b) from that lemma are needed here, stated as the following.

(a) $x \in G_n$, eventually almost surely.

(b) For all large enough k and t, for $(t+1)(k+g) \leq n < (t+2)(k+g)$, and for $y \in G_n(\gamma)$ there is at least one $s \in [0, k+g-1]$ for which there are at least $(1-\gamma)t$ indices j in the interval $[1, t]$ that are (γ, s, k, g)-splitting indices for y.

To complete the proof of the weak Bernoulli waiting-time theorem, it is enough to establish the upper bound,

$$\limsup_{k\to\infty} \frac{1}{k} \log W_k(x, y) \leq h, \quad \mu \times \mu - \text{a.s.} \tag{2}$$

For almost every $x \in A^\infty$ there exists $K(x)$ such that $x_1^k \in B_k,\ k \geq K(x)$. Fix such an x. For each k put $C_k(x) = \left\{y \colon x_1^k \notin \mathcal{U}_k(y_1^{n(k)})\right\}$, and note that

$$\frac{1}{k} \log W_k(x, y) > h + \epsilon \implies y \in C_k(x),$$

so that it is enough to show that $y \notin C_k(x)$, eventually almost surely.

Choose $k \geq K(x)$ and t so large that property (b) holds, and such that $(t+1)(k+g) \leq n(k) < (t+2)(k+g)$. The key bound is

$$\mu(C_k(x) \cap G_{n(k)}) \leq (k+g)2^{-2t\gamma \log \gamma}(1+\gamma)^t \left(1 - 2^{-k(h+\alpha)}\right)^{t(1-\gamma)}. \tag{3}$$

Indeed, if this holds, then α and γ can be chosen so small that the bound is summable in k, since $t \sim n/k$ and $n \sim 2^{k(h+\epsilon)}$. Since $y \in G_n$, eventually almost surely, it then follows that $y \notin C_k(x)$, eventually almost surely, which proves the theorem.

To establish the key bound (3), fix a set $J \subseteq [1, t]$ of cardinality at least $t(1-\gamma)$ and an integer $s \in [0, k+g-1]$ and let $G_{n(k)}(J, s)$ denote the set of all $y \in G_{n(k)}$ for which every $j \in J$ is a (γ, s, k, g)-splitting index for y. Since

$$G_{n(k)}(J, s) \subseteq \cap_{j \in J} B_{s,j},$$

the splitting-index product bound, (1), yields

$$\mu(\bigcap_{j \in J}[w(j)] \cap B_{s,j}) \leq (1+\gamma)^t \prod_{j \in J} \mu(w(j)),$$

for any $y \in G_{n(k)}(J, s)$. But if the blocks $w(j)$, $j \in J$ are treated as independent then, since $\mu(x_1^k) \geq 2^{-k(h+\alpha)}$ and since the cardinality of J is at least $t(1-\gamma)$, the probability that $w(j) \neq x_1^k$, for all $j \in J$, is upper bounded by

$$\left(1 - 2^{-k(h+\alpha)}\right)^{t(1-\gamma)}$$

and hence

$$\mu(C_k(x) \cap G_{n(k)}(J, s)) \leq (1+\gamma)^t \left(1 - 2^{-k(h+\alpha)}\right)^{t(1-\gamma)}.$$

This implies the bound (3), since there are $k+g$ ways to choose the shift s and at most $2^{-2t\gamma \log \gamma}$ ways to choose sets $J \subset [1, t]$ of cardinality at least $t(1-\gamma)$. This completes the proof of the exact-match waiting-time theorem, Theorem III.5.1. □

III.5.c Proof of the approximate-match theorem.

Let μ be a stationary coding of an i.i.d. process such that $h(\mu) = h$, and fix $\epsilon > 0$. The least n such that $k \leq (\log n)/(h+\epsilon)$ is denoted by $n(k)$, while $k(n)$ denotes the largest value of k for which $k \leq (\log n)/(h+\epsilon)$. Fix a positive number δ and define

$$G_n = \left\{y_1^n \colon \mu([\mathcal{U}_{k(n)}(y_1^n)]_{\delta/2}) > 1/2\right\}.$$

By the $\bar{d}$-admissibility theorem, Theorem III.3.1, $y_1^n \in G_n$, eventually almost surely. For almost every $y \in A^\infty$ there is an $N(y)$ such that $y_1^n \in G_n$, $n \geq N(y)$. Fix such a y. For each k, let

$$C_k(y) = \left\{x_1^k \colon x_1^k \notin [\mathcal{U}_k(y_1^{n(k)})]_\delta\right\},$$

so that, $(1/k) \log W_k(x, y, \delta) > h + \epsilon$ if and only if $x_1^k \in C_k(y)$, and hence it is enough to show that $x_1^k \notin C_k(y)$, eventually almost surely.

Let $\{B_k\}$ be the sequence of sets given by the almost blowing-up characterization theorem, Theorem III.4.3, and choose $\beta > 0$ and K such that if $k \geq K$ and $C \subset B_k$ is a set of k-sequences of measure at least $2^{-k\beta}$ then $\mu([C]_{\delta/2}) > 1/2$.

Suppose $k = k(n) \geq K$ and $n(k) \geq N(y)$, so that $y_1^{n(k)} \in G_{n(k)}$. If it were true that $\mu(C_k(y) \cap B_k) \geq 2^{-k\beta}$ then both $[C_k(y) \cap B_k]_{\delta/2}$ and $[\mathcal{U}_k(y_1^{n(k)})]_{\delta/2}$ each have measure larger than 1/2, hence intersect. But this implies that the sets $C_k(y)$ and $[\mathcal{U}_k(y_1^{n(k)})]_\delta$ intersect, which contradicts the definition of $C_k(y)$. Thus, $\mu(C_k(y) \cap B_k) < 2^{-k\beta}$, and hence

$$\sum_{k=1}^{\infty} \mu(C_k(y) \cap B_k) < \infty.$$

Since $x_1^k \in B_k$, eventually almost surely, it follows that $x_1^k \notin C_k(y)$, eventually almost surely, completing the proof of the approximate-match theorem. □

III.5.d An exact-match counterexample.

Let μ be the Kolmogorov measure of the i.i.d. process with binary alphabet $B = \{0, 1\}$ such that $\mu_1(0) = \mu_1(1) = 1/2$.. A stationary coding $\nu = \mu \circ F^{-1}$ will be constructed along with an increasing sequence $\{k(n)\}$ such that

$$\lim_{n\to\infty} P_{\nu\times\nu}\left(\frac{1}{k(n)} \log W_{k(n)}(y, \tilde{y}) \leq N\right) = 0, \tag{4}$$

for every N, where $P_{\nu\times\nu}$ denotes probability with respect to the product measure $\nu \times \nu$.

The classic example where waiting times are much longer than recurrence times is a "bursty" process, that is, one in which long blocks of 1's and 0's alternate. If $x_1 = 1$, say, then, $W_1(x, y)$ will be large for a set of y's of probability close to 1/2. If the process has a large enough alphabet, and sample paths cycle through long enough blocks of each symbol then $W_1(x, y)$ will be large for a set of probability close to 1.

The key to the construction is to force $W_k(y, \tilde{y})$ to be large, with high probability, with a code that makes only a *small* density of changes in sample paths, for then the construction can be iterated to produce infinitely many bad k. The initial step at each stage is to create markers by changing only a few places in a k-block, where k is large enough to guarantee that many different markers are possible. The same marker is used for a long succession of nonoverlapping k-blocks, then switches to another marker for a long time, cycling in this way through the possible markers to make sure that the time between blocks with the same marker is very large. By this mimicking of the bursty property of the classical example, $(1/k) \log W_k(y, \tilde{y})$ is forced to be large with high probability.

The marker construction idea is easier to carry out if the binary process μ is thought of as a measure on A^Z whose support is contained in B^Z, where $A = \{0, 1, 2, 3\}$, for then blocks of 2's and 3's can be used to form markers. This artificial expansion of the alphabet is not really essential, however, for with additional (but messy) effort one can create binary markers.

The first problem is the creation of a large number of block codes which don't change sequences by much, yet are pairwise separated at the k-block level. To make this precise, fix $G_L \subseteq A^L$. A function $C: G_L \mapsto A^L$ is said to be *δ-close to the identity* if points are moved no more than δ, that is, $\bar{d}_L(C(x_1^L), x_1^L) \leq \delta$, $x_1^L \in G_L$. A family $\{C_i\}$ of functions from G_L to A^L is δ-close to the identity if each member of the family is δ-close to the identity. A family $\{C_i\}$ of functions from G_L to A^L is said to be *pairwise k-separated,* if the k-block universes of the ranges of different functions are disjoint, that is

$$\mathcal{U}_k(y_1^L) \cap \mathcal{U}_k(\widehat{y}_1^L) = \emptyset,\ \ y_1^L \in C_i(x_1^L), \widehat{y}_1^L \in C_j(x_1^L), i \neq j.$$

The following lemma is stated in a form that allows the necessary iteration; its proof shows how a large number of pairwise separated codes close to the identity can be constructed using markers.

Lemma III.5.5

Suppose $k \geq (2g+3)/\delta$ and $L = mk$. Let G_L be the subset of $\{0, 1, 2, 3\}^L$ consisting of those sequences in which no block of 2's and 3's of length g occur. Then there is a pairwise k-separated family $\{C_i\colon\ 1 \leq i \leq 2^g\}$ of functions from G_L to A^L, which is δ-close to the identity, with the additional property that no $(g+1)$-block of 2's and 3's occurs in any concatenation of members of $\cup_i C_i(G_L)$.

Proof. Order the set $\{2, 3\}^g$ in some way and let w_1^g denote the i-th member. Define the i-th k-block encoder $\widetilde{c}_i(x_1^k) = y_1^k$ by setting

$$y_1 = y_{g+2} = y_{2g+3} = 0, \quad y_2^{g+1} = y_{g+3}^{2g+2} = w_1^g, \quad y_i = x_i, i > 2g+3,$$

that is, $\widetilde{c}_i(x_1^k)$ replaces the first $2g+3$ terms of x_1^k by the concatenation $0w_1^g0w_1^g0$, and leaves the remaining terms unchanged. The encoder C_i is defined by blocking x_1^L into k-blocks and applying $\widetilde{c}_i$ separately to each k-block, that is, $C_i(x_1^L) = y_1^L$, where

$$y_{jk+1}^{jk+k} = \widetilde{c}_i\left(x_{jk+1}^{jk+k}\right), \quad 0 \leq j < m.$$

Any block of length k in $C_i(x_1^L) = y_1^L$ must contain the $(g+2)$-block $0w_1^g0$, and since different w_1^g are used with different C_i, pairwise k-separation must hold. Since each $\widetilde{c}_i$ changes at most $2g+3 \leq \delta k$ coordinates in a k-block, the family is δ-close to the identity. Note also that the blocks of 2's and 3's introduced by the construction are separated from all else by 0's, hence no $(g+1)$-block of 2's and 3's can occur in any concatenation of members of $\cup_i C_i(G_L)$. Thus the lemma is established. □

The code family constructed by the preceding lemma is concatenated to produce a single code C of length $2^g L$. The block-to-stationary construction, Lemma I.8.5, is then used to construct a stationary coding for which the waiting time is long for one value of k. An iteration of the method, as done in the string matching construction of Section I.8.c, is used to produce the final process. The following lemma combines the codes of preceding lemma with the block-to-stationary construction in a form suitable for later iteration.

Lemma III.5.6

Let μ be an ergodic measure with alphabet $A = \{0, 1, 2, 3\}$ such that $\mu(\{2, 3\}^g) = 0$, where $g \geq 1$. Given $\epsilon > 0$, $\delta > 0$, and N, there is a $k \geq 2g+3$ and a stationary encoder $F\colon A^Z \mapsto A^Z$ such that for $\nu = \mu \circ F^{-1}$ the following hold.

(i) $P_{\nu \times \nu}\left(W_k(y, \tilde{y}) < 2^{kN}\right) \leq 2^{-g} + \epsilon.$

(ii) $\bar{d}(x, F(x)) \leq \delta$, *for all* x.

(iii) $\nu(\{2, 3\}^{g+1}) = 0.$

Proof. Choose $k \geq (2g+3)/\delta$, then choose $L \geq 2^{kN+2}/\epsilon$ so that L is divisible by k. Let L, G_L, and $\{C_i\colon\ 1 \leq i \leq 2^g\}$ be as in the preceding lemma for the given δ. Put

$M = 2^g L$ and define a code $C: A^M \mapsto A^M$ by setting $C(x_1^M) = x_1^M$, for $x_1^M \notin (G_L)^{2^g}$, and

$$y_{(i-1)L+1}^{iL} = C_i(x_{(i-1)L+1}^{iL}),\ 1 \le i \le 2^g,$$

for $x_1^M \in (G_L)^{2^g}$.

Next apply the block-to-stationary construction of Lemma I.8.5 to obtain a stationary code $F = F_C$ with the property that for almost every $x \in A^Z$ there is a partition, $Z = \cup_j I_j$, of the integers into subintervals of length no more than M, such that if J consists of those j for which I_j is shorter than M, then $\cup_{j\in J} I_j$ covers no more than a limiting $(\epsilon/4)$-fraction of Z, and such that $y = F(x)$ satisfies the following two conditions.

(a) If $I_j = [n_j + 1, n_j + M]$, then $y_{n_j+1}^{n_j+M} = C(x_{n_j+1}^{n_j+M})$.

(b) If $i \in \cup_{j\in J} I_j$, then $y_i = x_i$.

The intervals I_j that are shorter than M will be called *gaps*, while those of length M will be called *coding blocks*.

To show that $F = F_C$ has the desired properties, first note that the only changes made in x to produce $y = F(x)$ are made by the block codes C_i, each of which changes only the first $2g + 3$ places in a k-block. Thus $\bar{d}(x, F(x)) \le \delta$, for every x, since $k \ge (2g+3)/\delta$, and hence property (ii) holds. Furthermore, the changes produce blocks of 2's and 3's of length g which are enclosed between two 0's; hence, since μ had no blocks of 2's and 3's of length g, the encoded process can have no such blocks of length $g + 1$, which establishes property (iii).

To show that property (i) holds, note that if y and $\tilde{y}$ are chosen independently, and if $y = F(x)$ and $\tilde{y} = F(\tilde{x})$, then only two cases can occur.

Case 1. Either x_1 or $\tilde{x}_1$ is in a gap of the code F.

Case 2. Both x_1 and $\tilde{x}_1$ belong to coding blocks of F.

Case 1 occurs with probability at most $\epsilon/2$ since the limiting density of time in the gaps is upper bounded by $\epsilon/4$. In Case 2, since the M-block code C is a concatenation of the L-block codes, $\{C_i\}$, there will be integers n, m, i, and j such that

$$n \le 1 \le n + L - 1, \qquad m \le 1 \le m + L - 1,$$

and such that

$$y_n^{n+L-1} = C_i(x_n^{n+L-1}), \qquad \tilde{y}_m^{m+L-1} = C_j(\tilde{x}_m^{m+L-1}).$$

Three cases are now possible.

Case 2a. $i = j$.

Case 2b. Either $k > n + L - 1$ or $2^{kN} > m + L - 1$.

Case 2c. $i \ne j$, $k \le n + L - 1$, and $2^{kN} \le m + L - 1$.

Case 2a occurs with probability at most 2^{-g}, since y and $\tilde{y}$ were independently chosen and there are 2^g different kinds of L-blocks each of which is equally likely to be chosen. Case 2b occurs with probability at most $\epsilon/2$ since $L \ge 2^{kN+2}/\epsilon$. If Case 2c occurs then

$W_k(y, \tilde{y}) \geq 2^{kN}$, by the pairwise k-separation property. Cases 1, 2a, and 2b therefore imply that

$$P_{\nu \times \nu}\left(W_k(y, \tilde{y}) < 2^{kN}\right) \leq \epsilon/2 + 2^{-g} + \epsilon/2,$$

which completes the proof of Lemma III.5.6. □

The finish the proof it will be shown that there is a sequence of stationary encoders, $F_n: A^Z \mapsto A^Z$, $n = 1, 2, \ldots$, and an increasing sequence $\{k(n)\}$ such that for the composition $G_n = F_n \circ F_{n-1} \circ \ldots \circ F_1$ and measure $\nu^{(n)} = \mu \circ G_n^{-1}$, the following properties hold for each $n \geq 1$, where $G_0(x) \equiv x$.

(a) $P_{\nu^{(n)} \times \nu^{(n)}}\left(W_{k(i)}(y, \tilde{y}) \leq 2^{ik(i)}\right) \leq 2^{-i+1} + 2^{-i} + \ldots + 2^{-n+1} + 2^{-n}, 1 \leq i \leq n.$

(b) $\bar{d}\left(G_n(x), G_{n-1}(x)\right) \leq 2^{-n}$, almost surely.

(c) $\nu^{(n)}\left(\{2, 3\}^n\right) = 0.$

Condition (b) guarantees that eventually almost surely each coordinate is changed only finitely many times, so there will be a limit encoder F and measure $\nu = \mu \circ F^{-1}$. Condition (a) guarantees that the limit measure ν will have the desired property, that is,

$$\frac{1}{k(n)} \log W_{k(n)}(y, \tilde{y}) \to \infty, \text{ in } \nu \times \nu \text{ probability.}$$

Condition (c) is a technical condition which makes it possible to proceed from step n to step $n + 1$.

In summary, it is enough to show the existence of such a sequence of encoders $\{F_n\}$ and integers $\{k(n)\}$. This will be done by induction.

The first step is to apply Lemma III.5.6 with $g = 0$ to select $k(1) \geq 3$ and F_1 such that (a), (b), and (c) hold for $n = 1$, with $G_1 = F_1$. Assume $F_1, F_2, \ldots, F_n$ have been constructed so that (a), (b), and (c) hold for n.

Let $\delta \leq 2^{-(n+1)}$ be a positive number to be specified later. Apply Lemma III.5.6 with $g = n$ and μ replaced by $\nu^{(n)}$ to select $k(n + 1)$ larger than both $2n + 2$ and $k(n)$ and to select an encoder $F_{n+1}: A^Z \mapsto A^Z$ such that for $\nu^{(n+1)} = \nu^{(n)} \circ F_{n+1}^{-1}$, the properties (i)-(iii) of Lemma III.5.6 hold. These properties imply that conditions (b) and (c) hold for $n + 1$, since $\nu^{(n+1)} = \mu \circ G_{n+1}^{-1}$, while condition (a) holds for the case $i = n + 1$, since it was assumed that $\delta \leq 2^{-(n+1)}$. If δ is small enough, however, then $\nu^{(n+1)}$ will be so close to $\nu^{(n)}$ that condition (a), with n replaced by $n + 1$, will hold for $i < n + 1$. This completes the construction of the counterexample.

Remark III.5.7

The number ϵ can be made arbitrarily small in the preceding argument, and hence one can make $P(x_0 \neq F(x)_0)$ as small as desired. In particular, the entropy of the encoded process ν can be forced to be arbitrarily close to $\log 2$, the entropy of the coin-tossing process μ.

III.5.e An approximate-match counterexample.

Let α be a positive number smaller than 1/2. The goal is to show that there is an ergodic process μ for which

$$\lim_{k\to\infty} \text{Prob}\left(\frac{1}{k}\log W_k(x, y, \alpha) \leq N\right) = 0, \ \forall N. \tag{5}$$

Such a μ can be produced merely by suitably choosing the parameters in the strong-nonadmissibility example constructed in III.3.b. A weaker subsequence result is discussed first as it illustrates the ideas in simpler form.

First some definitions and results from III.3.b will be recalled. Two sets $C, D \subseteq A^k$ are α-separated if $C \cap [D]_\alpha = \emptyset$, where $[D]_\alpha$ denotes the α-blowup of D. The full k-block universe of $\mathcal{S} \subseteq A^\ell$ is the set k-blocks that appear anywhere in any concatentation of members of $\mathcal{S}$. Two subsets $\mathcal{S}, \mathcal{S}' \subset A^\ell$ are (α, K)-*strongly-separated* if their full k-block universes are α-separated for any $k \geq K$. A merging of a collection $\{\mathcal{S}_j \colon j \in J\}$ is a product of the form

$$\mathcal{S} = \prod_{m=1}^{M} \mathcal{S}_{\phi(m)},$$

where J divides M and $\phi\colon [1, M] \mapsto [1, J]$ is such that $|\phi^{-1}(j)| = M/J$, for each $j \in [1, J]$.

Given an increasing unbounded sequence $\{N_m\}$, and a decreasing sequence $\{\alpha_m\} \subset (\alpha, 1/2)$, inductive application of the merging/separation lemma, Lemma III.3.5, produces a sequence $\{\mathcal{S}(m) \subset A^{\ell(m)} \colon m \geq 1\}$ and an increasing sequence $\{k_m\}$, such that the following properties hold for each m.

- **(a)** $\mathcal{S}(m)$ is a disjoint union of a collection $\{\mathcal{S}_j(m)\colon j \leq J_m\}$ of pairwise (α_m, k_m)-strongly-separated sets of the same cardinality.
- **(b)** Each $\mathcal{S}_j(m)$ is a merging of $\{\mathcal{S}_j(m-1)\colon j \leq J_{m-1}\}$.
- **(c)** $2^{k_m N_m} \leq \ell(m)/m$.

The only new fact here is property (c). Once k_m is determined, however, $\mathcal{S}(m)$ can be replaced by $(\mathcal{S}(m))^n$ for any positive integer n without disturbing properties (a) and (b), and hence property (c) can also be assumed to hold, see Remark III.3.6.

Property (b) guarantees that the measure μ defined by $\{\mathcal{S}(m)\}$ is ergodic, while property (c) implies that

$$\lim_{m\to\infty} \text{Prob}\left(W_{k_m}(x, y, \alpha) \leq 2^{k_m N}\right) = 0, \tag{6}$$

for any integer N. This is just the subsequence form of the desired result (5). Indeed, if sample paths sample paths x and y are picked independently at random, then with probability at least $(1-1/J_m)^2(1-1/m)^2$, their starting positions x_1 and y_1 will belong to $\ell(m)$-blocks that belong to different $\mathcal{S}_j(m)$ and lie at least $2^{k_m N_m}$-indices below the end of each $\ell(m)$-block. Thus,

$$\text{Prob}\left(W_{k_m}(x, y, \alpha) \geq 2^{k_m N_m}\right) \geq (1-1/J_m)^2(1-1/m)^2,$$

which proves (6), since N_m is unbounded.

The stronger form, discussed in III.3.b.2, controls what happens in the interval $k_m \leq k \leq k_{m+1}$. Further independent cutting and stacking can always be applied at any stage without losing separation properties already gained, and hence, as in the preceding discussion, the column structures at any stage can be made so long that bad waiting-time behavior is guaranteed for the entire range $k_m \leq k \leq k_{m+1}$. This leads to an example satisfying the stronger result, (5). The reader is referred to [76] for a complete discussion of this final argument.

Chapter IV

B-processes.

Section IV.1 Almost block-independence.

The focus in this chapter is on B-processes, that is, finite alphabet processes that are stationary codings of i.i.d. processes. Various other names have been used, each arising from a different characterization, including almost block-independent processes, finitely determined processes, and very weak Bernoulli processes. This terminology and many of the ideas to be discussed here are rooted in Ornstein's fundamental work on the much harder problem of characterizing the *invertible* stationary codings of i.i.d. processes, the so-called isomorphism problem in ergodic theory, [46]. Invertibility is a basic concept in the abstract study of transformations, but is of little interest in stationary process theory where the focus is on the joint distributions rather than on the particular space on which the random variables are defined. This is fortunate, for while the theory of stationary codings of i.i.d. processes is still a complex theory it becomes considerably simpler when the invertibility requirement is dropped, as will be done here.

A natural and useful characterization of stationary codings of i.i.d. processes, the almost block-independence property, will be discussed in this first section. The almost block-independence property, like other characterizations of B-processes, is expressed in terms of the $\bar{d}$-metric (or some equivalent metric.) Either measure or random variable notation will be used for the $\bar{d}$-distance, for example, $\bar{d}_n(X_1^n, Y_1^n)$ will often be used in place of $\bar{d}_n(\mu, \nu)$, if μ is the distribution of X_1^n and ν the distribution of Y_1^n.

A *block-independent process* is formed by extending a measure μ_n on A^n to a product measure on $(A^n)^\infty$, then transporting this to a T^n-invariant measure $\widetilde{\mu}$ on A^∞. In other words, $\widetilde{\mu}$ is the measure on A^∞ defined by the formula

$$\widetilde{\mu}(x_1^{mn}) = \prod_{j=1}^{m} \mu_n\left(x_{(j-1)n+1}^{(j-1)n+n}\right), \quad x_1^{mn} \in A^{mn}, m \geq 1,$$

and the requirement that

$$\widetilde{\mu}(x_i^j) = \sum_{x_1^{i-1}} \sum_{x_{j+1}^{mn}} \widetilde{\mu}(x_1^{mn}),$$

for all $i \leq j \leq mn$ and all x_i^j. Note that $\widetilde{\mu}$ is T^n-invariant, though it is not, in general, stationary. As in earlier chapters, randomizing the start produces a stationary process, called the concatenated-block process defined by μ_n. The theory to be developed in this chapter could be stated in terms of approximation by concatenated-block processes, but it is generally easier to use the simpler block-independence ideas.

The *independent n-blocking* of an ergodic process μ is the block-independent process defined by the restriction μ_n of μ to A^n. It will be denoted by $\widetilde{\mu}(n)$, or by $\widetilde{\mu}$ if n is understood. In random variable language, the independent n-blocking of $\{X_i\}$ is the T^n-invariant process $\{Y_j\}$ defined by the following two conditions.

(a) $Y_{(j-1)n+1}^{(j-1)n+n}$ and X_1^n have the same distribution, for each $j \geq 1$.

(b) $Y_{(j-1)n+1}^{(j-1)n+n}$ is independent of $\{Y_i,\ i \leq (j-1)n\}$, for each $j \geq 1$.

An ergodic process μ is *almost block-independent (ABI)* if given $\epsilon > 0$, there is an N such that if $n \geq N$ and $\widetilde{\mu}$ is the independent n-blocking of μ then $\bar{d}(\mu, \widetilde{\mu}) < \epsilon$. An i.i.d. process is clearly almost block-independent, since the ABI condition holds for every $n \geq 1$ and every $\epsilon > 0$. The almost block-independence property is preserved under stationary coding and, in fact, characterizes the class of stationary codings of i.i.d. processes. This result and the fact that mixing Markov chains are almost block-independent are the principal results of this section.

Theorem IV.1.1 (The almost block-independence theorem.)
An ergodic process is almost block-independent if and only if it is a stationary coding of an i.i.d. process.

Theorem IV.1.2
A mixing Markov chain is almost block-independent.

Note, in particular, that the two theorems together imply that a mixing Markov chain is a stationary coding of an i.i.d. process, which is not at all obvious.

The fact that stationary codings of i.i.d. processes are almost block-independent is established by first proving it for finite codings, then by showing that the ABI property is preserved under the passage to $\bar{d}$-limits. Both of these are quite easy to prove.

It is not as easy to show that an almost block-independent process μ is a stationary coding of an i.i.d. process, for this requires the construction of a stationary coding from some i.i.d. process ν onto the given almost block-independent process μ. This will be carried out by showing how to code a suitable i.i.d. process onto a process $\bar{d}$-close to μ, then how to how to make a small density of changes in the code to produce a process even closer in $\bar{d}$. The fact that only a small density of changes are needed insures that an iteration of the method produces a limit coding equal to μ. Of course, all this requires that $h(\nu) \geq h(\mu)$, since stationary coding cannot increase entropy. In fact it is possible to show that an almost block-independent process μ is a stationary coding of *any* i.i.d. process ν for which $h(\nu) \geq h(\mu)$. The construction is much simpler, however, and is sufficient for the purposes of this book, if it is assumed that the i.i.d. process has infinite alphabet with continuous distribution for then any n-block code can be represented as a function of the first n-coordinates of the process and $\bar{d}$-joinings can be used to modify such codes.

IV.1.a B-processes are ABI processes.

First it will be shown that finite codings of i.i.d. processes are almost block-independent. This is because blocks separated by twice the window half-width are independent. Let $\{Y_i\}$ be a finite coding the i.i.d. process $\{X_i\}$ with window half-width w, and let $\{Z_i\}$ be the independent n-blocking of $\{Y_i\}$. If the first and last w terms of each n-block are omitted, then $mn\bar{d}_{nm}(Z_1^{nm}, Y_1^{nm})$ is upper bounded by

$$m(n-2w)\bar{d}_{m(n-2w)}(Z_{w+1}^{n-w-1}\cdots Z_{(m-1)n+w+1}^{mn-w-1}, Y_{w+1}^{n-w-1}\cdots Y_{(m-1)n+w+1}^{mn-w-1}) + 2mw,$$

by property (c) of the $\bar{d}$-property list, Lemma I.9.11. The successive $n-2w$ blocks

$$Z_{w+1}^{n-w-1}, \ldots, Z_{(m-1)n+w+1}^{mn-w-1}$$

are independent with the distribution of Y_{w+1}^{n-w-1}, by the definition of independent n-blocking, while the successive $n-2w$ blocks

$$Y_{w+1}^{n-w-1}, \ldots, Y_{(m-1)n+w+1}^{mn-w-1}$$

are independent with the distribution of Y_{w+1}^{n-w-1} by the definition of window half-width and the assumption that $\{X_i\}$ is i.i.d., so that property (f) of the $\bar{d}$-property list, Lemma I.9.11, yields

$$\bar{d}_{m(n-2w)}(Z_{w+1}^{n-w-1}\cdots Z_{(m-1)n+w+1}^{mn-w-1}, Y_{w+1}^{n-w-1}\cdots Y_{(m-1)n+w+1}^{mn-w-1}) = 0.$$

Thus

$$\bar{d}_{nm}(Z_1^{nm}, Y_1^{nm}) \leq \frac{2w}{n}.$$

Since this is less than ϵ, for any $n \geq w/\epsilon$, it follows that finite codings of i.i.d. processes are almost block-independent.

Next assume μ is the $\bar{d}$-limit of almost block-independent processes. Given $\epsilon > 0$ choose an almost block-independent process ν such that $\bar{d}(\mu, \nu) < \epsilon/3$, and hence,

$$\bar{d}_n(\mu, \nu) < \epsilon/3, \tag{1}$$

for all n. Since ν is almost block-independent there is an N such that if $n \geq N$ and $\tilde{\nu}$ is the independent n-blocking of ν then $\bar{d}(\nu, \tilde{\nu}) < \epsilon/3$. Fix such an n and let $\tilde{\mu}$ be the independent n-blocking of μ. The fact that both $\tilde{\nu}$ and $\tilde{\mu}$ are i.i.d. when thought of as A^n-valued processes implies that

$$\bar{d}(\tilde{\nu}, \tilde{\mu}) = \bar{d}_n(\tilde{\nu}, \tilde{\mu}) = \bar{d}_n(\nu, \mu) < \epsilon/3,$$

by (1). The triangle inequality then gives

$$\bar{d}(\mu, \tilde{\mu}) \leq \bar{d}(\mu, \nu) + \bar{d}(\nu, \tilde{\nu}) + \bar{d}(\tilde{\nu}, \tilde{\mu}) < \epsilon.$$

This proves that the class of ABI processes is $\bar{d}$-closed.

In summary, finite codings of i.i.d. processes and $\bar{d}$-limits of ABI processes are almost block-independent, hence stationary codings of i.i.d. processes are almost block-independent, since a stationary coding is a $\bar{d}$-limit of finite codings. Furthermore, the preceding argument applies to stationary codings of infinite alphabet i.i.d. processes onto finite alphabet processes, hence such coded processes are also almost block-independent. This completes the proof that B-processes are almost block-independent. □

IV.1.b ABI processes are B-processes.

Let $\{X_i\}$ be an almost block-independent process with finite alphabet A and Kolmogorov measure μ. The goal is to show that μ is a stationary coding of an infinite alphabet i.i.d. process with continuous distribution. An i.i.d. process $\{Z_i\}$ is *continuously distributed* if $\text{Prob}(Z_0 = b) = 0$, for all alphabet symbols b. It will be shown that there is a continuously distributed i.i.d. process $\{Z_i\}$ with Kolmogorov measure λ for which there is a sequence of stationary codes $\{F_t\}$ such that the following hold.

(a) $\bar{d}(\mu, \lambda \circ F_t^{-1}) \to 0$, as $t \to \infty$.

(b) $\sum_t \text{Prob}((F_{t+1}(z))_0 \neq (F_t(z))_0) < \infty$.

The second condition is important, for it means that, almost surely, each coordinate $F_t(z)_i$ changes only finitely often as $t \to \infty$. Therefore, there will almost surely exist a limit code $F(z)_i = \lim_t F_t(z)_i$ which is stationary and for which $\bar{d}(\mu, \lambda \circ F^{-1}) = 0$, that is, $\mu = \lambda \circ F^{-1}$.

The exact representation of $\{Z_i\}$ is unimportant, since any random variable with continuous distribution is a function of any other random variable with continuous distribution, and hence any i.i.d. process with continuous distribution is a stationary coding, with window width 1, of any other i.i.d. process with continuous distribution. This fact, which allows one to choose whatever continuously distributed i.i.d. process is convenient at a given stage of the construction, will be used in the sequel.

The key to the construction is the following lemma, which allows the creation of arbitrarily better codes by making small changes in a good code.

Lemma IV.1.3 (Stationary code improvement.)
If $\bar{d}(\mu, \lambda \circ G^{-1}) \leq \epsilon$, where μ is almost block-independent and G is a finite stationary coding of a continuously distributed i.i.d. process $\{Z_i\}$ with Kolmogorov measure λ, and if $\{V_i\}$ is a binary equiprobable i.i.d. process independent of $\{Z_i\}$, then given $\delta > 0$ there is a finite stationary coding F of $\{(Z_i, V_i)\}$ such that the following two properties hold.

(i) $\bar{d}(\mu, (\lambda \times \nu) \circ F^{-1}) \leq \delta$.

(ii) $(\lambda \times \nu)(\{(z, v) \colon G(z)_0 \neq F(z, v)_0\}) \leq 4\epsilon$.

The lemma immediately yields the desired result for one can start with the vector valued, countable component, i.i.d. process

$$\{(V(0)_i, V(1)_i, V(2)_i, \ldots V(t)_i, \ldots)\} \tag{2}$$

with independent components, where $V(0)_0$ is uniform on $[0, 1)$ and, for $t \geq 1$, $V(t)_0$ is binary and equiprobable, together with an arbitrary initial coding F_0 of $\{V(0)_i\}$. The lemma can then be applied sequentially with

$$Z(t)_i = (V(0)_i, V(1)_i, \ldots, V(t-1)_i),$$

and $\epsilon_t = 2^{-t+1}$ and $\delta_t = 2^{-t}$, to obtain a sequence of stationary codes $\{F_t\}$ that satisfy the desired properties (a) and (b), producing thereby a limit coding F from the vector-valued i.i.d. process (2) onto the almost block-independent process μ.

The idea of the proof of Lemma IV.1.3 is to first truncate G to a block code. A block code version of the lemma is then used to change to an arbitrarily better block code.

Finally, the better block code is converted to a stationary code by a modification of the block-to-stationary method of I.8.b, using the auxiliary process $\{V_i\}$ to tell when to apply the block code. The i.i.d. nature of the auxiliary process $\{V_i\}$ and its independence of the $\{Z_i\}$ process guarantee that the final coded process satisfies an extension of almost block-independence in which occasional spacing between blocks is allowed, an extension which will be shown to hold for ABI processes.

A finite stationary code is truncated to a block code by using it except when within a window half-width of the end of a block.

Lemma IV.1.4 (Code truncation.)
Let F be a finite stationary coding which carries the stationary process ν onto the stationary process $\widetilde{\nu}$. Given $\delta > 0$ there is an N such that if $n \geq N$ then there is an n-block code G, called the (δ, n)-truncation of F, such that $d_n(F(u)_1^n, G(u_1^n)) < \delta$, for ν-almost every sample path u.

Proof. Let w be the window half-width of F and let $N = 2w/\delta$. For $n \geq N$ define $G(u_1^n) = bF(\tilde{u})_{1+w}^{n-w}c$, where $\tilde{u}$ is any sample path for which $\tilde{u}_1^n = u_1^n$, and b and c are fixed words of length w. This is well-defined since $F(\tilde{u})_{1+w}^{n-w}$ depends only on the values of $\tilde{u}_1^n$. Clearly $d_n(F(u)_1^n, G(u_1^n)) < \delta$, for ν-almost every sample path u, since the only disagreements can occur in the first or last w places. □

The block code version of the stationary code improvement lemma is stated as the following lemma.

Lemma IV.1.5 (Block code improvement.)
Let (Y, ν) be a nonatomic probability space and suppose $\psi: Y \mapsto A^n$ is a measurable mapping such that $\bar{d}_n(\mu_n, \nu \circ \psi^{-1}) < \delta$. Then there is a measurable mapping $\phi: Y \mapsto A^n$ such that $\mu_n = \nu \circ \phi^{-1}$ and such that $E_\nu(d_n(\phi(y), \psi(y))) < \delta$.

Proof. This is really just the mapping form of the idea that $\bar{d}$-joinings are created by cutting up nonatomic spaces, see Exercise 12 in Section I.9. Indeed, a joining has the form $\lambda(a_1^n, b_1^n) = \nu(\phi^{-1}(a_1^n) \cap \psi^{-1}(b_1^n))$ for $\phi: Y \mapsto A^n$ such that $\mu_n = \nu \circ \phi^{-1}$, so that $E_\lambda(d_n(a_1^n, b_1^n)) = E_\nu(d_n(\phi(y), \psi(y)))$. □

The almost block-independence property can be strengthened to allow gaps between the blocks, while retaining the property that the blocks occur independently. The following terminology will be used. A binary, ergodic process $\{R_i\}$ will be called a (δ, n)-*blocking process,* if the waiting time between 1's is never more than n and is exactly equal to n with probability at least $1 - \delta$. A block R_i^{i+n-1} of length n will be called a *coding block* if $R_j = 0$, for $i \leq j < i + n - 1$ and $R_{i+n-1} = 1$, that is, $R_i^{i+n-1} = 0^{n-1}1$.

An ergodic pair process $\{(W_i, R_i)\}$ will be called a (δ, n)-*independent process* if $\{R_i\}$ is a (δ, n)-blocking process and if W_1^n is independent of $\{(W_i, R_i): i \leq 0\}$, given that R_1^n is a coding block. An ergodic process $\{W_i\}$ is called a (δ, n)-*independent extension of a measure μ_n on A^n, with associated (δ, n)-blocking process* $\{R_i\}$, if $\{(W_i, R_i)\}$ is a (δ, n)-independent process and

$$\text{Prob}(W_1^n = a_1^n | R_1^n \text{ is a coding block}) = \mu_n(a_1^n),\ a_1^n \in A^n.$$

The following lemma supplies the desired stronger form of the ABI property.

Lemma IV.1.6 (Independent-extension.)

Let $\{X_i\}$ be an almost block-independent process with Kolmogorov measure μ. Given $\epsilon > 0$ there is an N and a $\delta > 0$ such that if $n \geq N$ then $\bar{d}(\mu, \widetilde{\mu}) < \epsilon$, for any (δ, n)-independent extension $\widetilde{\mu}$ of μ_n.

Proof. In outline, here is the idea of the proof. It is enough to show how an independent N_1-blocking that is close to μ can be well-approximated by a (δ, n)-independent extension, provided only that δ is small enough and n is large enough. The obstacle to a good $\bar{d}$-fit that must be overcome is the possible nonsynchronization between n-blocks and multiples of N_1-blocks. This is solved by noting that conditioned on starting at a coding n-block, the starting position, say i, of an n-block may not be an exact multiple of N_1, but $i + j$ will be for some $j < N_1$. Thus after omitting at most N_1 places at the beginning and end of the n-block it becomes synchronized with the independent N_1-blocking, hence all that is required is that $2N_1$ be negligible relative to n, and that δ be so small that very little is outside coding blocks.

To make this outline precise, first choose N_1 so large that $\bar{d}(\mu, \nu) < \epsilon/3$, where ν is the Kolmogorov measure of the independent N_1-blocking $\{Y_i\}$ of μ. Since ν is not stationary it need not be true that $\bar{d}_m(\mu_m, \nu_m) < \epsilon/3$, for all m, but at least there will be an M_1 such that

$$\bar{d}_m(\mu_m, \nu_m) < \epsilon/3, \ m \geq M_1. \tag{3}$$

Put $N_2 = M_1 + 2N_1$ and fix $n \geq N_2$. Let δ be a positive number to be specified later and let $\widetilde{\mu}$ be the Kolmogorov measure of a (δ, n)-independent extension $\{W_i\}$ of μ_n, with associated (δ, n)-blocking process $\{R_i\}$.

The goal is to show that if M_1 is large enough and δ small enough then $\bar{d}(\nu, \widetilde{\mu}) < 2\epsilon/3$, and hence $\bar{d}(\mu, \widetilde{\mu}) < \epsilon$. Towards this end, let $\{\widehat{W}_i\}$ denote the W-process conditioned on a realization $\{r_i\}$ of the associated (δ, n)-blocking process $\{R_i\}$. Let $\{n_k\}$ be the successive indices of the locations of the 1's in $\{r_i\}$. For each k for which $n_{k+1} - n_k = n$ choose the least integer t_k and greatest integer s_k such that

$$t_k N_1 \geq n_k, \ s_k N_1 \leq n_{k+1}.$$

Note that $m = s_k N_1 - t_k N_1$ is at least as large as M_1, and that $\widehat{W}_{t_k N_1 + 1}^{s_k N_1}$ has the same distribution as $X_{t_k N_1 + 1}^{s_k N_1}$, so that (3) yields

$$\bar{d}_m(Y_{t_k N_1 + 1}^{s_k N_1}, \widehat{W}_{t_k N_1 + 1}^{s_k N_1}) < \epsilon/3. \tag{4}$$

The processes $\{Y_i\}$ and $\{\widehat{W}_i\}$ are now joined by using the joining that yields (4) on the blocks $[t_k N_1 + 1, s_k N_1]$, and an arbitrary joining on the other parts. If M_1 is large enough so that $t_k N_1 - n_k + n_{k+1} - t_k N_1$ is negligible compared to M_1 and if δ is small enough, then, for almost every realization $\{r_k\}$, the $\bar{d}$-distance between the processes $\{Y_i\}$ and $\{\widehat{W}_i\}$ will be less than $2\epsilon/3$. An application of property (g) in the list of $\bar{d}$-properties, Lemma I.9.11, yields the desired result, $\bar{d}(\nu, \widetilde{\mu}) < 2\epsilon/3$, completing the proof of Lemma IV.1.6. □

Proof of Lemma IV.1.3.

Let $\{Z_i\}$ be a continuously distributed i.i.d. process and $\{V_i\}$ a binary, equiprobable i.i.d. process, independent of $\{Z_i\}$, with respective Kolmogorov measures λ and ν, and let $\lambda \times \nu$ denote the Kolmogorov measure of the direct product $\{(Z_i, V_i)\}$. Let μ be an almost block-independent process with alphabet A, and let G be a finite stationary coding such that $\bar{d}(\mu, \lambda \circ G^{-1}) \leq \epsilon$.

Fix $\delta > 0$. The goal is to construct a finite stationary coding F of $\{(Z_i, V_i)\}$ such that the following hold.

(i) $\bar{d}(\mu, (\lambda \times \nu) \circ F^{-1}) \leq \delta$.

(ii) $(\lambda \times \nu)(\{(z, v): G(z)_0 \neq F(z, v)_0\}) \leq 4\epsilon$.

Towards this end, first choose n and $\bar{\delta} < \epsilon$ so that if $\widetilde{\mu}$ is a $(\bar{\delta}, n)$-independent extension of μ then $\bar{d}(\mu, \widetilde{\mu}) < \delta$, and so that there is an (ϵ, n)-truncation $\widetilde{G}$ of G. In particular, this means that

$$\bar{d}_n(\mu_n, \lambda_n \circ \widetilde{G}^{-1}) \leq 2\epsilon,$$

and hence the block code improvement lemma provides an n-block code ϕ such that $\mu_n = \lambda_n \circ \phi^{-1}$ and

$$E(d_n(\widetilde{G}(Z_1^n), \phi(Z_1^n))) \leq 2\epsilon. \tag{5}$$

Next, let g be a finite stationary coding of $\{V_i\}$ onto a $(\bar{\delta}, n)$-blocking process $\{R_i\}$ and lift this to the stationary coding g^* of $\{(Z_i, V_i)\}$ onto $\{(Z_i, R_i)\}$ defined by $g^*(z, v) = (z, g(v))$. Let $F(z, r)$ be the (stationary) code defined by applying ϕ whenever the waiting time between 1's is exactly n, and coding to some fixed letter whenever this does not occur. In other words, fix $a \in A$, let $\{\gamma_k\}$ denote the (random) sequence of indices that mark the starting positions of the successive coding blocks in a (random) realization r of $\{R_i\}$, define

$$F(z, r)_{\gamma_k}^{\gamma_k+n-1} = \phi(z_{\gamma_k}^{\gamma_k+n-1}),$$

for all k, and define $F(z, r)_i = a$, $i \notin \cup_k[\gamma_k, \gamma_k + n - 1]$. Clearly F maps onto a $(\bar{\delta}, n)$-independent extension of μ, hence property (i) holds.

The see why property (ii) holds, first note that $\{Z_j\}$ and $\{R_j\}$ are independent, and hence the process defined by $W_k = Z_{\gamma_k}^{\gamma_k+n-1}$ is i.i.d. with W_0 distributed as Z_1^n. Furthermore,

$$E(d_n(\widetilde{G}(Z_1^n), \phi(Z_1^n))) = E(d_n(\widetilde{G}(Z_1^n), \phi(Z_1^n)) \mid R_1^n \text{ is a coding block})$$

so the ergodic theorem applied to $\{W_k\}$, along with the fitting bound, (5), implies that the concatenations satisfy

$$\lim_{k\to\infty} d_{nk}(\phi(W_1)\cdots\phi(W_k), \widetilde{G}(W_1)\cdots\widetilde{G}(W_k)) < 2\epsilon, \tag{6}$$

almost surely. By definition, however, $F(u, v)_{\gamma_k}^{\gamma_k+n-1} = \phi(W_k)$, if $g(v) = r$, and $d_n(G(z)_{\gamma_k}^{\gamma_k+n-1}, \widetilde{G}(z_{\gamma_k}^{\gamma_k+n-1})) < \epsilon$. Since, by the ergodic theorem applied to $\{R_i\}$, the limiting density of indices i for which $i \notin \cup_k[\gamma_k, \gamma_k + n - 1]$ is almost surely at most $\bar{\delta}$, it follows from (6) that

$$(\lambda \times \nu)(\{(z, v): G(z)_0 \neq F(z, v)_0\}) < 2\epsilon + \bar{\delta} + \epsilon \leq 4\epsilon,$$

thereby establishing property (ii). This completes the proof of the lemma, and hence the proof that an ABI process is a stationary coding of an i.i.d. process. □

Remark IV.1.7

The existence of block codes with a given distribution as well as the independence of the blocking processes used to convert block codes to stationary codes onto independent extensions are easy to obtain for i.i.d. processes with continuous distribution. With considerably more effort, suitable approximate forms of these ideas can be obtained for any finite alphabet i.i.d. process whose entropy is at least that of μ. These results, which are the basis for Ornstein's isomorphism theory, are discussed in detail in his book, [46], see also [63, 42].

IV.1.c Mixing Markov and almost block-independence.

The key to the fact that mixing Markov implies ABI, as well as several later results about mixing Markov chains is a simple form of coupling, a special type of joining frequently used in probability theory, see [35]. Let $\{X_n\}$ be a (stationary) mixing Markov process and let $\{Y_n\}$ be the nonstationary Markov chain with the same transition matrix as $\{X_n\}$, but which is conditioned to start with $Y_0 = a$. A joining $\{(X_n, Y_n)\}$ is obtained by running the two processes independently until they agree, then running them together. The Markov property guarantees that this is indeed a joining of the two processes, with the very strong additional property that a sample path is always paired with a sample path which agrees with it ever after as soon as they agree once. In particular, this guarantees that $n\bar{d}_n(X_n, Y_n)$ cannot exceed the expected time until the two processes agree.

The precise formulation of the coupling idea needed here follows. Let μ and ν be the Kolmogorov measures of $\{X_n\}$ and $\{Y_n\}$, respectively. For each $n \geq 1$, the *coupling function* is defined by the formula

$$w_n(a_1^n, b_1^n) = \begin{cases} \min\{i \in [1, n-1]: a_i = b_i\}, & \text{if } \{i \in [1, n-1]: a_i = b_i\} \neq \emptyset \\ n, & \text{otherwise.} \end{cases}$$

The *coupling measure* is the measure λ_n on $A^n \times A^n$ defined by

$$\text{(7)} \quad \lambda_n(a_1^n, b_1^n) = \begin{cases} \nu(a_1^w)\mu(b_1^n), & \text{if } w_n(a_1^n, b_1^n) = w < n, \text{ and } a_{w+1}^n = b_{w+1}^n \\ \nu(a_1^n)\mu(b_1^n) & \text{if } w_n(a_1^n, b_1^n) = n \\ 0 & \text{otherwise.} \end{cases}$$

A direct calculation, making use of the Markov properties for both measures, shows that λ_n is a probability measure with ν_n and μ_n as marginals.

Lemma IV.1.8 (The Markov coupling lemma.)

The coupling function satisfies the bound

$$\text{(8)} \qquad \bar{d}_n(\mu, \nu) \leq \frac{E_{\lambda_n}(w_n)}{n},$$

In particular, $\bar{d}_n(\mu, \nu) \to 0, \ n \to \infty$.

Proof. The inequality (8) follows from the fact that the λ_n defined by (7) is a joining of μ_n and ν_n, and is concentrated on sequences that agree ever after, once they agree. Furthermore, the expected value of the coupling function is bounded, independent of n and a, see Exercise 2, below, so that $\bar{d}_n(\mu, \nu)$ indeed goes to 0. This establishes the lemma. □

To prove that a mixing Markov chain is $\bar{d}$-close to its independent n-blocking, for all large enough n, the idea to construct a joining by matching successive n-blocks, conditioned on the past. This will produce the desired $\bar{d}$-closeness, because the distribution of X_{kn+1}^{kn+n} conditioned on the past depends only on X_{kn}, by the Markov property, and the distribution of X_{kn+1}^{kn+n} conditioned on X_{kn} is $\bar{d}$-close to the unconditioned distribution of X_{kn+1}^{kn+n}, by the coupling lemma, provided only that n is large enough.

Fix a mixing Markov chain $\{X_i\}$ with Kolmogorov measure μ and fix $\epsilon > 0$. The notation $\bar{d}_n(X_{kn+1}^{kn+n}, X_{kn+1}^{kn+n}/x_{kn})$ will be used to denote the $\bar{d}_n$-distance between the unconditioned distribution of X_{kn+1}^{kn+n} and the conditional distribution of X_{kn+1}^{kn+n}, given $X_{kn} = x_{kn}$. The coupling lemma implies that

$$\bar{d}_n(X_{kn+1}^{kn+n}, X_{kn+1}^{kn+n}/x_{kn}) \leq \epsilon, \tag{9}$$

for all sufficiently large n.

Fix n for which (9) holds and let $\{Z_i\}$ be the independent n-blocking of $\{X_i\}$, with Kolmogorov measure $\widetilde{\mu}$. For each positive integer m, a joining λ_{mn} of μ_{mn} with $\widetilde{\mu}_{mn}$ will be constructed such that

$$E_{\lambda_{mn}}(d_{mn}(x_1^{mn}, z_1^{mn})) \leq \epsilon,$$

which, of course, proves that $\bar{d}(\mu, \widetilde{\mu}) \leq \epsilon$, and hence that $\{X_i\}$ is almost block-independent.

To construct λ_{mn}, first choose, for each x_{kn}, a joining $\lambda_{/x_{kn}}$ of the unconditioned distribution of X_{kn+1}^{kn+n} and the conditional distribution of X_{kn+1}^{kn+n}, given $X_{kn} = x_{kn}$, that realizes the $\bar{d}_n$-distance between them. For each positive integer m, let λ_{mn} be the measure on $A^{mn} \times A^{mn}$ defined by

$$\lambda_{mn}(x_1^{mn}, z_1^{mn}) = \mu(x_1^n)\mu(z_1^n)\prod_{k=1}^{m-1} \lambda_{/x_{kn}}(x_{kn+1}^{kn+n}, z_{kn+1}^{kn+n}). \tag{10}$$

A direct calculation, using the Markov property of $\{X_i\}$ plus the fact that Z_{kn+1}^{kn+n} is independent of Z_j, $j \leq kn$, shows that λ_{mn} is a joining of μ_{mn} with $\widetilde{\mu}_{mn}$. Likewise, summing over the successive n-blocks and using the bound (9) yields the bound $E_{\lambda_{mn}}(d_{mn}(x_1^{mn}, z_1^{mn})) \leq \epsilon$. Since the proof extends to Markov chains of arbitrary order, this completes the proof that mixing Markov chains are almost block-independent. □

The almost block-independent processes are, in fact, exactly the almost mixing Markov processes, in the sense that

Theorem IV.1.9

An almost block-independent process is the $\bar{d}$-limit of mixing Markov processes.

Proof. By the independent-extension lemma, Lemma IV.1.6, it is enough to show that for any $\delta > 0$, a measure μ on A^n has a (δ, n)-independent extension $\widetilde{\mu}$ which is mixing Markov of some order. The concatenated-block process defined by μ is a function of an irreducible Markov chain, see Example I.1.11. A suitable random insertion of spacers between the blocks produces a (δ, n)-independent extension which is a function of a mixing Markov chain. (See Exercise 13, Section I.2.) The only problem, therefore, is to produce a (δ, n)-independent extension which is actually a mixing Markov chain, not just a function of a mixing Markov chain. This is easy to accomplish by marking long concatenations of blocks with a symbol not in the alphabet, then randomly inserting an occasional repeat of this symbol.

To describe the preceding in a rigorous manner, let $C \subseteq A^n$ consist of those a_1^n for which $\mu(a_1^n) > 0$ and let s be a symbol not in the alphabet A. For each positive integer m, let sC^m denote the set of all concatenations of the form $sw(1)\cdots w(m)$, where $w(i) \in C$, and let $\hat{\mu}$ be the measure on sC^m defined by the formula

$$\hat{\mu}(sw(1)\cdots w(m)) = \prod_{i=1}^{m} \mu(w(i)).$$

Fix $0 < p < 1$, to be specified later, and let $\{Y_j\}$ be the stationary, irreducible, aperiodic Markov chain with state space $sC^m \times \{0, 1, \ldots, nm+1\}$ and transition matrix defined by the rules

(a) If $i < nm+1$, (sx_1^{mn}, i) can only go to $(sx_1^{mn}, i+1)$.

(b) $(sx_1^{mn}, nm+1)$ goes to $(sy_1^{mn}, 0)$ with probability $p\hat{\mu}(sy_1^{mn})$, and to $(sy_1^{mn}, 1)$ with probability $(1-p)\hat{\mu}(sy_1^{mn})$, for each $sy_1^{mn} \in sC^m$.

Next, define $f(sx_1^{mn}, i)$ to be x_i if $1 < i \leq mn+1$, and s otherwise. The process $\{Z_j\}$ defined by $Z_j = f(Y_j)$ is easily seen to be $(mn+2)$-order Markov, since once the marker s is located in the past all future probabilities are known. As a function of the mixing Markov chain $\{Y_j\}$, it is mixing. Furthermore, $\{Z_j\}$ is clearly a (δ, n)-independent extension of μ, provided m is large enough and p is small enough (so that only a δ-fraction of indices are spacers s.)

This completes the proof of Theorem IV.1.9. □

Remark IV.1.10

The use of a marker symbol not in the original alphabet in the above construction was only to simplify the proof that the final process $\{Z_j\}$ is Markov. A proof that does not require alphabet expansion is outlined in Exercise 6c, below.

Remark IV.1.11

The proof that the B-processes are exactly the almost block-independent processes is based on the original proof, [66]. The coupling proof that mixing Markov chains are almost block-independent is modeled after the proof of a conditional form given in [41]. The proof that ABI processes are $\bar{d}$-limits of mixing Markov processes is new.

IV.1.d Exercises.

1. Show that (7) defines a joining of μ_n and ν_n.

2. Show the expected value of the coupling function is bounded, independent of n.

3. Show that the λ_{mn} defined by (10) is indeed a joining of μ_{mn} with $\widetilde{\mu}_{mn}$, such that $E_{\lambda_{mn}}(d_{mn}(x_1^{mn}, z_1^{mn})) \leq \epsilon$.

4. Show that if $\{W_i\}$ and $\{\widehat{W}_i\}$ are (δ, n)-independent extensions of μ_n and $\hat{\mu}_n$, respectively, with the same associated (δ, n)-blocking process $\{R_i\}$, then $\bar{d}(\nu, \hat{\nu}) = \bar{d}_n(\mu_n, \hat{\mu}_n)$, where ν and $\hat{\nu}$ are the Kolmogorov measures of $\{W_i\}$ and $\{\widehat{W}_i\}$, respectively.

5. Show that even if the marker s belongs to the original alphabet A, the process $\{Z_j\}$ defined in the proof of Theorem IV.1.9 is a (δ, n)-independent extension of μ. (Hint: use Exercise 4.)

6. Let μ be the Kolmogorov measure of an ABI process for which $0 < \mu_1(a) < 1$, for some $a \in A$.

   (a) Show that μ must be mixing.

   (b) Show that $\mu(a^n) \to 0$ as $n \to \infty$, where a^n denotes the concatenation of a with itself n times. (See Exercise 4c, Section I.5.)

   (c) Prove Theorem IV.1.9 without extending the alphabet. (Hint: if n is large enough, then by altering probabilities slightly, and using Exercise 6b and Exercise 4, it can be assumed that $\mu(a^n) = 0$ for some a. The sequence $ba^{2n}b$ can be used in place of s to mark the beginning of a long concatenation of n-blocks.)

7. Some special codings of i.i.d. processes are Markov. Let $\{X_n\}$ be binary i.i.d. and define Y_n to be 1 if $X_n \neq X_{n-1}$, and 0, otherwise. Show that $\{Y_n\}$ is Markov, and not i.i.d. unless $X_n = 0$ with probability 1/2.

Section IV.2 The finitely determined property.

The most important property of a stationary coding of an i.i.d. process is the finitely determined property, for it allows $\bar{d}$-approximation of the process merely by approximating its entropy and a suitable finite joint distribution.

A stationary process μ is *finitely determined (FD)* if any sequence of stationary processes which converges to it weakly and in entropy also converges to it in $\bar{d}$. In other words, a stationary μ is finitely determined if, given $\epsilon > 0$, there is a $\delta > 0$ and a positive integer k, such that any stationary process ν which satisfies the two conditions,

(i) $|\mu_k - \nu_k| < \delta$

(ii) $|H(\mu) - H(\nu)| < \delta$

must also satisfy

(iii) $\bar{d}(\mu, \nu) < \epsilon$.

Some standard constructions produce sequences of processes converging weakly and in entropy to a given process. Properties of such approximating processes that are preserved under $\bar{d}$-limits are automatically inherited by any finitely determined limit process. This principle is used in the following theorem to establish ergodicity and almost block-independence for finitely determined processes.

Theorem IV.2.1

A finitely determined process is ergodic, mixing, and almost block-independent. In particular, a finitely determined process is a stationary coding of an i.i.d. process.

Proof. Let $\mu^{(n)}$ be the concatenated-block process defined by the restriction μ_n of a stationary measure μ to A^n. If k fixed and n is large relative to k, then the distribution of k-blocks in n-blocks is almost the same as the μ_k-distribution, the only error in this approximation being the end effect caused by the fact that the final k symbols of the n-block have to be ignored. Thus $\mu_k^{(n)} \to \mu_k$, as $n \to \infty$. The entropy of $\mu^{(n)}$ is equal to $H(\mu_n)/n$ which converges to $H(\mu)$, so that $\{\mu^{(n)}\}$ converges weakly and in entropy to μ. Since each $\mu^{(n)}$ is ergodic and since ergodicity is preserved by passage to $\bar{d}$-limits, it follows that a finitely determined process is ergodic.

Concatenated-block processes can be modified by randomly inserting spacers between the blocks. For each n, let $\hat{\mu}^{(n)}$ be a (δ_n, n)-independent extension of μ_n (such processes were defined as part of the independent-extension lemma, Lemma IV.1.6.) If δ_n is small enough then $\hat{\mu}^{(n)}$ and $\mu^{(n)}$ have almost the same n-block distribution, hence almost the same k-block distribution for any $k \leq n$, and furthermore, $\hat{\mu}^{(n)}$ and $\mu^{(n)}$ have almost the same entropy. Thus if δ_n goes to 0 suitably fast, $\{\hat{\mu}^{(n)}\}$ will converge weakly and in entropy to μ, hence in $\bar{d}$ if μ is assumed to be finitely determined. Since $\hat{\mu}^{(n)}$ can be chosen to be a function of a mixing Markov chain, see Exercise 5, Section IV.1, and hence almost block-independent, and since $\bar{d}$-limits of ABI processes are ABI, a finitely determined process must be almost block-independent. Also, since functions of mixing processes are mixing and $\bar{d}$-limits of mixing processes are mixing, a finitely determined process must be mixing. This completes the proof of Theorem IV.2.1. □

The converse result, that a stationary coding of an i.i.d. process is finitely determined is much deeper and will be established later.

Also useful in applications is a form of the finitely determined property which refers only to finite distributions. If ν is a measure on A^n, and $k \leq n$, *the ν-distribution of k-blocks in n-blocks* is the measure $\phi = \phi(k, \nu)$ on A^k defined by

$$\phi(a_1^k) = \frac{1}{n-k+1} \sum_{i=0}^{n-k} \sum_{u \in A^i} \sum_{v \in A^{n-k-i}} \nu(u a_1^k v),$$

that is, the average of the ν-probability of a_1^k over the first $n-k+1$ starting positions in sequences of length n. A direct calculation shows that

$$\phi(a_1^k) = \sum_{x_1^n} p_k(a_1^k | x_1^n) \nu(x_1^n) = E_\nu(p_k(a_1^k | X_1^n)),$$

the expected value of the empirical k-block distribution with respect to ν. Note also that if $\widehat{\nu}$ is the concatenated-block process defined by ν, then

$$|\phi(k, \nu) - \widehat{\nu}_k| \leq 2(k-1)/n, \tag{1}$$

since the only difference between the two measures is that $\widehat{\nu}_k$ includes in its average the ways k-blocks can overlap two n-blocks.

Theorem IV.2.2 (The FD property: finite form.)
A stationary process μ is finitely determined if and only if given $\epsilon > 0$, there is a $\delta > 0$ and positive integers k and N such that if $n \geq N$ then any measure ν on A^n for which $|\mu_k - \phi(k, \nu)| < \delta$ and $|H(\mu_n) - H(\nu)| < n\delta$ must also satisfy $\bar{d}_n(\mu, \nu) < \epsilon$.

Proof. One direction is easy for if ν_n is the projection onto A^n of a stationary measure ν, then $\phi(k, \nu_n) = \nu_k$, for any $k \leq n$, and $(1/n)H(\nu_n) \to H(\nu)$. Thus, if μ satisfies the finite form of the FD property stated in the theorem, then μ must be finitely determined.

To prove the converse assume μ is finitely determined and let ϵ be a given positive number. The definition of finitely determined yields a $\delta > 0$ and a k, such that any stationary process ν that satisfies $|\nu_k - \mu_k| < \delta$ and $|H(\nu) - H(\mu)| < \delta$ must also satisfy $\bar{d}(\nu, \mu) < \epsilon$.

Choose N so large that if $n \geq N$ then $|(1/n)H(\mu_n) - H(\mu)| < \delta/2$. By making N enough larger relative to k to take care of end effects, that is, using (1), it can also be supposed that if $n \geq N$ and if $\widehat{\nu}$ is the concatenated-block process defined by a measure ν on A^n, then $|\phi(k, \nu) - \widehat{\nu}_k| < \delta/2$.

Fix $n \geq N$ and let ν be a measure on A^n for which $|\phi(k, \nu) - \mu_k| < \delta/2$ and $|H(\nu) - H(\mu_n)| < n\delta/2$ and let $\widehat{\nu}$ be the concatenated-block process defined by ν. The definition of N guarantees that

$$|\widehat{\nu}_k - \mu_k| \leq |\widehat{\nu}_k - \phi(k, \nu)| + |\phi(k, \nu) - \mu_k| < \delta.$$

Furthermore, $|H(\widehat{\nu}) - H(\mu)| < \delta$, since $H(\widehat{\nu}) = (1/n)H(\nu)$, which is within $\delta/2$ of $(1/n)H(\mu_n)$, which is, in turn, within $\delta/2$ of $H(\mu)$. Thus, the definitions of k and δ imply that $\bar{d}(\mu, \widehat{\nu}) < \epsilon$. A finitely determined process is mixing, by Theorem IV.2.1, hence is totally ergodic, so Exercise 2 implies that $\bar{d}_n(\mu_n, \nu) \leq \bar{d}(\mu, \widehat{\nu}) < \epsilon$. This completes the proof of Theorem IV.2.2. □

IV.2.a ABI implies FD.

The proof that almost block-independence implies finitely determined will be given in three stages. First it will be shown that an i.i.d. process is finitely determined. This proof will then be extended to establish that mixing Markov processes are finitely determined. Finally, it will be shown that the class of finitely determined processes is closed in the $\bar{d}$-metric. These results immediately show that ABI implies FD, for an ABI process is the $\bar{d}$-limit of mixing Markov processes by Theorem IV.1.9 of the preceding section.

IV.2.a.1 I.i.d. processes are finitely determined.

The i.i.d. proof will be carried out by induction. It is easy to get started for closeness in first-order distribution means that first-order distributions can be $\bar{d}_1$-well-fitted. The idea is to extend a good $\bar{d}_n$-fitting of the distributions of X_1^n and Y_1^n by fitting the conditional distribution of X_{n+1}, given X_1^n, as well as possible to the conditional distribution of Y_1^n, given Y_1^n. The conditional distribution of X_{n+1}, given X_1^n, does not depend on X_1^n, if independence is assumed, and hence the method works, provided closeness in distribution and entropy implies that the conditional distribution of Y_1^n, given Y_1^n, is almost independent of Y_1^n.

To make this sketch into a real proof, an approximate independence concept will be used. Let $p(x, y) = p_{X,Y}(x, y) =$ Prob$(X = x, Y = y)$ denote the joint distribution of the pair (X, Y) of finite-valued random variables, and let $p(x) = p_X(x) = \sum_y p(x, y)$ and $p(y) = p_Y(y) = \sum_x p(x, y)$ denote the marginal distributions. The random variables are said to be *ϵ-independent* if

$$\sum_{x,y} |p(x, y) - p(x)p(y)| < \epsilon.$$

If $H(X) = H(X|Y)$, then X and Y are independent. An approximate form of this is stated as the following lemma.

Lemma IV.2.3 (The ϵ-independence lemma.)

There is a positive constant c such that if $H(X) - H(X|Y) < c\epsilon^2$, then X and Y are ϵ-independent.

Proof. The entropy difference may be expressed as the divergence of the joint distribution $p_{X,Y}$ from the product distribution $p_X p_Y$, that is,

$$\begin{aligned} H(X) - H(X|Y) &= -\sum_x p(x)\log p(x) + \sum_{x,y} p(x,y)\log\frac{p(x,y)}{p(y)} \\ &= \sum_{x,y} p(x,y)\log\frac{p(x,y)}{p(x)p(y)} \\ &= D(p_{X,Y}|p_X p_Y). \end{aligned}$$

Pinsker's inequality, Exercise 6 of Section I.6, yields,

$$D(p_{X,Y}|p_X p_Y) \geq c\left(\sum_{x,y}|p(x,y) - p(x)p(y)|\right)^2,$$

where c is a positive constant, independent of X and Y. This establishes the lemma. □

A stationary process $\{Y_i\}$ is said to be *an ϵ-independent process* if $(Y_1, \ldots, Y_{j-1})$ and Y_j are ϵ-independent for each $j > 1$. An i.i.d. process is, of course, ϵ-independent for every $\epsilon > 0$. The next lemma asserts that a process is ϵ-independent if it is close enough in entropy and first-order distribution to an i.i.d. process.

Lemma IV.2.4

Let $\{X_i\}$ be i.i.d. and $\{Y_i\}$ be stationary with respective Kolmogorov measures μ and ν. Given $\epsilon > 0$, there is a $\delta > 0$ such that if $|\mu_1 - \nu_1| < \delta$ and $|H(\mu) - H(\nu)| < \delta$, then $\{Y_i\}$ is an ϵ-independent process.

Proof. If first-order distributions are close enough then first-order entropies will be close. Thus if μ_1 is close enough to ν_1, then $H(X_1)$ will be close to $H(Y_1)$. The i.i.d. property, however, implies that $H(\mu) = H(X_1)$, so that if $H(\nu) = \lim_m H(Y_1|Y_{-m}^0)$ is close enough to $H(\mu)$, then $H(Y_1) - H(Y_1|Y_{-m}^0)$ will be small, for all $m \geq 0$, since $H(Y_1|Y_{-m}^0)$ is decreasing in m. The lemma now follows from the preceding lemma. □

A conditioning formula useful in the i.i.d. proof as well as in later results is stated here, without proof, as the following lemma. In its statement, $p(x, y)$ denotes the joint distribution of a pair (X, Y) of finite-valued random variables, with first marginal $p(x) = \sum_y p(x, y)$ and conditional distribution $p(y|x) = p(x, y)/p(x)$.

Lemma IV.2.5

If $f(x, y) = g(x) + h(y)$, then

$$\sum_{x,y} f(x,y)p(x,y) = \sum_x g(x)p(x) + \sum_x p(x)\sum_y h(y)p(y|x).$$

Use will also be made of the fact that first-order $\bar{d}$-distance is just one-half of variational distance, that is, $\bar{d}_1(\mu, \nu) = |\mu_1 - \nu_1|/2$, which is just the equality part of property (a) in the $\bar{d}$-properties lemma, Lemma I.9.11.

Theorem IV.2.6

An i.i.d. process is finitely determined.

Proof. The following notation and terminology will be used. The random variable X_{n+1}, conditioned on $X_1^n = x_1^n$ is denoted by X_{n+1}/x_1^n, and $\bar{d}_1(X_{n+1}/x_1^n, Y_{n+1}/y_1^n)$ denotes the $\bar{d}$-distance between the corresponding conditional distributions.

Fix an i.i.d. process $\{X_i\}$ and $\epsilon > 0$. Lemma IV.2.4 provides a positive number δ so that if $|\mu_1 - \nu_1| < \delta$ and $|H(\mu) - H(\nu)| < \delta$, then the process $\{Y_i\}$ defined by ν is an ϵ-independent process. Without loss of generality it can be supposed that $\delta < \epsilon$. Fix such a $\{Y_i\}$. It will be shown by induction that $\bar{d}_n(X_1^n, Y_1^n) < \epsilon$.

Getting started is easy, since $\bar{d}_1(X_1, Y_1) = (1/2)|\mu_1 - \nu_1| < \delta/2 < \epsilon/2$. Assume it has been shown that $\bar{d}_n(X_1^n, Y_1^n) < \epsilon$, and let λ_n realize $\bar{d}_n(X_1^n, Y_1^n)$. The strategy is to extend by fitting the conditional distributions X^{n+1}/x_1^n and Y_{n+1}/y_1^n as well as possible. This strategy is successful because the distribution of X^{n+1}/x_1^n is equal to the distribution of X_{n+1}, by independence, and the distribution of Y^{n+1}/y_1^n is, by ϵ-independence, close on the average to that of Y_{n+1}, which is in turn close to the distribution of X_{n+1}.

For each x_1^n, y_1^n, let $\lambda_{x_1^n, y_1^n}$ realize $\bar{d}_1(X_{n+1}/x_1^n, Y_{n+1}/y_1^n)$. The measure λ_{n+1} defined by

$$\lambda_{n+1}(x_1^{n+1}, y_1^{n+1}) = \lambda_n(x_1^n, y_1^n)\lambda_{x_1^n, y_1^n}(x_{n+1}, y_{n+1}),$$

is certainly a joining of μ_{n+1} and ν_{n+1}. Furthermore, since

$$(n+1)d_{n+1}(x_1^{n+1}, y_1^{n+1}) = nd_n(x_1^n, y_1^n) + d_1(x_{n+1}, y_{n+1}),$$

the conditional formula, Lemma IV.2.5, yields

$$\begin{aligned} &(n+1)E_{\lambda_{n+1}}(d_{n+1}) \\ &\quad = nE_{\lambda_n}(d_n) + \sum_{x_1^{n+1}, y_1^{n+1}} \lambda_n(x_1^n, y_1^n) d_1(x_{n+1}, y_{n+1}) \lambda_{x_1^n, y_1^n}(x_{n+1}, y_{n+1}). \end{aligned} \tag{2}$$

The first term is upper bounded by $n\epsilon$, since it was assumed that λ_n realizes $\bar{d}_n(X_1^n, Y_1^n)$, while the second sum is equal to

$$\sum_{x_1^n, y_1^n} \lambda_n(x_1^n, y_1^n)\bar{d}_1(X_{n+1}/x_1^n, Y_{n+1}/y_1^n), \tag{3}$$

since it was assumed that $\lambda_{x_1^n, y_1^n}$ realizes $\bar{d}_1(X_{n+1}/x_1^n, Y_{n+1}/y_1^n)$.

The triangle inequality yields

$$\begin{aligned} &\bar{d}_1(X_{n+1}/x_1^n, Y_{n+1}/y_1^n) \\ &\quad \leq \bar{d}_1(X_{n+1}/x_1^n, X_{n+1}) + \bar{d}_1(X_{n+1}, Y_{n+1}) + \bar{d}_1(Y_{n+1}, Y_{n+1}/y_1^n) \end{aligned}$$

The first term is 0, by independence, while, by stationarity, the second term is equal to $\bar{d}_1(X_1, Y_1)$, which is at most $\epsilon/2$. The third term is just (1/2) the variational distance between the distributions of the unconditioned random variable Y_{n+1} and the conditioned random variable Y_{n+1}/y_1^n, whose expected value (with respect to y_1^n) is at most $\epsilon/2$, since Y_{n+1} and Y_1^n are ϵ-independent. Thus the second sum in (2) is at most $\epsilon/2 + \epsilon/2$, which along with the bound $nE_{\lambda_n}(d_n) < n\epsilon$ and the fact that λ_{n+1} is a joining of μ_{n+1} and ν_{n+1}, produces the inequality

$$(n+1)\bar{d}_{n+1}(\mu, \nu) \leq (n+1)E_{\lambda_{n+1}}(d_{n+1}) \leq n\bar{d}_n(\mu, \nu) + \epsilon < (n+1)\epsilon,$$

thereby establishing the induction step.

This completes the proof of Theorem IV.2.6. □

IV.2.a.2 Mixing Markov processes are finitely determined.

The Markov result is based on a generalization of the i.i.d. proof. In that proof, the fitting was done one step at a time. A good match up to stage n was continued to stage $n+1$ by using the fact that X_{n+1} is independent of X_1^n for the i.i.d. process, and Y_{n+1} is almost independent of Y_1^n, provided the Y-process is close enough to the X-process in distribution and entropy.

To extend the i.i.d. proof to the mixing Markov case, two properties will be used, the Markov property and the Markov coupling lemma, Lemma IV.1.8. The Markov property guarantees that for any $n \geq 1$, a future block X_{n+1}^{n+m} depends on the immediate past X_n, but on no previous values. The Markov coupling lemma guarantees that for mixing Markov chains even this dependence on X_n dies off in the $\bar{d}$-sense, as m grows. The key is to show that approximate versions of both properties hold for any process close enough to the Markov process in entropy and in joint distribution for a long enough time, for then a good match after n steps can be carried forward by fitting future m-blocks, for suitable choice of m.

A conditional form of ϵ-independence will be needed. The random variables X and Y are *conditionally ϵ-independent, given Z*, if

$$\sum_{z}\sum_{x,y}|p(x,y|z)-p(x|z)p(y|z)|p(z)<\epsilon,$$

where $p(x,y|z)$ denotes the conditional joint distribution and $p(x|z)$ and $p(y|z)$ denote the respective marginals. The conditional form of the ϵ-independence lemma, Lemma IV.2.3, extends to the following result, whose proof is left to the reader.

Lemma IV.2.7 (The conditional ϵ-independence lemma.)
Given $\epsilon > 0$, there is a $\gamma = \gamma(\epsilon) > 0$ such that if $H(X|Z) - H(X|Y,Z) < \gamma$, then X and Y are conditionally ϵ-independent, given Z.

The next lemma provides the desired approximate form of the Markov property.

Lemma IV.2.8
Let $\{X_i\}$ be an ergodic Markov process with Kolmogorov measure μ and let $\{Y_i\}$ be a stationary process with Kolmogorov measure ν. Given $\epsilon > 0$ and a positive integer m, there is a $\delta > 0$ such that if $|\mu_{m+1} - \nu_{m+1}| < \delta$ and $|H(\mu) - H(\nu)| < \delta$, then, for every $n \geq 1$, Y_1^m and Y_{-n}^{-1} are conditionally ϵ-independent, given Y_0.

Proof. Choose $\gamma = \gamma(\epsilon)$ from preceding lemma, then choose $\delta > 0$ so small that if $|\mu_{m+1} - \nu_{m+1}| < \delta$ then

$$|H(Y_1^m|Y_0) - H(X_1^m|X_0)| < \gamma/2,$$

which is possible since conditional entropy is continuous in the variational distance. Fix m. By decreasing δ if necessary it can be assumed that if $|H(\mu) - H(\nu)| < \delta$ then

$$|mH(\mu) - mH(\nu)| < \gamma/2.$$

Fix a stationary process ν for which $|\mu_{m+1} - \nu_{m+1}| < \delta$ and $|H(\mu) - H(\nu)| < \delta$. The choice of δ and the fact that $H(Y_1^m|Y_{-n}^0)$ decreases to $mH(\nu)$ as $n \to \infty$, then guarantee that

$$H(Y_1^m|Y_0) - H(Y_1^m|Y_{-n}^0) < \gamma,\ n \geq 1,$$

which, in turn, implies that Y_1^m and Y_{-n}^{-1} are conditionally ϵ-independent, given Y_0, for all $n \geq 1$, by the choice of δ_1. This proves Lemma IV.2.8. □

Theorem IV.2.9

A mixing finite-order Markov process is finitely determined.

Proof. In the proof, X_{n+1}^{n+m}/x_1^n will denote the random vector X_{n+1}^{n+m}, conditioned on $X_1^n = x_1^n$, and $\bar{d}(X_{n+1}^{n+m}/x_1^n, Y_{n+1}^{n+m}/y_1^n)$ will denote the $\bar{d}$-distance between the distributions of X_{n+1}^{n+m}/x_1^n and Y_{n+1}^{n+m}/y_1^n.

Only the first-order proof will be given; the extension to arbitrary finite order is left to the reader. Fix a mixing Markov process $\{X_n\}$ with Kolmogorov measure μ and fix $\epsilon > 0$. The Markov coupling lemma, Lemma IV.1.8, provides an m such that

$$\bar{d}_m(X_1^m, X_1^m/x_0) < \epsilon/4, \quad x_0 \in A. \tag{4}$$

Since this inequality depends only on μ_{m+1} there is a $\delta > 0$ so that if ν is a stationary measure for which $|\mu_{m+1} - \nu_{m+1}| < \delta$ then

$$\bar{d}_m(Y_1^m, Y_1^m/y_0) < \epsilon/4, \quad y_0 \in A. \tag{5}$$

Furthermore, since $\bar{d}_m(\mu, \nu)$ is upper bounded by $|\mu_m - \nu_m|/2$, it can also be assumed that δ is small enough to guarantee that

$$\bar{d}_m(X_1^m, Y_1^m) < \epsilon/4. \tag{6}$$

By making δ smaller, if necessary, it can also be assumed from Lemma IV.2.8 that

(7) $\qquad Y_1^m$ and Y_{-n}^1 are conditionally $(\epsilon/2)$-independent, given Y_0,

for any stationary process $\{Y_i\}$ with Kolmogorov measure ν for which $|\mu_{m+1} - \nu_{m+1}| < \delta$ and $|H(\mu) - H(\mu)| < \delta$.

Fix a stationary ν for which (5), (6), and (7) all hold. To complete the proof that mixing Markov implies finitely determined it is enough to show that $\bar{d}(\mu, \nu) \leq \epsilon$.

The $\bar{d}$-fitting of μ and ν is carried out m steps at a time, using (6) to get started. As in the proof for the i.i.d. case, it is enough to show that

$$\sum_{x_1^n, y_1^n} \lambda_n(x_1^n, y_1^n)\bar{d}_m(X_{n+1}^{n+m}/x_1^n, Y_{n+1}^{n+m}/y_1^n) \leq \epsilon, \tag{8}$$

where λ_n realizes $\bar{d}_n(\mu, \nu)$. The distribution of X_{n+1}^{n+m}/x_1^n is the same as the distribution of X_{n+1}^{n+m}, conditioned on $X_n = x_n$, and hence the triangle inequality yields,

$$\begin{aligned} \bar{d}_m(X_{n+1}^{n+m}/x_1^n, Y_{n+1}^{n+m}/y_1^n) \leq\ & \bar{d}_m(X_{n+1}^{n+m}/x_n, X_{n+1}^{n+m}) + \bar{d}_m(X_{n+1}^{n+m}, Y_{n+1}^{n+m}) \\ & + \bar{d}_m(Y_{n+1}^{n+m}, Y_{n+1}^{n+m}/y_n) + \bar{d}_m(Y_{n+1}^{n+m}/y_n, Y_{n+1}^{n+m}/y_1^n). \end{aligned}$$

The first three terms contribute less than $3\epsilon/4$, by (4), (6) and (5). The fourth term is upper bounded by half of the variational distance between the distribution of Y_{n+1}^{n+m}/y_n and the distribution of Y_{n+1}^{n+m}/y_1^n. But the expected value of this variational distance is less than $\epsilon/2$, since Y_{n+1}^{n+m} and Y_1^{n-1} are conditionally $(\epsilon/2)$-independent, given Y_n, by (7). Thus the expected value of the fourth term is at most $(1/2)\epsilon/2 = \epsilon/4$, and hence the sum in (8) is less than ϵ.

This completes the proof of Theorem IV.2.9. □

IV.2.a.3 The finitely determined processes are $\bar{d}$-closed.

The final step in the proof that almost block-independence implies finitely determined is stated as the following theorem, a result also useful in other contexts.

Theorem IV.2.10 (The $\bar{d}$-closure theorem.)
The $\bar{d}$-limit of finitely determined processes is finitely determined.

The theorem is an immediate consequence of the following lemma, which guarantees that any process $\bar{d}$-close enough to a finitely determined process must have an approximate form of the finitely determined property.

Lemma IV.2.11
If $\bar{d}(\mu, \nu) < \epsilon$ and ν is finitely determined, then there is a $\delta > 0$ and a k such that

$$\bar{d}(\mu, \tilde{\mu}) < 3\epsilon,$$

for any stationary process $\tilde{\mu}$ for which $|\mu_k - \tilde{\mu}_k| < \delta$ and $|H(\mu) - H(\tilde{\mu}| < \delta$.

The basic idea for the proof is as follows. Let λ be a stationary joining of μ and ν such that $E_\lambda(d(x_1, y_1)) = \bar{d}(\mu, \nu)$. It will be shown that if $\tilde{\mu}$ is close enough to μ in distribution and entropy there will be a stationary $\tilde{\lambda}$ with first marginal $\tilde{\mu}$ such that

(a) $\tilde{\lambda}$ is close to λ in distribution and entropy.

(b) The second marginal $\tilde{\nu}$ of $\tilde{\lambda}$ is close to ν in distribution and entropy.

If the second marginal, $\tilde{\nu}$ is close enough to ν in distribution and entropy then the finitely determined property of ν guarantees that $\bar{d}(\nu, \tilde{\nu})$ will be small. The fact that $\tilde{\lambda}$ is a stationary joining of $\tilde{\mu}$ and $\tilde{\nu}$ guarantees that

$$\bar{d}(\tilde{\mu}, \tilde{\nu}) \leq E_{\tilde{\lambda}}(d(x_1, y_1)),$$

which is however, close to $E_\lambda(d(x_1, y_1)) = \bar{d}(\mu, \nu)$, by property (a). The triangle inequality

$$\bar{d}(\mu, \tilde{\mu}) \leq \bar{d}(\mu, \nu) + \bar{d}(\nu, \tilde{\nu}) + \bar{d}(\tilde{\nu}, \tilde{\mu})$$

then implies that $\bar{d}(\mu, \tilde{\mu})$ is also small.

A simple language for making the preceding sketch into a rigorous proof is the language of channels, borrowed from information theory. (None of the deeper ideas from channel theory will be used, only its suggestive language.)

Fix n, and for each a_1^n, let $\lambda_n(\cdot|a_1^n)$ be the conditional measure on A^n defined by the formula

$$\lambda_n(b_1^n|a_1^n) = \frac{\lambda_n(a_1^n, b_1^n)}{\mu(a_1^n)}, \quad a_1^n, b_1^n \in A^n. \tag{9}$$

The family $\{\lambda_n(\cdot|a_1^n)\}$ of conditional measures can thought of as the (noisy) channel (or black box),

$$a_1^n \rightarrow \square \rightarrow b_1^n,$$

which, given the input a_1^n, outputs b_1^n with probability $\lambda_n(b_1^n|a_1^n)$. Such a finite-sequence channel extends to an infinite-sequence channel

$$x \rightarrow \square \rightarrow y, \tag{10}$$

which outputs an infinite sequence y, given an infinite input sequence x, by breaking x into blocks of length n and applying the n-block channel to each block separately, that is, y_{jn+1}^{jn+n} has the value b_1^n, with probability $\lambda_n(b_1^n|x_{jn+1}^{jn+n})$, independent of $\{x_i : i \leq jn\}$ and $\{y_i : i \leq jn\}$.

Given an input measure α on A^∞ the infinite-sequence channel (10) defined by the conditional probabilities (9) determines a joining $\Lambda^* = \Lambda^*(\lambda_n, \alpha)$ of α with an output measure $\beta^* = \beta^*(\lambda_n, \alpha)$ on A^∞. The joining Λ^*, called the *joint input-output measure* defined by the channel and the input measure α, is the measure on $A^\infty \times A^\infty$ defined for $m \geq 1$ by the formula

$$\Lambda^*(a_1^{mn}, b_1^{mn}) = \alpha(a_1^{mn}) \prod_{j=0}^{m-1} \lambda_n(b_{jn+1}^{jn+n}|a_{jn+1}^{jn+n}).$$

The projection of Λ^* onto its first factor is clearly the input measure α. The output measure β^* is the projection of Λ^* onto its second factor, that is, the measure on A^∞ defined for $m \geq 1$ by the formula

$$\beta^*(b_1^{mn}) = \sum_{a_1^{mn}} \Lambda^*(a_1^{mn}, b_1^{mn}).$$

If α is stationary then neither $\Lambda^*(\lambda_n, \alpha)$ nor $\beta^*(\lambda_n, \alpha)$ need be stationary, though both are certainly n-stationary, and hence stationary measures $\Lambda = \Lambda(\lambda_n, \alpha)$, and $\beta = \beta(\lambda_n, \alpha)$ are obtained by randomizing the start. A direct calculation shows that Λ is a joining of α and β. The measures Λ and β are, respectively, called the *stationary joint input-output measure* and *stationary output measure defined by* λ_n and the input measure α.

The next two lemmas contain the facts that will be needed about the continuity of $\Lambda(\lambda_n, \alpha)$ and $\beta(\lambda_n, \alpha)$, in n and in α, with respect to weak convergence and convergence in entropy.

Lemma IV.2.12 (Continuity in n.)
If λ is a stationary joining of the two stationary processes μ and ν, then $\Lambda(\lambda_n, \mu)$ converges weakly and in entropy to λ and $\beta(\lambda_n, \mu)$ converges weakly and in entropy to ν, as $n \to \infty$.

Proof. Fix (a_1^k, b_1^k), let $n > k$, and put $\Lambda^* = \Lambda^*(\lambda_n, \mu)$ and $\Lambda = \Lambda(\lambda_n, \mu)$. The measure $\Lambda(a_1^k, b_1^k)$ is an average of the n measures $\Lambda^*(a(i)_{i+1}^{i+k}, b(i)_{i+1}^{i+k})$, $0 \leq i < n$, where $a(i)_{i+s} = a_s$ and $b(i)_{i+s} = b_s$, for $1 \leq s \leq k$. But, for $i < n - k$

$$\Lambda^*(a(i)_{i+1}^{i+k}, b(i)_{i+1}^{i+k}) = \lambda(a(i)_{i+1}^{i+k}, b(i)_{i+1}^{i+k}) = \lambda(a_1^k, b_1^k),$$

so that, $\Lambda(a_1^k, b_1^k)$ converges to $\lambda(a_1^k, b_1^k)$ as $n \to \infty$. Likewise, the averaged output $\beta(\lambda_n, \mu)(b_1^k) = \sum_{a_1^k} \Lambda(a_1^k, b_1^k)$ converges to $\sum_{a_1^k} \lambda(a_1^k, b_1^k) = \nu(b_1^k)$ as $n \to \infty$.

To establish convergence in entropy, first note that if (X_1^n, Y_1^n) is a random vector with distribution λ_n then $H(X_1^n, Y_1^n) = H(X_1^n) + H(Y_1^n|X_1^n)$, and hence

$$\lim_{n\to\infty} \frac{1}{n} H(Y_1^n|X_1^n) = H(\lambda) - H(\mu).$$

On the other hand, if (X_1^{mn}, Y_1^{mn}) is a random vector with distribution $\Lambda^*(\lambda_n, \mu)_{mn}$, then

$$H(X_1^{mn}, Y_1^{mn}) \quad = \quad H(X_1^{mn}) + H(Y_1^{mn}|X_1^{mn})$$

$$= H(X_1^{mn}) + \sum_{i=1}^{m} H(Y_{(i-1)n+1}^{in} | X_{(i-1)n+1}^{in})$$
$$= H(X_1^{mn}) + mH(Y_1^n | X_1^n),$$

since the channel treats input n-block independently. Dividing by mn and letting $m \to \infty$ gives

$$H(\Lambda^*(\lambda_n, \mu)) = H(\mu) + \frac{1}{n} H(Y_1^n | X_1^n),$$

so that $H(\Lambda^*(\lambda_n, \mu)) \to H(\lambda)$, as $n \to \infty$. Randomizing the start then produces the desired result, $H(\Lambda(\lambda_n, \mu)) \to H(\lambda)$, since the entropy of an n-stationary process is the same as the entropy of the average of its first n-shifts, see Exercise 5, Section I.6. Furthermore, randomizing the start also yields $H(\beta(\lambda_n, \mu)) \to H(\nu)$, since $\widetilde{\nu}_n = \nu_n$, if $\widetilde{\nu} = \beta^*(\lambda_n, \mu)$. This completes the proof of Lemma IV.2.12. □

Lemma IV.2.13 (Continuity in input distribution.)

Fix $n \geq 1$, and a stationary measure λ on $A^\infty \times A^\infty$. If a sequence of stationary processes $\{\alpha(m)\}$ converges weakly and in entropy to a stationary process α as $m \to \infty$, then,

(i) *$\{\Lambda(\lambda_n, \alpha^{(m)})\}$ converges weakly and in entropy to $\Lambda(\lambda_n, \alpha)$, as $m \to \infty$.*

(ii) *$\{\beta(\lambda_n, \alpha^{(m)})\}$ converges weakly and in entropy to $\beta(\lambda_n, \alpha)$, as $m \to \infty$.*

Proof. For simplicity assume $n = 1$, so that both Λ^* and β^* are stationary for any stationary input. Fix a stationary input measure α and put $\Lambda = \Lambda(\lambda_1, \alpha)$. By definition,

$$\Lambda_m(a_1^m, b_1^m) = \alpha(a_1^m) \prod_{i=1}^{m} \lambda(b_i | a_i),$$

which is weakly continuous in α. Furthermore, if (X_1^m, Y_1^m) is a random vector with distribution Λ_m then $H(X_1^m, Y_1^m) = H(X_1^m) + mH(Y_1 | X_1)$, since the channel treats input symbols independently. Thus, dividing by m and letting m go to ∞ yields $H(\Lambda) = H(\alpha) + H(Y_1 | X_1)$, which depends continuously on $H(\alpha)$ and the distribution of (X_1, Y_1). Thus (i) holds.

Likewise, $\beta = \beta(\lambda_1, \alpha)$ is weakly continuous in α, and $H(\beta) = H(\Lambda) - H(X_1 | Y_1)$ is continuous in $H(\Lambda)$ and the distribution of (X_1, Y_1), so (ii) holds. The extension to $n > 1$ is straightforward, and hence Lemma IV.2.13 is established. □

Proof of Lemma IV.2.11.

Assume $\bar{d}(\mu, \nu) < \epsilon$, assume ν is finitely determined, and let λ be a stationary measure realizing $\bar{d}(\mu, \nu)$, so that, in particular, $E_\lambda(d(x_1, y_1)) < \epsilon$. The finitely determined property of ν provides γ and ℓ such that any stationary process $\widetilde{\nu}$ for which $|\nu_\ell - \widetilde{\nu}_\ell| < \gamma$ and $|H(\nu) - H(\widetilde{\nu})| < \gamma$, must also satisfy $\bar{d}(\nu, \widetilde{\nu}) < \epsilon$.

The continuity-in-n lemma, Lemma IV.2.12, provides an n such that the stationary input-output measure $\Lambda = \Lambda(\lambda_n, \mu)$ satisfies

$$E_\Lambda(d(x_1, y_1)) < E_\lambda(d(x_1, y_1)) + \epsilon, \tag{11}$$

and the stationary output measure $\beta = \beta(\lambda_n, \mu)$ satisfies

$$|\beta_\ell - \nu_\ell| < \gamma/2, \text{ and } |H(\beta) - H(\nu)| < \gamma/2.$$

The continuity-in-input lemma, Lemma IV.2.13, provides δ and k such that if $\widetilde{\mu}$ is any stationary process for which $|\widetilde{\mu}_k - \mu_k| < \delta$ and $|H(\widetilde{\mu}) - H(\mu)| < \delta$, then the joint input-output distribution $\widetilde{\Lambda} = \Lambda(\lambda_n, \widetilde{\mu})$ satisfies

$$E_{\widetilde{\Lambda}}(d(x_1, y_1)) < E_{\Lambda}(d(x_1, y_1)) + \epsilon \tag{12}$$

and the output distribution $\widetilde{\nu} = \beta(\lambda_n, \widetilde{\mu})$ satisfies

$$|\widetilde{\nu}_\ell - \beta_\ell| < \gamma/2, \text{ and } |H(\widetilde{\nu}) - H(\beta)| < \gamma/2.$$

Given such an input $\widetilde{\mu}$ it follows that the output $\widetilde{\nu}$ satisfies

$$|\widetilde{\nu}_\ell - \nu_\ell| \le |\widetilde{\nu}_\ell - \beta_\ell| + |\beta_\ell - \nu_\ell| < \gamma,$$

and

$$|H(\widetilde{\nu}) - H(\nu)| \le |H(\widetilde{\nu}) - H(\beta)| + |H(\beta) - H(\nu)| < \gamma,$$

so the definitions of ℓ and γ imply that $\bar{d}(\widetilde{\nu}, \nu) < \epsilon$. Furthermore,

$$\bar{d}(\widetilde{\mu}, \widetilde{\nu}) \le E_{\widetilde{\Lambda}}(d(x_1, y_1)),$$

since $\widetilde{\Lambda} = \Lambda(\lambda_n, \widetilde{\mu})$ is a joining of the input $\widetilde{\mu}$ with the output $\widetilde{\nu}$ and

$$\begin{aligned} E_{\widetilde{\Lambda}}(d(x_1, y_1)) &< E_{\Lambda}(d(x_1, y_1)) + \epsilon \\ &< E_{\lambda}(d(x_1, y_1)) + 2\epsilon < 3\epsilon, \end{aligned}$$

by the inequalities (11) and (12), and the fact that λ realizes $\bar{d}(\mu, \nu)$ which was assumed to be less than ϵ.

This completes the proof of Lemma IV.2.11, thereby completing the proof that $\bar{d}$-limits of finitely determined processes are finitely determined, and hence the proof that almost block-independent processes are finitely determined. □

Remark IV.2.14

The proof that i.i.d. processes are finitely determined is based on Ornstein's original proof, [46], while the proof that mixing Markov chains are finitely determined is based on the Friedman-Ornstein proof, [13]. The fact that $\bar{d}$-limits of finitely determined processes are finitely determined is due to Ornstein, [46]. The observation that the ideas of that proof could be expressed in the simple channel language used here is from [83]. The fact that almost block-independence is equivalent to finitely determined first appeared in [67]; the proof given here that ABI processes are finitely determined is new.

IV.2.b Exercises.

1. The n-th order Markovization of a stationary process μ is the n-th order Markov process with transition probabilities $\nu(x_{n+1}|x_1^n) = \mu(x_1^{n+1})/\mu(x_1^n)$. Show that a finitely determined process is the $\bar{d}$-limit of its n-th order Markovizations. (Hint: ν is close in distribution and entropy to μ.)

2. Show that if μ is totally ergodic and ν is any probability measure on A^n, then $\bar{d}_n(\mu_n, \nu) \leq \bar{d}(\mu, \widehat{\nu})$, where $\widehat{\nu}$ denotes the concatenated-block process defined by ν. (Hint: $\bar{d}(\mu, \widehat{\nu}) = \bar{d}(x, y)$, for some y for which the limiting nonoverlapping k-block distribution of $T^i y$ is ν, for some $i < n$, and for some x all of whose shifts have limiting nonoverlapping n-block distribution equal to μ_n, see Exercise 1a in Section I.4. Hence any of the limiting nonoverlapping n-block distributions determined by $(T^i x, T^i y)$ must be a joining of μ_n and ν.)

3. Show that the class of B-processes is $\bar{d}$-separable.

4. Prove the conditional ϵ-independence lemma, Lemma IV.2.7.

5. Show that a process built by repeated independent cutting and stacking is a B-process if the initial structure has two columns with heights differing by 1.

Section IV.3 Other B-process characterizations.

A more careful look at the proof that mixing Markov chains are finitely determined shows that all that is really needed for a process to be finitely determined is a weaker form of the Markov coupling property. This form, called the very weak Bernoulli property, is actually equivalent to the finitely determined property, hence serves as another characterization of B-processes. Equivalence will be established by showing that very weak Bernoulli implies finitely determined and then that almost blowing-up implies very weak Bernoulli. Since it has already been shown that finitely determined implies almost blowing-up, see Section III.4, it follows that almost block-independence, almost blowing-up, very weak Bernoulli, and finitely determined are, indeed, equivalent to being a stationary coding of an i.i.d. process.

IV.3.a The very weak Bernoulli and weak Bernoulli properties.

As in earlier discussions, either measure or random variable notation will be used for the $\bar{d}$-distance, and X_1^n/x_{-k}^0 will denote the random vector X_1^n, conditioned on the past values x_{-k}^0.

A stationary process $\{X_i\}$ is *very weak Bernoulli (VWB)* if given $\epsilon > 0$ there is an n such that for any $k \geq 0$,

$$E_{X_{-k}^0}(\bar{d}_n(X_1^n/X_{-k}^0, X_1^n)) \leq \epsilon, \tag{1}$$

where $E_{X_{-k}^0}$ denotes expectation with respect to the random past X_{-k}^0. Informally stated, a process is VWB if the past has little effect in the $\bar{d}$-sense on the future.

The significance of very weak Bernoulli is that it is equivalent to finitely determined and that many physically interesting processes, such as geodesic flow on a manifold of constant negative curvature and various processes associated with billiards, can be shown to be very weak Bernoulli by exploiting their natural expanding and contracting foliation structures. The reader is referred to [54] for a discussion of such applications.

Theorem IV.3.1 (The very weak Bernoulli characterization.)
A process is very weak Bernoulli if and only if it is finitely determined.

The proof that very weak Bernoulli implies finitely determined will be given in the next subsection, while the proof of the converse will be given later after some discussion of the almost blowing-up property.

A somewhat stronger property, called weak Bernoulli, is obtained by using variational distance with a gap in place of the $\bar{d}$-distance. A stationary process $\{X_i\}$ is *weak Bernoulli (WB)* or absolutely regular, if past and future become ϵ-independent if separated by a gap g, that is, given $\epsilon > 0$ there is a gap g such that for any $k \geq 0$ and $m > 0$, the random vectors X_g^{g+m} and X_{-k}^0 are ϵ-independent.

The class of weak Bernoulli processes includes the mixing Markov processes and the large class of mixing regenerative processes, see Exercise 2 and Exercise 3. Furthermore, as noted in earlier theorems, Theorems III.2.3 and III.5.1, weak Bernoulli processes have nice empirical distribution and waiting-time properties. Their importance here is that weak Bernoulli processes are very weak Bernoulli. To see why, first note that if X_g^{g+m} and X_{-k}^0 are ϵ-independent then

$$E_{X_{-k}^0}\left(\bar{d}_m(X_{g+1}^{g+m}/X_{-k}^0, X_{g+1}^{g+m})\right) \leq \epsilon/2,$$

since $\bar{d}_m$-distance is upper bounded by one-half the variational distance. If this is true for all m and k, then one can take $n = g + m$, with m so large that $g/(g+m) \leq \epsilon/2$, and use the fact that

$$\bar{d}_n(U_1^n, V_1^n) \leq \bar{d}_{n-g}(U_{g+1}^n, V_{g+1}^n) + g/n,$$

for any $n \geq g$ and any random vector (U_1^n, V_1^n), to obtain $E_{X_{-k}^0}(\bar{d}_n(X_1^n/X_{-k}^0, X_1^n)) \leq \epsilon$. Thus weak Bernoulli indeed implies very weak Bernoulli.

In a sense made precise in Exercises 4 and 5, weak Bernoulli requires that with high probability, the conditional measures on different infinite pasts can be joined in the future so that, with high probability, names agree after some point, while very weak Bernoulli only requires that the density of disagreements be small. This property was the key to the example constructed in [65] of a very weak Bernoulli process that is not weak Bernoulli, a result established by another method in [78].

IV.3.b Very weak Bernoulli implies finitely determined.

The proof models the earlier argument that mixing Markov implies finitely determined. Entropy is used to guarantee approximate independence from the distant past, conditioned on intermediate values, while very weak Bernoulli is used to guarantee that even such conditional dependence dies off in the $\bar{d}$-sense as block length grows. Since approximate versions of both properties hold for any process close enough to $\{X_n\}$ in entropy and in joint distribution for a long enough time, a good fitting can be carried forward by fitting future m-blocks, for suitable choice of m.

As in the earlier proofs, the key is to show that

$$\sum_{x_1^n, y_1^n} \lambda(x_1^n, y_1^n)\bar{d}_m(X_{n+1}^{n+m}/x_1^n, Y_{n+1}^{n+m}/y_1^n)$$

is small for some fixed m and *every* n, given only that the processes are close enough in entropy and k-th order distribution, for some fixed $k \geq m$. The triangle inequality gives

$$\begin{aligned}&\bar{d}_m(X_{n+1}^{n+m}/x_1^n, Y_{n+1}^{n+m}/y_1^n)\\ &\quad\leq \bar{d}_m(X_{n+1}^{n+m}, Y_{n+1}^{n+m}) + \bar{d}_m(X_{n+1}^{n+m}, X_{n+1}^{n+m}/x_1^n) + \bar{d}_m(Y_{n+1}^{n+m}, Y_{n+1}^{n+m}/y_1^n)\end{aligned}$$

By stationarity, the first term is equal $\bar{d}_m(X_1^m, Y_1^m)$, which is small since $m \leq k$, while the expected value of the second term is small for large enough m, uniformly in n, by the very weak Bernoulli property. Thus, once such an m is determined, all that remains to be shown is that the expected value of $\bar{d}_m(Y_{n+1}^{n+m}, Y_{n+1}^{n+m}/y_1^n)$ is small, for all n, provided only that $\{Y_i\}$ is close enough to $\{X_i\}$ in k-order distribution for some large enough $k \geq m$, as well as close enough in entropy.

The details of the preceding sketch are carried out as follows. Fix a very weak Bernoulli process $\{X_n\}$ with Kolmogorov measure μ and fix $\epsilon > 0$. The very weak Bernoulli property provides an m so that

$$E_{X_{-t}^0}\left(\bar{d}_m(X_1^m/X_{-t}^0, X_1^m)\right) < \epsilon/8, \quad t \geq 0. \tag{2}$$

The goal is to show that a process sufficiently close to $\{X_i\}$ in distribution and entropy satisfies almost the same bound. Towards this end, let γ be a positive number to be specified later. For m fixed, $H(X_1^m|X_{-t}^0)$ converges to $mH(\mu)$, as $t \to \infty$, and hence there is a K such that

$$H(X_1^m|X_{-K}^0) < mH(\mu) + \gamma.$$

Fix $k = m + K + 1$. The quantity $E_{X_{-K}^0}(\bar{d}_m(X_1^m/X_{-K}^0, X_1^m))$ depends continuously on μ_k, so if δ is small enough and $\{Y_i\}$ is a stationary process with Kolmogorov measure ν such that $|\mu_k - \nu_k| < \delta$, then

$$E_{Y_{-K}^0}\left(\bar{d}_m(Y_1^m/Y_{-K}^0, Y_1^m)\right) < \epsilon/4. \tag{3}$$

Furthermore, $H(X_1^m|X_{-K}^0)$ also depends continuously on μ_k, so that if it also assumed that $|H(\nu) - H(\mu)| < \delta$, and δ is small enough, then $H(Y_1^m|Y_{-K}^0) < mH(\nu) + 2\gamma$ holds. But, since $H(Y_1^m|Y_{-t}^0)$ decreases to $mH(\nu)$ this means that

$$H(Y_1^m|Y_{-K}^0) - H(Y_1^m|Y_{-K-j}^0) < 2\gamma, \quad j \geq 0.$$

If γ is small enough, the conditional ϵ-independence lemma, Lemma IV.2.7, implies that Y_1^m and Y_{-K-j}^{-K-1} are conditionally $(\epsilon/2)$-independent, given Y_{-K}^0. Since $\bar{d}$-distance is upper bounded by one-half the variational distance, this means that

$$E_{Y_{-K-j}^0}(\bar{d}_m(Y_1^m/Y_{-K}^0, Y_1^m/Y_{-K-j}^0)) < \epsilon/4, \quad j \geq 1.$$

This result, along with the triangle inequality and the earlier bound (3), yields the result that will be needed, namely,

$$E_{Y_{-t}^0}(\bar{d}_m(Y_1^m, Y_1^m/Y_{-t}^0)) < \epsilon/2, \quad t \geq 1. \tag{4}$$

In summary, since it can also be supposed that $\delta < \epsilon/2$, and hence that $\bar{d}_m(X_1^m, Y_1^m) < \epsilon/4$, it follows from the triangle inequality that there is a $\delta > 0$ such that

$$\sum_{x_1^n, y_1^n} \lambda(x_1^n, y_1^n)\bar{d}_m(X_{n+1}^{n+m}/x_1^n, Y_{n+1}^{n+m}/y_1^n) < \epsilon,$$

for all n, for any stationary process $\{Y_i\}$ with Kolmogorov measure ν for which $|\mu_k - \nu_k| < \delta$ and $|H(\mu) - H(\nu)| < \delta$, which completes the proof that very weak Bernoulli implies finitely determined. □

IV.3.c The almost blowing-up characterization.

The almost blowing-up property (ABUP), was introduced in Section III.4. The ϵ-blowup $[C]_\epsilon$ of a set $C \subseteq A^n$ is its ϵ-neighborhood relative to the d_n-metric, that is,

$$[C]_\epsilon = \left\{b_1^n \colon d_n(a_1^n, b_1^n) \le \epsilon, \text{ for some } a_1^n \in C\right\}.$$

A set $B \subset A^n$ has the (δ, ϵ)-blowing-up property if $\mu([C]_\epsilon) \ge 1 - \epsilon$, for any subset $C \subset B$ for which $\mu(C) \ge 2^{-k\delta}$. A stationary process has the almost blowing-up property if for each n there is a $B_n \subset A^n$ such that $x_1^n \in B_n$, eventually almost surely, and for any $\epsilon > 0$, there is a $\delta > 0$ and an N such that B_n has the (δ, ϵ)-blowing-up property for $n \ge N$.

It was shown in Section III.4.c that a finitely determined process has the almost blowing-up property. In this section it will be shown that almost blowing-up implies very weak Bernoulli. Since very weak Bernoulli implies finitely determined, this completes the proof that almost block-independence, finitely determined, very weak Bernoulli, and almost blowing-up are all equivalent ways of saying that a process is a stationary coding of an i.i.d. process.

The proof is based on the following lemma, which, in a more general setting and stronger form, is due to Strassen, [81].

Lemma IV.3.2

Let μ and ν be probability measures on A^n. If $\nu([C]_\epsilon) \ge 1-\epsilon$, whenever $\mu(C) \ge \epsilon$, then $\bar{d}_n(\mu, \nu) \le 2\epsilon$.

Proof. The ideas of the proof will be described first, after which the details will be given. The goal is to construct a joining λ of μ and ν for which $d_k(x_1^n, y_1^n) \le \epsilon$, except on a set of (x_1^n, y_1^n) of λ-measure less than ϵ. A joining can be thought of as a partitioning of the ν-mass of each y_1^n into parts and an assignment of these parts to various x_1^n, subject to the *joining requirement*, namely, that the total mass received by each x_1^n is equal to $\mu(x_1^n)$.

A simple strategy for beginning the construction of a good joining is to cut off an α-fraction of the ν-mass of y_1^n and assign it to some x_1^n for which $d_n(x_1^n, y_1^n) \le \epsilon$. A trivial argument shows that this is indeed possible for some positive α for those y_1^n that are within ϵ of some x_1^n of positive μ-mass. The set of such y_1^n is just the ϵ-blowup of the support of μ, which, by hypothesis, has ν-measure at least $1 - \epsilon$.

The key to continuing is the observation that if the set of x_1^n whose μ-mass is not completely filled in the first stage has μ-measure at least ϵ then its blowup has large ν-measure and the simple strategy can be used again, namely, for most y_1^n a small fraction of the remaining ν-mass can be cut off and and assigned to the unfilled mass of a nearby x_1^n. This simple strategy can repeated as long as the set of x_1^n whose μ-mass is not yet completely filled has μ-measure at least ϵ. If the fraction is chosen to be largest possible at each stage then only a finite number of stages are needed to reach the point when the set of x_1^n whose μ-mass is not yet completely filled has μ-measure less than ϵ.

Some notation and terminology will assist in making the preceding sketch into a proof. A nonnegative function $\widetilde{\mu}$ is called a *mass function* if it has nonempty support and $\sum \widetilde{\mu}(x_1^n) \le 1$. A function $\phi \colon [S]_\epsilon \mapsto S$, is called an *$\epsilon$-matching over S* if $d_k(y_1^n, \phi(y_1^n)) \le \epsilon$, for all $y_1^n \in [S]_\epsilon$. If the domain S of such a ϕ is contained in the support of a mass function $\widetilde{\mu}$, then the number $\alpha = \alpha(\nu, \widetilde{\mu}, \phi)$ defined by

$$\alpha = \min\left\{1, \inf_{\widetilde{\mu}(x_1^n)>0} \frac{\widetilde{\mu}(x_1^n)}{\widetilde{\mu}(\phi^{-1}(x_1^n))}\right\} \tag{5}$$

is positive. It is called the *maximal* $(\nu, \widetilde{\mu}, \phi)$*-stuffing fraction,* for it is the largest number $\alpha \leq 1$ for which

$$\alpha\nu(\phi^{-1}(x_1^n)) \leq \widetilde{\mu}(x_1^n), \quad x_1^n \in S,$$

that is, for which an α-fraction of the ν-mass of each $y_1^n \in [S]_\epsilon$ can be assigned to the $\widetilde{\mu}$-mass of $x_1^n = \phi(y_1^n)$.

The desired joining λ is constructed by induction. To get started let $\mu^{(1)} = \mu$, let S_1 be the support of $\mu^{(1)}$, let ϕ_1 be any ϵ-matching over S_1 (such a ϕ_1 exists by the definition of ϵ-blowup, as long as $S_1 \neq \emptyset$) and let α_1 be the maximal $(\nu, \mu^{(1)}, \phi_1)$-stuffing fraction. Having defined $S_i, \mu^{(i)}, \phi_i$ and α_i, let $\mu^{(i+1)}$ be the mass function defined by

$$\mu^{(i+1)}(x_1^n) = \mu^{(i)}(x_1^n) - \alpha_i\nu(\phi_i^{-1}(x_1^n)).$$

The set S_{i+1} is then taken to be the support of $\mu^{(i+1)}$, the function ϕ_{i+1} is taken to be any ϵ-matching over S_{i+1}, and α_{i+1} is taken to be the maximal $(\nu, \mu^{(i+1)}, \phi_{i+1})$-stuffing fraction.

If $\alpha_i < 1$ then, by the definition of maximal stuffing fraction, there is an $x_1^n \in S_i$ for which $\mu^{(i)}(x_1^n) = \alpha_i\nu(\phi_i^{-1}(x_1^n))$ and therefore $\mu(S_{i+1}) < \mu(S_i)$. Hence there is first i, say i^*, for which $\alpha_i = 1$ or for which $\mu(S_{i+1}) \leq \epsilon$. The construction can be stopped after i^*, cutting up and assigning the remaining unassigned ν-mass in any way consistent with the joining requirement that the total mass received by each x_1^n is equal to $\mu(x_1^n)$.

If $\alpha_{i^*} = 1$, then

$$\mu^{(i^*)}(S_{i^*}) \geq \nu([S_{i^*}]_\epsilon) \geq 1 - \epsilon,$$

otherwise $\mu(S_{i^*+1}) \leq \epsilon$. In either case, $d_k(x_1^n, y_1^n) \leq \epsilon$, except on a set of (x_1^n, y_1^n) of λ-measure less than ϵ, and the proof of Lemma IV.3.2 is finished. □

With Lemma IV.3.2 in hand, only one simple entropy fact about conditional measures will be needed to prove that almost blowing-up implies very weak Bernoulli. This is the fact that, with high probability, the conditional probability $\mu(x_1^n|x_{-\infty}^0)$ has approximately the same exponential size as the unconditioned probability $\mu(x_1^n)$, provided only that n is large enough. Let $\Sigma(X_{-\infty}^n)$ denote the σ-algebra generated by the collection of cylinder sets $\{[x_{-k}^n]: k \geq 0\}$. Also, for $F_n \in \Sigma(X_{-\infty}^n)$, the notation $x_{-\infty}^n \in F_n$ will be used as shorthand for the statement that $\cap_k[x_{-k}^n] \subset F_n$.

Lemma IV.3.3 (The conditional entropy lemma.)

Let μ be an ergodic process of entropy H and let α be a given positive number. There is an $N = N(\alpha) \geq 0$ such that if $n \geq N$ then there is a set $F_n \in \Sigma(X_{-\infty}^n)$ such that $\mu(F_n) \geq 1 - \alpha$ and so that if $x_{-\infty}^n \in F_n$ then

$$2^{-\alpha n}\mu(x_1^n) \leq \mu(x_1^n|x_{-\infty}^0) \leq 2^{\alpha n}\mu(x_1^n).$$

Proof. By the ergodic theorem, $-(1/n)\log\mu(x_1^n|x_{-\infty}^0) \to H$, almost surely, while, by the entropy theorem, $-(1/n)\log\mu(x_1^n) \to H$, almost surely. These two facts together imply the lemma. □

Only the following consequence of the upper bound will be needed.

Lemma IV.3.4

Let $N = N(\alpha)$ be as in the preceding lemma. If $n \geq N$, there is a set $D_n \in \Sigma(X_{-\infty}^0)$ of measure at least $1 - \sqrt{\alpha}$ such that if $x_{-\infty}^0 \in D_n$ and $C \subseteq A^n$, then

$$\mu(C|x_{-\infty}^0) \leq 2^{\alpha n}\mu(C) + \sqrt{\alpha}.$$

Proof. For $n \geq N$, the Markov inequality yields a set $D_n \in \Sigma(X_{-\infty}^0)$ of measure at least $1 - \sqrt{\alpha}$ and, for each $x_{-\infty}^0 \in D_n$, a set $F_n \subset A^n$ such that

(a) $\mu(F_n|x_{-\infty}^0) > 1 - \sqrt{\alpha}$.

(b) $\mu(x_1^n|x_{-\infty}^0) \leq 2^{\alpha n}\mu(x_1^n),\ x_1^n \in F_n$.

If $x_{-\infty}^0 \in D_n$ and $C \subseteq A^n$ then (a) and (b) give

$$\begin{aligned}\mu(C|x_{-\infty}^0) &\leq \mu(C \cap F_n|x_{-\infty}^0) + \sqrt{\alpha} \\ &\leq 2^{\alpha n}\mu(C \cap F_n) + \sqrt{\alpha} \\ &\leq 2^{\alpha n}\mu(C) + \sqrt{\alpha},\end{aligned}$$

which establishes the lemma. □

Now the main theorem of this section will be proved.

Theorem IV.3.5

Almost blowing-up implies very weak Bernoulli.

Proof. Let μ be a stationary process with the almost blowing-up property. The goal is to show that μ is very weak Bernoulli, that is, to show that for each $\epsilon > 0$, there is an n and a set $V \in \Sigma(X_{-\infty}^0)$ such that $\mu(V) > 1 - \epsilon$ and such that

$$\bar{d}_n(\mu, \mu(\cdot|x_{-\infty}^0)) < \epsilon, \quad x_{-\infty}^0 \in V.$$

By Lemma IV.3.2, it is enough to show how to find n and $V \in \Sigma(X_{-\infty}^0)$ such that $\mu(V) > 1 - \epsilon$ and such that the following holds for $x_{-\infty}^0 \in V$.

(6) If $C \subset A^n$ and $\mu(C|x_{-\infty}^0) \geq \epsilon$ then $\mu([C]_\epsilon) \geq 1 - \epsilon$.

Towards this end, let $B_n \subset A^n, n \geq 1$, be such that $x_1^n \in B_n$, eventually almost surely, and such that for any $\epsilon > 0$, there is a $\delta > 0$ and an N such that B_n has the (δ, ϵ)-blowing-up property for $n \geq N$.

The quantity $\mu(B_n)$ is an average of the quantities $\mu(B_n|x_{-\infty}^0)$, so that if n is so large that $\mu(B_n) \geq 1 - \epsilon^2/4$, then, by the Markov inequality, there is a set $D_n^* \in \Sigma(X_{-\infty}^0)$ such that $\mu(D_n^*) \geq 1 - \epsilon/2$ and such that $\mu(B_n|x_{-\infty}^0) \geq 1 - \epsilon/2$, for $x_{-\infty}^0 \in D_n^*$. Combining this with Lemma IV.3.4 it follows that if n is large enough and $\alpha \leq \epsilon^2/4$ then the set $V = D_n \cap D_n^* \in \Sigma(X_{-\infty}^0)$, has measure at least $1 - \epsilon$, and for $x_{-\infty}^0 \in V$ and $C \subseteq A^n$,

$$\begin{aligned}\mu(C|x_{-\infty}^0) &\leq \mu(C \cap B_n|x_{-\infty}^0) + \epsilon/2 \\ &\leq 2^{\alpha n}\mu(C \cap B_n) + \sqrt{\alpha} + \epsilon/2,\end{aligned}$$

which, if $\mu(C|x_{-\infty}^0) \geq \epsilon$ can be rewritten as

$$\mu(C \cap B_n) \geq 2^{-\alpha n}(\epsilon - \sqrt{\alpha} - \epsilon/2).$$

If α is chosen to be less than both δ and $\epsilon^2/8$, then $2^{-\alpha n}(\epsilon - \sqrt{\alpha} - \epsilon/2) \geq 2^{-\delta n}$, for all sufficiently large n, so that the blowing-up property of B_n yields

$$\mu([C]_\epsilon) \geq \mu([C \cap B_n)]_\epsilon) \geq 1 - \epsilon,$$

provided only that n is large enough. This establishes the desired result (6) and completes the proof that almost blowing-up implies very weak Bernoulli. □

Remark IV.3.6

The proof of Theorem IV.3.5 appeared in [39]. The proof of Strassen's lemma given here uses a construction suggested by Ornstein and Weiss which appeared in [37]. A more sophisticated result, using a marriage lemma to pick the ϵ-matchings ϕ_i, yields the stronger result that the $\bar{d}_n$-metric is equivalent to the metric $d_n^*(\mu, \nu)$, defined as the minimum $\epsilon > 0$ for which $\mu(C) \leq \nu([C]_\epsilon + \epsilon$, for all $C \subset A^n$. This and related results are discussed in Pollard's book, [59, Example 26, pp. 79-80].

IV.3.d Exercises.

1. Show that a stationary process is very weak Bernoulli if and only if for each $\epsilon > 0$ there is an m such that for each $k \geq 1$,

$$\bar{d}_n(X_1^n / x_{-k}^0, X_1^n) \leq \epsilon,$$

   except for a set of x_{-k}^0 of measure at most ϵ.

2. Show using coupling that a mixing Markov chain is weak Bernoulli.

3. Show that a mixing regenerative process is weak Bernoulli. (Hint: coupling.)

4. Show by using the martingale theorem that a stationary process μ is very weak Bernoulli if given $\epsilon > 0$ there is a positive integer n and a measurable set $G \in \Sigma(X_{-\infty}^0)$ of measure at least $1 - \epsilon$, such that if $x_{-\infty}^0, \tilde{x}_{-\infty}^0 \in G$ then there is a measurable mapping $\phi: [x_{-\infty}^0] \mapsto [\tilde{x}_{-\infty}^0]$ which maps the conditional measure $\mu(\cdot|x_{-\infty}^0)$ onto the conditional measure $\mu(\cdot|\tilde{x}_{-\infty}^0)$, and a measurable set $B \subset [x_{-\infty}^0]$ such that $\mu(B|x_{-\infty}^0) \leq \epsilon$, with the property that for all $x \notin B$, $\phi(x)_i = x_i$ for all except at most ϵn indices $i \in [1, n]$. .

5. Show by using the martingale theorem that a stationary process μ is weak Bernoulli if given $\epsilon > 0$ there is a positive integer K such that for every $n \geq K$ there is a measurable set $G \in \Sigma(X_{-\infty}^0)$ of measure at least $1 - \epsilon$, such that if $x_{-\infty}^0, \tilde{x}_{-\infty}^0 \in G$ then there is a measurable mapping $\phi: [x_{-\infty}^0] \mapsto [\tilde{x}_{-\infty}^0]$ which maps the conditional measure $\mu(\cdot|x_{-\infty}^0)$ onto the conditional measure $\mu(\cdot|\tilde{x}_{-\infty}^0)$, and a measurable set $B \subset [x_{-\infty}^0]$ such that $\mu(B|x_{-\infty}^0) \leq \epsilon$, with the property that for all $x \notin B$, $\phi(x)_i = x_i$ for $i \in [K, n]$.

Bibliography

[1] Arnold, V.I. and Avez, A., *Ergodic problems of classical mechanics.* W.A. Benjamin, Inc., New York, 1968.

[2] R. Arratia and M. Waterman, "The Erdős-Rényi strong law for pattern matching with a given proportion of mismatches," Ann. Probab., 17(1989), 1152-1169.

[3] A. Barron, "Logically smooth density estimation," Ph. D. Thesis, Dept. of Elec. Eng., Stanford Univ., 1985.

[4] P. Billingsley, *Ergodic theory and information.* John Wiley and Sons, New York, 1965.

[5] D. C. Kohn, *Measure Theory.* Birkhauser, New York, 1980.

[6] T. M. Cover and J. A. Thomas, *Elements of information theory.* John Wiley and Sons, New York, 1991.

[7] I. Csiszár and J. Körner, *Information Theory. Coding theorems for discrete memoryless systems.* Akadémiai Kiadó, Budapest, 1981.

[8] P.Elias, "Universal codeword sets and representations of the integers," IEEE Trans. Info. Th., IT-21(1975), 194-203.

[9] P. Erdős and A. Rényi, "On a new law of large numbers," J. d'analyse, 23(1970), 103-111.

[10] J. Feldman, "r-Entropy, equipartition, and Ornstein's isomorphism theorem,", Israel J. of Math., 36(1980), 321-343.

[11] W. Feller, *An introduction to probability theory and its applications.* Volume II (Second Edition)., Wiley, New York, 1971.

[12] N. Friedman, *Introduction to ergodic theory.* Van Nostrand Reinhold, New York, NY, 1970.

[13] N. Friedman and D. Ornstein, "On isomorphism of weak Bernoulli transformations," Advances in Math., 5(1970), 365-394.

[14] H. Furstenberg, *Recurrence in ergodic theory and combinatorial number theory.* Princeton Univ. Press, Princeton, NJ, 1981.

[15] R. Gallager, *Information theory and reliable communication.* Wiley, New York, NY, 1968.

[16] P. Grassberger, "Estimating the information content of symbol sequences and efficient codes," IEEE Trans. Inform. Theory, IT-35(1989), 669-675.

[17] R. Gray, *Probability, random processes, and ergodic properties.* Springer-Verlag, New York, NY, 1988.

[18] R. Gray, *Entropy and information theory.* Springer-Verlag, New York, NY, 1990.

[19] R. Gray and L. D. Davisson, "Source coding without the ergodic assumption," IEEE Trans. Info. Th., IT-20(1975), 502-516.

[20] P. Halmos, *Measure theory.* D. van Nostrand Co., Princeton, NJ, 1950.

[21] P. Halmos, *Lectures on ergodic theory.* Chelsea Publishing Co., New York, 1956.

[22] W. Hoeffding, "Asymptotically optimal tests for multinomial distributions," Ann. of Math. Statist., 36(1965), 369-400.

[23] M. Kac, "On the notion of recurrence in discrete stochastic processes," Ann. of Math. Statist., 53(1947), 1002-1010.

[24] S. Kakutani, "Induced measure-preserving transformations," Proc. Japan Acad., 19(1943), 635-641.

[25] T. Kamae, "A simple proof of the ergodic theorem using nonstandard analysis," Israel J. Math., 42(1982), 284-290.

[26] S. Karlin and G. Ghandour, "Comparative statistics for DNA and protein sequences - single sequence analysis," Proc. Nat. Acad. Sci., U.S.A., 82(1985), 5800-5804.

[27] I. Katznelson and B. Weiss, "A simple proof of some ergodic theorems," Israel J. Math., 42(1982), 291-296.

[28] M. Keane and M. Smorodinsky, " Finitary isomorphism of irreducible Markov shifts," Israel J. Math., 34(1979) 281-286.

[29] J. G. Kemeny and J. L. Snell, *Finite Markov chains.* Van Nostrand Reinhold, Princeton, New Jersey, 1960.

[30] J. Kieffer, "Sample converses in source coding theory," IEEE Trans. Info. Th., IT-37(1991), 263-268.

[31] I. Kontoyiannis and Y. M. Suhov, "Prefixes and the entropy rate for long-range sources," *Probability, statistics, and optimization.* (F. P. Kelly, ed.) Wiley, New York, 1993.

[32] U. Krengel, *Ergodic theorems.* W. de Gruyter, Berlin, 1985.

[33] R. Jones, "New proofs of the maximal ergodic theorem and the Hardy-Littlewood maximal inequality," Proc. AMS, 87(1983), 681-4.

[34] D. Lind and B. Marcus, *An Introduction to Symbolic Dynamics and Coding.* Cambridge Univ. Press, Cambridge, 1995.

[35] T. Lindvall, *Lectures on the coupling method.* John Wiley and Sons, New York, 1992.

[36] K. Marton, "A simple proof of the blowing-up lemma," IEEE Trans. Info. Th., IT-42(1986), 445-447.

[37] K. Marton and P. Shields, "The positive-divergence and blowing-up properties," Israel J. Math., 86(1994), 331-348.

[38] K. Marton and P. Shields, "Entropy and the consistent estimation of joint distributions," Ann. Probab., 22(1994), 960-977. Correction: Ann. Probab., to appear.

[39] K. Marton and P. Shields, "Almost sure waiting time results for weak and very weak Bernoulli processes," Ergodic Th. and Dynam. Sys., 15(1995), 951-960.

[40] D. Neuhoff and P. Shields, "Block and sliding-block source coding," IEEE Trans. Info. Th., IT-23(1977), 211-215.

[41] D. Neuhoff and P. Shields, "Indecomposable finite state channels and primitive approximation," IEEE Trans. Inform. Th., IT-28(1982), 11-18.

[42] D. Neuhoff and P. Shields, "Channel entropy and primitive approximation," Ann. Probab., 10(1982), 188-198.

[43] D. Neuhoff and P. Shields, "A very simplistic, universal, lossless code," IEEE Workshop on Information Theory, Rydzyna, Poland, June, 1995.

[44] A. Nobel and A. Wyner, "A recurrence theorem for dependent processes with applications to data compression," IEEE Trans. Inform. Th., IT-38(1992), 1561-1563.

[45] D. Ornstein, "An application of ergodic theory to probability theory", Ann. Probab., 1(1973), 43-58.

[46] D. Ornstein, *Ergodic theory, randomness, and dynamical systems.* Yale Mathematical Monographs 5, Yale Univ. Press, New Haven, CT, 1974.

[47] D. Ornstein, D. Rudolph, and B. Weiss, "Equivalence of measure preserving transformations," Memoirs of the AMS, 262(1982).

[48] D. Ornstein and P. Shields, "An uncountable family of K-Automorphisms", Advances in Math., 10(1973), 63-88.

[49] D. Ornstein and P. Shields, "Universal almost sure data compression," Ann. Probab., 18(1990), 441-452.

[50] D. Ornstein and P. Shields, "The $\bar{d}$-recognition of processes." Advances in Math., 104(1994), 182-224.

[51] D. Ornstein and B. Weiss, "The Shannon-McMillan-Breiman theorem for amenable groups," Israel J. Math., 44(1983), 53-60.

[52] D. Ornstein and B. Weiss, "How sampling reveals a process," Ann. Probab., 18(1990), 905-930.

[53] D. Ornstein and B. Weiss, "Entropy and data compression," IEEE Trans. Inform. Th., IT-39(1993), 78-83.

[54] D. Ornstein and B. Weiss, "On the Bernoulli nature of systems with some hyperbolic structure," Ergodic Th. and Dynam. Sys., to appear.

[55] W. Parry, *Topics in ergodic theory.* Cambridge Univ. Press, Cambridge, 1981.

[56] K. Petersen, *Ergodic theory.* Cambridge Univ. Press, Cambridge, 1983.

[57] Phelps, R., *Lectures on Choquet's theorem.* Van Nostrand, Princeton, N.J., 1966

[58] M. Pinsker, *Information and information stability of random variables and processes.* (In Russian) Vol. 7 of the series *Problemy Peredači Informacii*, AN SSSR. Moscow, 1960. English translation: Holden-Day, San Francisco, 1964.

[59] D. Pollard, *Convergence of stochastic processes.* Springer-Verlag, New York, 1984.

[60] V. Rohlin, "A 'general' measure-preserving transformation is not mixing," Dokl. Akad. Nauk SSSR, 60(1948), 349-351.

[61] D. Rudolph, "If a two-point extension of a Bernoulli shift has an ergodic square, then it is Bernoulli," Israel J. of Math., 30(1978), 159-180.

[62] I. Sanov, "On the probability of large deviations of random variables," (in Russian), Mat. Sbornik, 42(1957), 11-44. English translation: Select. Transl. Math. Statist. and Probability, 1(1961), 213-244.

[63] P. Shields, *The theory of Bernoulli shifts.* Univ. of Chicago Press, Chicago, 1973.

[64] P. Shields, "Cutting and independent stacking of intervals," Mathematical Systems Theory, 7 (1973), 1-4.

[65] P. Shields, "Weak and very weak Bernoulli partitions," Monatshefte für Mathematik, 84 (1977), 133-142.

[66] P. Shields, "Stationary coding of processes," IEEE Trans. Inform. Th., IT-25(1979), 283-291.

[67] P. Shields, "Almost block independence," Z. fur Wahr., 49(1979), 119-123.

[68] P. Shields, "The ergodic and entropy theorems revisited," IEEE Trans. Inform. Th., IT-33(1987), 263-6.

[69] P. Shields, "Universal almost sure data compression using Markov types," Problems of Control and Information Theory, 19(1990), 269-277.

[70] P. Shields, "The entropy theorem via coding bounds," IEEE Trans. Inform. Th., IT-37(1991), 1645-1647.

[71] P. Shields, "String matching - the general ergodic case," Ann. Probab., 20(1992), 1199-1203.

[72] P. Shields, "Entropy and prefixes," Ann. Probab., 20(1992), 403-409.

[73] P. Shields, "Cutting and stacking. A method for constructing stationary processes," IEEE Trans. Inform. Th., IT-37(1991), 1605-1617.

[74] P. Shields, "Universal redundancy rates don't exist," IEEE Trans. Inform. Th., IT-39(1993), 520-524.

[75] P. Shields, "Two divergence-rate counterexamples," J. of Theor. Prob., 6(1993), 521-545.

[76] P. Shields, "Waiting times: positive and negative results on the Wyner-Ziv problem," J. of Theor. Prob., 6(1993), 499-519.

[77] P. Shields and J.-P. Thouvenot, " Entropy zero $\times$ Bernoulli processes are closed in the $\bar{d}$-metric," Ann. Probab., 3(1975), 732-6.

[78] M. Smorodinsky, " A partition on a Bernoulli shift which is not 'weak Bernoulli'," Math. Syst. Th., 5(1971), 210-203.

[79] M. Smorodinsky, "Finitary isomorphism of m-dependent processes," Contemporary Math., vol. 135(1992) 373-376.

[80] M. Steele, "Kingman's subadditive ergodic theorem," Ann. Inst. Henri Poincaré, 25(1989), 93-98.

[81] V. Strassen, "The existence of probability measures with given marginals," Ann. of Math. Statist., 36(1965), 423-439.

[82] W. Szpankowski, Asymptotic properties of data compression and suffix trees," IEEE Trans. Inform. Th., IT-39(1993), 1647-1659.

[83] J. Moser, E. Phillips, and S.Varadhan, *Ergodic theory : a seminar.* Courant Inst. of Math. Sciences, NYU, New York, 1975.

[84] P. Walters, *An introduction to ergodic theory.* Springer-Verlag, New York, 1982.

[85] F. M. J. Willems, "Universal data compression and repetition times," IEEE Trans. Inform. Th., IT-35(1989), 54-58.

[86] A. Wyner and J. Ziv, "Some asymptotic properties of the entropy of a stationary ergodic data source with applications to data compression," IEEE Trans. Inform. Th., IT-35(1989), 125-1258.

[87] A. Wyner and J. Ziv, "Fixed data base version of the Lempel-ziv data compression algorithm," IEEE Trans. Inform. Th., IT-37(1991), 878-880.

[88] A. Wyner and J. Ziv, "The sliding-window Lempel-Ziv algorithm is asymptotically optimal," Proc. of the IEEE, 82(1994), 872-877.

[89] S. Xu, "An ergodic process of zero divergence-distance from the class of all stationary processes," J. of Theor. Prob., submitted.

[90] J. Ziv, "Coding theorems for individual sequences," IEEE Trans. Inform. Th., IT-24(1978), 405-412.

[91] J. Ziv and A. Lempel, "A universal algorithm for sequential data compression," IEEE Trans. Info. Th., IT-23(1977), 337-343.

[92] J. Ziv, "Coding of sources with unknown statistics-Part I: Probability of encoding error; Part II: Distortion relative to a fidelity criterion", IEEE Trans. Info. Th., IT-18(1972), 384-394.

Index